Contributions to Statistics

The series **Contributions to Statistics** contains publications in theoretical and applied statistics, including for example applications in medical statistics, biometrics, econometrics and computational statistics. These publications are primarily monographs and multiple author works containing new research results, but conference and congress reports are also considered.
Apart from the contribution to scientific progress presented, it is a notable characteristic of the series that publishing time is very short, permitting authors and editors to present their results without delay.

More information about this series at http://www.springer.com/series/2912

Ignacio Rojas · Héctor Pomares
Olga Valenzuela
Editors

Time Series Analysis and Forecasting

Selected Contributions from ITISE 2017

Springer

Editors
Ignacio Rojas
ETSIIT
University of Granada
Granada, Spain

Olga Valenzuela
Faculty of Sciences
University of Granada
Granada, Spain

Héctor Pomares
CITIC-UGR and ETSIIT
University of Granada
Granada, Spain

ISSN 1431-1968
Contributions to Statistics
ISBN 978-3-030-07276-6 ISBN 978-3-319-96944-2 (eBook)
https://doi.org/10.1007/978-3-319-96944-2

Mathematics Subject Classification (2010): 62-XX, 68-XX, 60-XX, 58-XX, 37-XX

This Springer imprint is published by the registered company Springer Nature Switzerland AG
The registered company address is: Gewerbestrasse 11, 6330 Cham, Switzerland

Preface

George Santayana, a Spanish philosopher and writer of the nineteenth century, once said: "Those who do not remember the past are condemned to repeat it".[1] And this is, in a way, what time series analysis is about: we analyze the past (actually, samples obtained usually at regular intervals from a system of our interest) so that we can obtain valuable information with which we can analyze what has happened in the past and, if used correctly, try to avoid disastrous (or at least non-desirable) outcomes in the future. This analysis is not easy though. These sampled data may have very complex dependencies on other external variables (exogenous variables) or on other sampled data (multivariate time series), and it can contain a certain amount of noise due to inaccurate sensor readings or even some samples can be completely incorrect (outliers). To make things worse, the amount of data available may be so huge that the computational complexity of even the more simple methods may need such a long time that only parallelism and big data infrastructures can be used to solve the problem. The applications are, however, also huge. We can find time series analysis, modeling, and forecasting methods in the literature applied in so diverse fields such as business economics, statistics, engineering, environmental sciences, or physical sciences, just to mention a few of them. As a consequence, we can state that time series-related research is arguably one the most important research fields at the present time (more than 14,000 research papers only in 2017 according to Web of Science from Thomson Reuters).

The origin of this book stems from the International work-conference on Time Series, ITISE 2017, held in Granada (Spain) in September 2017. Our aim with the organization of ITISE 2017 was to create a friendly discussion forum for scientists, engineers, educators, and students about the latest ideas and realizations in the foundations, theory, models, and applications for interdisciplinary and multidisciplinary research encompassing disciplines of statistics, mathematical models, econometrics, engineering, and computer science in the field of time series analysis and forecasting.

[1]The Essential Santayana. Selected Writings Edited by the Santayana Edition, Compiled and with an introduction by Martin A. Coleman. Bloomington: Indiana University Press. 2009.

The list of topics in the successive Call for Papers has also evolved, resulting in the following list for the last edition:

1. **Time Series Analysis and Forecasting.**
   - Nonparametric and functional methods.
   - Vector processes.
   - Probabilistic approach to modeling macroeconomic uncertainties.
   - Uncertainties in forecasting processes.
   - Nonstationarity.
   - Forecasting with many models. Model integration.
   - Forecasting theory and adjustment.
   - Ensemble forecasting.
   - Forecasting performance evaluation.
   - Interval forecasting.
   - Econometric models.
   - Econometric forecasting.
   - Data preprocessing methods: data decomposition, seasonal adjustment, singular spectrum analysis, and detrending methods.
2. **Advanced Methods and Online Learning in Time Series.**
   - Adaptivity for stochastic models.
   - Online machine learning for forecasting.
   - Aggregation of predictors.
   - Hierarchical forecasting.
   - Forecasting with computational intelligence.
   - Time series analysis with computational intelligence.
   - Integration of system dynamics and forecasting models.
3. **High Dimension and Complex/Big Data.**
   - Local versus global forecast.
   - Techniques for dimension reduction.
   - Multiscaling.
   - Forecasting from Complex/Big data.
4. **Forecasting in Real Problems.**
   - Health forecasting.
   - Telecommunication forecasting.
   - Modeling and forecasting in power markets.
   - Energy forecasting.
   - Financial forecasting and risk analysis.
   - Forecasting electricity load and prices.
   - Forecasting and planning systems.
   - Real-time macroeconomic monitoring and forecasting.
   - Applications in other disciplines.

After a careful peer review and evaluation process (each submission was reviewed by at least 2, and on the average 2.9, program committee members or additional reviewer), 121 contributions are presenting in this proceedings (accepted for oral, poster, or virtual presentation), according to the recommendations of reviewers and the authors' preferences.

High-quality candidate papers (22 contributions) were invited to submit an extended version of their conference paper to be considered for this special publication in the book series of Springer: Contributions to Statistics. For the selection procedure, the information/evaluation of the chairman of every session, in conjunction with the review comments and the summary of reviews, was taken into account.

So, now we are pleased to have reached the end of the whole process and present the readers with these final contributions that we hope, will provide a clear overview of the thematic areas covered by the ITISE 2017 conference, ranging from theoretical aspects to real-world applications of Time Series Analysis and Forecasting.

It is important to note that for the sake of consistency and readability of the book, the presented papers have been classified into the following parts:

- **Part I: Advanced Mathematical Methodologies in Time Series.**
  The main objective of this part is to present mathematical and logical structures, methodologies, and theories that could be used with time series and also that has been used so far. It also aims to bring into existence recent and becoming developments in computational mathematics that could be used in the field of time series. In particular, six contributions have been selected for this part. Two of them shed some light on the problem of robustly estimating model parameters in the presence of highly noisy data and outliers. Another contribution applies to time series analysis techniques to demonstrate the weakness of some cryptographic systems used in some bank's cash dispensers or credit cards. There are two additional contributions which deal with more complex mathematical models: one of them proposes a way to estimate the probability distribution of the next observation of a time series using a closed solution to a Fokker–Planck equation model of the time series, which is a partial differential equation well known in statistical mechanics. The other proposes an extension to problems with exogenous variables of an existing methodology to predict high-dimensional time series using the Koopman operator framework. To that end, the authors assume that the time series are generated by some underlying unknown dynamical system whose inputs are precisely these exogenous variables. Finally, the last contribution selected for this part studies the limit of the empirical distribution function of the eigenvalues of a symmetric matrix which emulates a covariance matrix of several AR(1) processes. The conclusion of the paper is that this empirical distribution of eigenvalues converges almost surely to a nonrandom limit function given by the Marchenko–Pastur distribution.

- **Part II: Computational Intelligence Methods for Time Series.**
  Although time series analysis can be considered a discipline originated within the statistical area, in the last decades many computational intelligence methods or machine learning approaches have been proposed to solve time series-related problems. In fact, new and further computational intelligence approaches, their efficiency and their comparison to statistical methods and other fact-checked computational intelligence methods are significant topics in academic and professional projects. It is not uncommon the existence of time series forecasting competitions which try to elucidate which of the two main research streams is better. For instance, the M4-Competition for the first time makes explicit mention to machine learning forecasting methods. Within this topic, two contributions have been selected for this book. The first one uses neural networks to detect anomalies from time series, in this case, from datasets corresponding to Internet traffic. The second one uses support vector regression with different methods for scaling the data which are taken from a competition, in this case, the Knowledge Discovery and Data Mining Cup which took place in 2017.
- **Part III: Dimensionality Reduction and Similarity Measures in Time Series.**
  With the arrival of the Internet of Things, supported by the cheapening of sensors, storing devices, and wireless communications alike, time series data are becoming even bigger. One fundamental way to tackle with the computational complexity of this problem is trying to reduce the dimensionality of the problem but still maintaining an efficient representation of the data or to study similarity measures in the data so we can classify the data into several clusters, thereby reducing its complexity as we subsequently work with the clusters instead of with the data. Four contributions have been selected within this topic. The first one deals with the automatic search of segments within a time series whose trend can be linearized. The second one proposes an efficient algorithm for anomaly detection of quasi-periodic time series data, efficient enough to operate in real time on computationally weak platforms like smartphones. The third one is about the definition of a similarity measure for comparing a special kind of time series, time interval datasets, which consist of a start point in time, an end point in time and any amount of metadata. Finally, the last contribution selected for this part deals with the task of selecting the right comparison measure when trying to obtain groupings or clusters from time series or from features obtained from them under the perspective of a pure classification problem.

  After these first mainly theoretical parts, we have dedicated the last three parts of the book to the practical applications of time series analysis, modeling, and forecasting. Part IV will deal with finance applications, part V with energy-related applications, and finally, the last part will include other real scenarios where this field can be applied.
- **Part IV: Econometric Models.**
  One of the most prominent applications of time series modeling and forecasting lies within the field of econometrics. Playing a little with the quote given at the beginning of this preface, Jeremy Grantham, a well-known British

econometrician, once said: "Remember that history always repeats itself. Every great bubble in history has broken. There are no exceptions". This part aims at presenting some recent developments of time series research applied to financial and futures data with the original idea of focusing on studies that develop and apply recent nonlinear econometric models to reproduce financial market dynamics and to capture financial data properties with the hope of eventually predict the next economic bubble. Three contributions have been selected to that end. The first one tries to contribute to finding an answer to this question: how can we, without the knowledge (or an estimation) of the variance of the observations, test whether a change in the dynamics of a time series has occurred? A particular example is provided for the case of the Prague Stock Exchange Index. The second one investigates the presence of spatial differences in the dynamic link between unemployment rate variation and GDP growth in some European Union countries by providing a novel distance measure for evaluating the closeness of two vector autoregressive moving average models. The third and last contribution selected for this part makes use of copulas, which are well-known multivariate probability distributions for which the marginal probability distribution of each variable is uniform, to analyze the dependence between inflation and US/Euro exchange rates in the Euro area, during different periods with very interesting conclusions.

- **Part V: Energy Time Series Forecasting.**
  This part makes particular emphasis on the application of time series analysis, modeling, and forecasting applied to energy-related data. By energy, we refer to any kind of energy, such as electrical, solar, microwave, wind, and so on. The first of the contributions selected for this part uses computational intelligence methods, particularly a neural network trained with a genetic algorithm, to estimate aircraft fuel consumption with the purpose of reducing civil aviation carbon emissions. The next contribution uses NARX models, recursive least squares, and genetic algorithms to predict the energy consumption of a given access point within a Wi-Fi Infrastructure, so that better decisions so as to the placement of the access point could be made in order to alleviate energy expenditure. Finally, the last paper uses an ARMA-X-GARCH-X modeling approach to investigate the impact of wind energy and photovoltaic feed-in on electricity spot price level in Germany and its volatility with very interesting findings.

- **Part VI: Forecasting in Real Problems.**
  This last part is dedicated to other real applications of time series analysis, modeling, and forecasting different from those especially mentioned before (financial and energy-related). Four contributions were finally selected to that end. The first one proposes a methodology based on multivariate Mahalanobis distance calculation combined with surrogate time series testing, which is especially interesting for short time series extracted from so a diverse set of real problems such as seismological, meteorological, physiological, and economic datasets. The second one describes and analyzes the association between respiratory diseases and air pollution concentrations by handling the

multicollinearity and serial dependence between these time series using a hybrid model, which is a combination of a generalized additive model with principal component analysis and a vector autoregressive model. The third selected contribution deals with a special class of time series measured from black-hole systems and whose analysis can lead us to derive information about the geometrical structure of astronomical objects. These time series are especially difficult to manipulate due to their sparseness but the authors manage to do it here using a fully Bayesian-based method based on a state-space model. Finally, we conclude with a contribution whose objective is to model transient oscillations measured by a high-accuracy accelerometer in the presence of outliers. For that purpose, the authors define an observation time series model consisting of a linear regression model with time-variable autoregressive errors, where each coefficient is described by a second linear regression model throughout time and where the white noise components follow a scaled t-distribution with unknown degree of freedom.

Last but not least, we would like to point out that this edition of ITISE was organized by the University of Granada together with the Spanish chapter of the IEEE Computational Intelligence Society and the Spanish Network on Time Series (RESeT). The Guest Editors would also like to express their gratitude to all the people who supported them in the compilation of this book, and especially to the contributing authors for their submissions, the chairmen of the different sessions, and to the anonymous reviewers for their comments and useful suggestions in order to improve the quality of the papers.

We wish to thank our main sponsors as well: the Department of Computer Architecture and Computer Technology, the Faculty of Science of the University of Granada, the Research Centre for Information and Communications Technologies (CITIC-UGR), and the Ministry of Science and Innovation for their support and grants. Finally, we wish also to thank Prof. Alfred Hofmann, Vice President Publishing—Computer Science, Springer-Verlag and Dr. Veronika Rosteck Springer, Associate Editor, for their interest in editing a book series of Springer based on the best papers of ITISE 2017.

We hope the readers of this book can make the most of these selected contributions.

Granada, Spain
March 2018

Ignacio Rojas
Héctor Pomares
Olga Valenzuela

Contents

Part VI Forecasting in Real Problems

Part I
Advanced Mathematical Methodologies in Time Series

Forecasting via Fokker–Planck Using Conditional Probabilities

Chris Montagnon

Abstract Using a closed solution to a Fokker–Planck equation model of a time series, a probability distribution for the next observation is developed. This pdf has one free parameter, b. Various approaches to selecting this parameter have been explored: most recent value, weighted moving average, etc. Here, we explore using a conditional probability distribution for this parameter b, based upon the most recent observation. These methods are tested against some real-world product sales for both a one-step ahead and a two-step ahead forecast. Significant reduction in safety stock levels is found versus an ARMA approach, without a significant increase in out-of-stocks.

Keywords Forecasting · Fokker-Planck · Parameter distribution · Conditional probability · Stock control

1 Introduction

When forecasting a time series $\{X_t, t = 1, \ldots, N\}$, rather than the "best" (e.g., minimum squared error) prediction of a single value $\hat{X}_{N+1}$, the expected value of the next point in the time series, one often requires a probability distribution of the possible values of $\hat{X}_{N+1}$. Kantz and Schreiber [1] proposed tackling this through a Fokker–Planck equation [2] but did not take this further because of difficulty estimating the parameters. Several more recent papers (e.g., Refs. [3–5]) have sought to use forecasting methods based upon a diffusion model leading to a Fokker–Planck equation but the solutions have been numerical. References [6–8] also report difficulties in estimating the parameters in a Fokker–Planck model. In this paper, we use a conditional probability approach to estimate these parameters.

In Ref. [9], we modeled a time series using a drift coefficient $D^{(1)} = -\gamma x$ and diffusion coefficient $D^{(2)} = c - bx^2$ in a Fokker–Planck equation:

C. Montagnon (✉)
Department of Mathematics, Imperial College, London SW7 2AZ, UK
e-mail: chrismontagnon@compuserve.com

I. Rojas et al. (eds.), *Time Series Analysis and Forecasting*,
Contributions to Statistics, https://doi.org/10.1007/978-3-319-96944-2_1

$$\frac{\partial W}{\partial t} = -\frac{\partial}{\partial x}(D^{(1)}W) + \frac{\partial^2}{\partial x^2}(D^{(2)}W)$$

where $W(x, t|x_N = X_N)dx$ is the probability of finding the actual X_{N+t} in $(x, x + dx)$ when the value X_N has been observed for x_N.

This lead to a differential equation in W:

$$(c - bx^2)W_{xx} + (\gamma - 4b)xW_x + (\gamma - 2b)W = W_t \tag{1a}$$

The diffusion coefficient should be positive, so $c \geq bX^2_{\text{max}}$, and one can show that the variance of $W(x, t)$ increases with increasing c, so c should be as small as possible, giving

$$c = bX^2_{\text{max}}, \text{ where } X_{\text{max}} = \max(|X_t|, t = 1, \ldots, N) \tag{1b}$$

In order to reach a closed solution in Ref. [9], we needed the constraint:

$$\gamma = 3b \tag{1c}$$

This lead to the solution:

$$W_b(x, t|x_n = X_N) = \frac{e^{bt}}{2\sqrt{\pi t b}(X^2_{\text{max}} - X^2_N)^{\frac{1}{2}}} \exp\left\{-\frac{1}{4tb}\left(\sin^{-1}\left(\frac{x}{X_{\text{max}}}\right) - \sin^{-1}\left(\frac{X_N}{X_{\text{max}}}\right)\right)^2\right\} \tag{2}$$

The two constraints (1b) and (1c) on the parameters γ, c, and b mean that this distribution is dependent only on one free parameter, which in this paper we choose to be b. This paper explores how our knowledge of previous values of the time series $\{X_t, t = 1, \ldots, N\}$ helps select a value or a distribution for b.

2 Fokker–Planck Solution Versus $\mathcal{N}(\mu, \hat{s}_e)$

Conventionally, the distribution for the possible values of the next point in the time series is found by using $\mathcal{N}(\mu, \hat{s}_e)$ where μ is set equal to an ARMA forecast and $\hat{s}_e$ is found from the residuals of these past ARMA forecasts. It is relevant to see how this Fokker–Planck approach compares to that of $\mathcal{N}(\mu, \hat{s}_e)$.

The mean of $W_b(x, 1)$ is made the same as the mean μ of $\mathcal{N}(\mu, \hat{s}_e)$, so we compare the standard deviation of these two distributions. When $b = b^{(1)}$, the smallest b value, then as shown in Fig. 3 in the Appendix, the W graph is considerably tighter than the $\mathcal{N}$ graph. As we increase the b value used, to say $b^{(4)}$, we get Fig. 4 where again the W graph remains narrower than the $\mathcal{N}$ graph. In both these examples, the 95%

point for the W distribution will be substantially lower than that for the $\mathcal{N}(\mu, \hat{s}_e)$ distribution so in our retail example stock levels will be set lower using the $W_b(x, 1)$ distribution compared to using the $\mathcal{N}(\mu, \hat{s}_e)$ distribution.

3 Definition of the Past Values of Parameter *b*

We define $b_{\tau-j}$ to be the smallest b that makes the observed point $X_{\tau-j+1}$ just less than the 75% point of the distribution $W_b(x, 1|x_{\tau-j} = X_{\tau-j})$, i.e., $b_{\tau-j}$ is the solution to

$$\int^{X_{\tau-j+1}} W_{b_{\tau-j}}(x, 1|x_{\tau-j} = X_{\tau-j})dx = 0.75 \text{ solved for } b_{\tau-j} \quad (3)$$

For a given time series, up to point τ, this generates a set of values $(b_1, b_2, \ldots, b_{\tau-1})$. If we consider a discrete set of possible values for b, say $b^{(j)} : j = 1$ to 30, we obtain a distribution of these values similar to that shown in Fig. 1.

Note:
1. Whenever $X_{\tau+1}$ *is* $<$ *mean of* $W_{b_\tau}(x, 1|x_\tau = X_\tau)$, *then b is given the smallest value (in this case* 0.001) *in solving* (3). *Thus for at least* 50% *of the points this is the value selected.*
2. The final b value is the default value used when (3) *does not solve, so the final probability shown is really* $prob(b \geq 0.059)$.

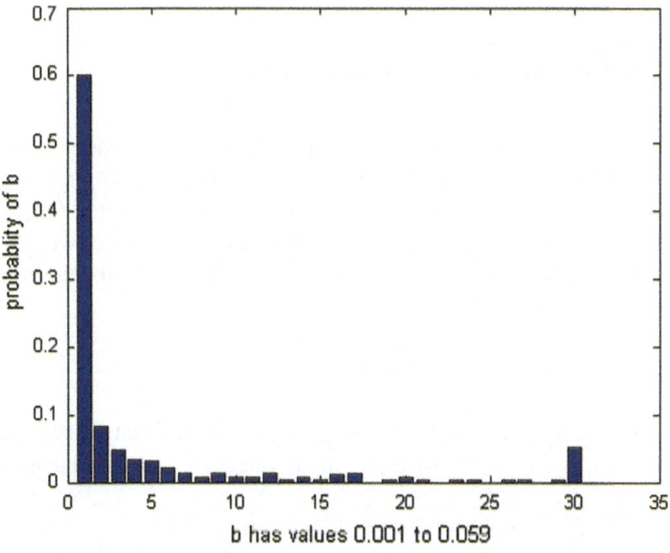

Fig. 1 Probability distribution for parameter *b* across all 300 points of the time series

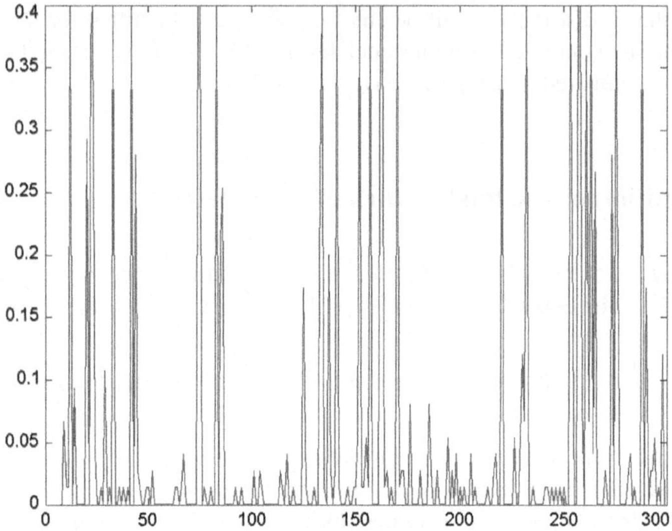

Fig. 2 Time series of the parameter b chosen to meet 75% point

Figure 2 shows the times series b_τ, so one can see that forecasting the values of b as though it was a normal time series (e.g., using an ARMA approach) is not practical, so in this paper we explore ways of choosing b_τ based upon what has just happened, i.e., based upon the observed actual for $b_{\tau-1}$ as defined in (3).

4 Method for One-Step Ahead Forecasts

We consider first the situation where we want a forecast for tomorrow (i.e., D + 1) when we are at the end of today (D + 0): so we might be placing an order to meet sales for tomorrow (D + 1). Given a time series of sufficient length (e.g., $N > 100$) and considering a discrete set of possible values for b, as above, we can form not only the overall probability distribution for b as in Fig. 1, but also a set of discrete conditional probability distributions:

$$p(b_\tau | b_{\tau-1} = b^{(k)}) \tag{4}$$

Thus, for example in an extreme situation, we might find that every time the value $b^{(1)}$ occurred it was always followed by the b value $b^{(31)}$, in which case we would have

$$\text{prob}(b_\tau | b_{\tau-1} = b^{(1)}) = \delta(b - b^{(31)})$$

For the test series we used, each of which had some 300 points, we found that in practice only $b^{(1)}, b^{(2)}, b^{(31)}$ occurred sufficiently often to build a conditional probability distribution as per (4), i.e., only for k $= 1, 2$, or 31 did we have enough (> ca. 20) points to generate any meaningful distribution.

With this information on the likely values of b_τ that follow a particular observed value for $b_{\tau-1}$, at least for $b_{\tau-1} = b^{(1)}, b^{(2)}$ or $b^{(31)}$, the method for generating $W_b(x, t|x_n = X_N)$ became

(i) If $b_{\tau-1} = b^{(1)}, b^{(2)}$ or $b^{(31)}$, then

$$W_b(x, 1|x_\tau = X_\tau) = \frac{1}{\mathcal{M}} \sum_{j=1}^{31} \text{prob}(b^{(j)}|b_{\tau-1}) W_{b^{(j)}}(x, 1|x_\tau = X_\tau) \tag{5}$$

where $b_{\tau-1} = b^{(k)}$ and $\mathcal{M}$ is a normalizing factor.

(ii) If $b_{\tau-1}$ not $= b^{(1)}, b^{(2)}$ or $b^{(31)}$, then

$$W_\tau(x, 1|x_\tau = X_\tau) = \frac{1}{28} \sum_{j=1}^{7} (8-j) W_{b_{\tau-j}}(x, 1|x_\tau = X_\tau) \tag{6}$$

i.e., as defined in Refs. [9, 10] where this was shown to be one of the better methods of defining $W_{b_\tau}(x, t|x_n = X_N)$ ($b_{\tau-j}$ is as defined in (3)).

(iii) In the situation of (ii) above, i.e., $b_{\tau-1}$ not $= b^{(1)}, b^{(2)}$ or $b^{(31)}$, we also tested using a probability distribution

$$W_b(x, 1|x_\tau = X_\tau) = \frac{1}{\mathcal{M}} \sum_{k=1}^{31} \text{prob}(b^{(k)}) W_{b^{(k)}}(x, t|x_\tau = X_\tau) \tag{7}$$

where prob$(b^{(k)})$ is the probability distribution of $b^{(k)}$ unconstrained by the value of $b_{\tau-1}$, i.e., similar to the discrete pdf in Fig. 1.

5 Results for One-Step Ahead Forecasts

We applied the above method to the ten test series a defined in Ref. [11]. We used 200 points for each series. This gave the results in Table 1.

The performance of all three versions of method 1 is compared to results using Normal distribution with mean equal to AR(7) forecast and with the variance calculated from the forecast residuals from past data.

Table 1 Results for one-step ahead forecasts from various conditional probability approaches to b_τ in $W_{b_\tau}(x, 1|x_\tau = X_\tau)$

	AR	Method 1a	Method 1b	Method 1c
Av stock	112	101	106	104
Stockout %	4.6	9.6	6.5	6.6

Key
AR Stock level is 95% pt of a Normal with mean = AR forecast, variance computed from past residuals
Method 1a Stock level is 95% pt. of $W_{b_\tau}(x, 1)$, b chosen as (6) above for all *b*
Method 1b Stock level is 95% pt. of $W_{b_\tau}(x, 1)$, b chosen as (5) and (6) above
Method 1c Stock level is 95% pt. of $W_{b_\tau}(x, 1)$, b chosen as (5) and (7) above

Using method 1a, i.e., stock level is set at 95% point of $W_b(x, 1|x_\tau = X\tau)$ which is chosen as (6) above—but for all values of $b_{\tau-1}$—gave average stock level of 101 which is lower than the AR reference of 112 but stockouts are higher at 9.6%.

Taking note of the value of $b_{\tau-1}$ that has just occurred and using the conditional probabilities as per (i) and (ii) above (i.e., method 1b), average stock level increased slightly over method 1a, at 106 but was still lower than the reference AR solution and stockouts were the lowest at 6.5%.

Applying a probability distribution for all $b_{\tau-1}$ using (5) and (7) above (i.e., method 1c) gave the lowest average stock level at 104 but stockouts rose slightly as compared to method 1b: 6.6% versus 6.5%.

Thus, from these results on this these test data series, we can conclude

- using method 1a (computing $W_b(x, 1|x_\tau = X\tau)$ from a weighted average of recent values) reduces stock by 10% versus a conventional AR method, but doubles the number of stockouts.
- introducing a "conditional probability" approach (methods 1b and 1c) still reduces stock versus the AR method by some 6% but now stockouts are only slightly over the 5% target.

So the conditional probability method of calculating b_τ is worth pursuing.

6 Method for Two-Step Ahead Forecasts

In many real-world situations, the reordering of stock to meet customer demand has to allow time for delivery from the supplying warehouse. Thus, at the end of day 0, one might calculate the stock that would need to be delivered at the end of day 1 in order to meet demand in day 2. We will call this situation a two-step ahead forecast.

In addition to making an estimate of demand in day 2, one needs to take a view as to what might have happened in day 1 so as to compute what stock might be available at the end of day1/start of day 2 before taking into account how much be added to this to meet demand in day 2. To do this, we need two pdfs:

$$W_{b'}(x, 1|x_\tau = X_\tau): \text{ the pdf made at the end of day } \tau \text{ for sales in day } \tau + 1.$$
$$W_{b''}(x, 2|x_\tau = X_\tau): \text{ the pdf made at the end of day } \tau \text{ for sales in day } \tau + 2.$$

Thus, if S is the stock available at the end of day τ (day 0) after, the delivery has been made that night (i.e., S is the total stock available for demand in day 1 (day $\tau + 1$)), and if $W_{b''}^{95}(2)$ is the stock required at the start of day 2 (i.e., the level that if achieved would meet 95% of demand in day 2, after the delivery that is to be made end day1 / start day 2), then the order to be delivered at end of day 1 is expected to be

$$\int_{-|X|_{\max}}^{+|X|_{\max}} \max\{[W_{b''}^{(95)}(2) - \max((S - x), 0)], 0\} \cdot W_{b'}(x, 1|x_\tau = X_\tau)dx \qquad (8)$$

7 Results for Two-Step Ahead Forecasts

In Table 2, we see the results of this two-step ordering process under various methods of choosing the values for b' and b''.

Table 2 Results for two-step ahead forecasts from various conditional probability approaches to b in $W_{b_\tau}(x, 2|x_\tau = X_\tau)$

	ARa	ARb	Method 2a	Method 2b	Method 2c
Av stock	112	160	114	112	109
Stockout %	8.5	3.5	11.4	8.4	8.2

Key
ARa pdf day 1, Normal (AR forecast day 1, variance from past), pdf day 2, Normal (AR forecast day 2, variance from past)
ARb pdf day 1, Delta (95% point of day 1), pdf day 2 Normal (AR forecast day 2, variance from past)
Method 2a (1) $W_{b_\tau}(x, 1)$, b as method 1b, (2) $W_{b_\tau}(x, 2)$, b as method 1a,
Method 2b (1) $W_{b_\tau}(x, 1)$, b as method 1b, (2) $W_{b_\tau}(x, 2)$, b as method 1b,
Method 1c (1) $W_{b_\tau}(x, 1)$, b as method 1b, (2) $W_{b_\tau}(x, 2)$, b as method 1c

These are two different AR solutions which are used as a reference. For the first solution (AR(a)), the pdf for sales in day 1 is taken as Normal with a mean (m1) = the AR one-step forecast and a variance (s_1^2) calculated from past forecast errors. Also, the pdf for sales in day 2 is taken as Normal with mean (m2) = the AR forecast from regression of X_t on $X_{t-2}, X_{t-3}, \ldots, X_{t-8}$ and variance (s_2^2) from the past errors in this two-step forecast. That is in (8) above

$$W_{b'}(x, 1|x_\tau = X_\tau) \text{ is replaced by Normal}(m1, s_1^2) \tag{9}$$

and to find $W_{b''}^{(95)}(2)$

$$W_{b''}(x, 2|x_\tau = X_\tau) \text{ is made } = \text{ Normal}(m2, s_2^2) \tag{10}$$

This reference AR(a) solution generates an average stock level of 112 with a stockouts at 8.5%.

The second AR solution (AR(b)) takes the pdf of sales in day 1 as

$$W_{b'}(x, 1|x_\tau = X_\tau) = \delta(W^{(95)}(1) - x).$$

where $W^{(95)}(1)$ is the 95% point of (9), i.e., we use a single value for our estimate of day 1 sales in calculating this order for delivery end day 1 / start day 2. $W_{b''}^{(95)}(2)$ is again the 95% point of the Normal distribution (10). This method gives an average stock level of 160 and stockouts at 3.5%.

In method 2a, the first application of our methods to this two-step ahead problem, we take

- the pdf for day 1, $W_{b'}(x, 1|x_\tau = X_\tau)$, where b' is calculated as in method 1b above, and
- the pfd for day 2, $W_{b''}(x, 2|x_\tau = X_\tau)$ is as method 1a and the b'' are as defined similar to (3) but of course the $b_{\tau-j}$ are redefined to reflect the "best" $b_{\tau-j}$ such that $X_{(\tau-j)+2}$ is at the 75% point of $W_{b_{\tau-j}}(x, 2|x_{\tau-j} = X_\tau)$.

This method, method 2a, gives an average stock level of 114 but stockouts of 11.4%.

In method 2b, we introduce the conditional probabilities $p(b_j|b_k)$ for $k = 1, 2,$ or 31, in order to calculate the pdf for day 2, i.e.,

(i) If $b_{\tau-1} = b^{(1)}, b^{(2)}$ or $b^{(31)}$ then

$$W(x, 2|x_\tau = X_\tau) = \frac{1}{\mathcal{M}} \sum_{j=1}^{31} \text{prob}(b^{(j)}|b_{\tau-1}) W_{b^j}(x, 2|x_\tau = X_\tau) \tag{11}$$

where $b_{\tau-1} = b^{(k)}$ and $\mathcal{M}$ is a normalizing factor.

(ii) If $b_{\tau-1}$ not $= b^{(1)}, b^{(2)}$ or $b^{(31)}$ then

$$W(x, 2|x_\tau = X_\tau) = \frac{1}{28}\sum_{j=1}^{7}(8-j)W_{b_{\tau-j}}(x, 2|x_\tau = X_\tau) \tag{12}$$

With this method 2b, we get the results shown in column 5 of Table 2: average stock level of 112 and stockouts at 8.4%.

Finally, in method 2c, we introduce a discrete probability distribution for all the b_j, i.e., not only $p(b_j|b_{\tau-1})$ where $b_{\tau-1} = b^{(k)}$ for $k = 1, 2,$ or 31, but $p(b_j) =$ unconditional $p(b_j)$ for all other k. As shown in column 6, this reduces the average stock level slightly further: now 109, but stockouts stay much the same at 8.2%.

8 Conclusion

Instead of setting reorder stock levels through a conventional approach with a forecast sales pdf Normal with mean equal to the AR forecast, we have used a solution to an appropriate Fokker–Planck equation to generate a pdf for sales $W_b(x, t|x_\tau = X_\tau)$ which has a free parameter b. Various methods (e.g., see Refs. [9, 10]) have been tried to generate a b that gives a forecast pdf resulting in a low stock level and a low number of stockouts. In this paper, we have used a conditional pdf for the value of b which depends on the value of the most recent b observed. Applying this method to over 2,000 points in a set of test series, first to a one-step ahead reordering system, reduces (when compared to an AR method) stock levels in these test series by some 7% although stockouts are still 1.6% above the target of 5%. When the reordering system requires orders to be placed at the end of day 0 for delivery at end of day 1 (and thus for use in day 2), using a conditional probability distribution to select the parameter b in the probability distribution for sales in day 2, $W_b(x, 2|x_\tau = X_\tau)$, gives an improvement of 3% in average stock level and also an improvement 0.3% points in stockouts, both compared to a conventional AR forecasting approach.

Thus, one may conclude that selecting b in $W_b(x, t|x_\tau = X_\tau)$ by a conditional probability approach is worthwhile.

Appendix

Fokker–Planck Distributions $\mathbf{W_b}$ Versus Normal (AR(7), $\hat{\mathrm{s}}_\mathrm{e}$)

See Figs. 3 and 4.

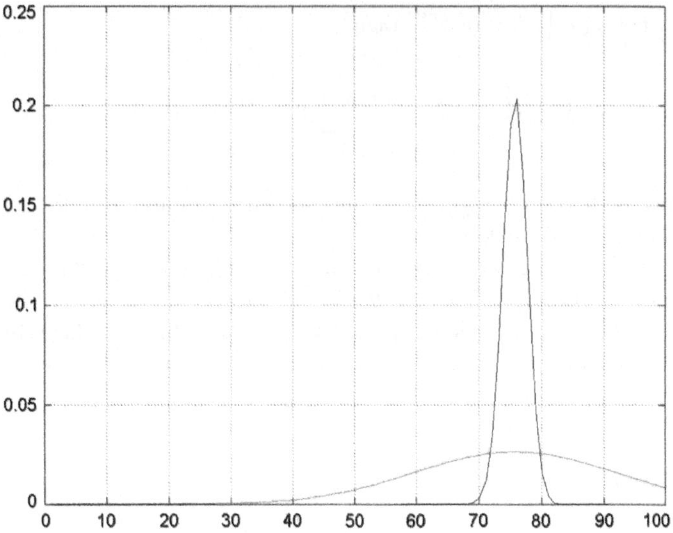

Fig. 3 Plot of (i) W_b and (ii) Normal (AR(7), $\hat{s}_e$)—(ii) is the wider graph b has its smallest value: b_1

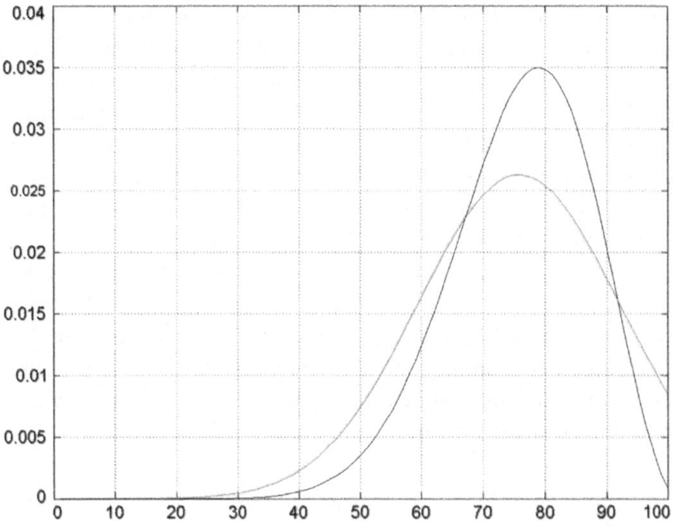

Fig. 4 Plot of (i) W_b and (ii) Normal (AR(7), $\hat{s}_e$)—(ii) is the wider graph b has its fourth value: b_4

References

1. Kantz, H., Schreiber, T.: Analysis, Nonlinear Time Series. Cambridge University Press (2003)
2. Risken, H.: The Fokker Planck Equation. Springer (1996)
3. Noble, P.L., Wheatland, M.S.: Modeling the sunspot number distribution with a Fokker Planck equation. Astrophys. J. (2011)
4. Medino, A.V., et al.: Generalized Langevin equation driven by Levy processes: a probabilistic, numerical and time series based approach. Physica A **391** (2012)
5. Czechowski, Z., Telesca, L.: Construction of a Langevin model from time series with a periodical correlation function: application to wind speed data. Physica A **392** (2013)
6. Sura, P., Barsugli, J.: A note on drift and diffusion parameters from time series. Phys. Lett. A **305** (2002)
7. Kantz, H., Olbrich, E.: Deterministic and probabilistic forecasting in reconstructed spaces
8. Gottschall, J., Peinke, J.: On the definition and handling of different drift and diffusion estimates. New J. Phys. **10** (2008)
9. Montagnon, C.: A closed solution to the Fokker-Planck equation applied to forecasting. Physica A (2014)
10. Montagnon, C.: Forecasting with the Fokker-Planck equation: Bayesian setting of parameters. Physica A (2017)
11. Montagnon, C.: Singular value decomposition and time series forecasting. Ph.D. Imperial College London (2011)

References

[illegible]

Cryptanalysis of a Random Number Generator Based on a Chaotic Ring Oscillator

Salih Ergün

Abstract This paper introduces cryptanalysis of a random number generator (RNG) based on a chaotic ring oscillator. An attack system is proposed to discover the security weaknesses of the chaos-based RNG. Convergence of the attack system is proved using master–slave synchronization scheme. Future evaluation of the RNG is obtained from a scalar time series where the only information available are the structure of the RNG and a scalar time series observed from the chaotic ring oscillator. Simulation and numerical results verifying the feasibility of the attack system are given. It is verified that deterministic chaos itself cannot be pointed out as the source of randomness.

Keywords Cryptanalysis · Random number generator · Chaotic ring oscillator Continuous-time chaos · Synchronization of chaotic systems

1 Introduction

Over the last decades, there has been an increasing emphasis on using tools of information secrecy. Certainly, random number generators (RNGs) have more prominently positioned into the focal point of research as the core component of the secure systems. Although many people are even unaware that they are using them, we use RNGs in our daily business. If we ever obtained money from a bank's cash dispenser, ordered goods over the Internet with a credit card, or watched pay TV we have used RNGs. Public/private key pairs for asymmetric algorithms, keys for symmetric and hybrid cryptosystems, one-time pad, nonces, and padding bytes are created by using RNGs [1].

Being aware of any knowledge on the design of the RNG should not provide a useful prediction about the output bit sequence. Even so, fulfilling the requirements for secrecy of cryptographic applications using the RNG dictates three secrecy criteria

S. Ergün (✉)
TÜBİTAK-Informatics and Information Security Research Center, PO Box 74, 41470 Gebze, Kocaeli, Turkey
e-mail: salih.ergun@tubitak.gov.tr; salih@ergun.com.tr

I. Rojas et al. (eds.), *Time Series Analysis and Forecasting*, Contributions to Statistics, https://doi.org/10.1007/978-3-319-96944-2_2

as a "must": 1. The output bit sequence of the RNG must pass all the statistical tests of randomness [2]; 2. The previous and the next random bit must be unpredictable [3] and; 3. The same output bit sequence of the RNG must not be able to be reproduced [4].

An important principle of modern cryptography is the Kerckhoff's assumption [2], which states that the overall security of any cryptographic system entirely depends on the security of the key, and assumes that all the other parameters of the system are publicly known. Cryptanalysis is complementary of cryptography. The interaction between these two branches of cryptology forms modern cryptography which has become strong only because of security analysis that reveals the weaknesses in the existing cryptographic systems.

There are four fundamental techniques for random number generation: 1. Amplification of a noise source [5, 6]; 2. Jittered oscillator sampling [1, 7]; 3. Discrete-time chaotic maps [8–10] and; 4. Continuous-time chaotic oscillators [11, 12]. Although the use of discrete-time chaotic maps in the realization of RNG has been widely accepted for a long period of time [8], it has been shown during the last decade that continuous-time chaotic oscillators can also be used to realize RNGs [11, 12]. In particular, a "true" RNG based on a chaotic ring oscillator has been proposed in [11]. In this paper, we target the RNG reported in [11] and further propose an attack system to discover the security weaknesses of the targeted system.

The strength of a cryptographic system almost depends on the strength of the key used or in other words on the difficulty for an attacker to predict the key. On the contrary to recent RNG design [12], where the effect of noise generated by circuit components was analyzed to address security issue, the target random number generation system [11] pointed out the deterministic chaos itself as the source of randomness.

The organization of the paper is as follows. In Sect. 2, the target RNG system is described in detail; In Sect. 3, an attack system is proposed to cryptanalyze the target system and its convergence is proved; Sect. 4 illustrates the numerical results with simulations which is followed by concluding remarks.

2 Target System

Chaotic systems are categorized into two groups: discrete time or continuous time, respectively regarding on the evolution of the dynamical systems. In target random number generation system [11], a simple continuous-time chaotic circuit is utilized as the core of the RNG. This chaotic system is derived from two ring oscillators coupled by diodes [11].

Using the normalized quantities: $x_n = v_n/V_{th}$, $y_d = i_d R_d/V_{th}$, $t = T/RC$, $\alpha = G_m R$, $\beta = C/(C + C_1)$, $\gamma = R/R_1$, $\delta = R/R_d$, and $\varepsilon = R/R_2$, the equations of the chaotic circuit transform into Eq. 1:

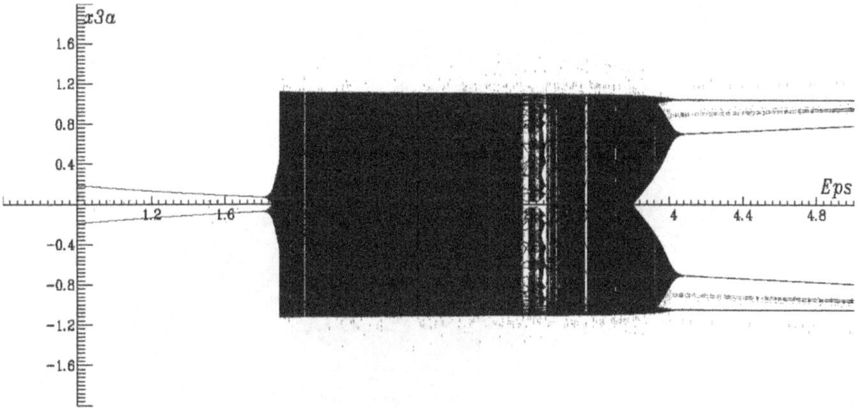

Fig. 1 Bifurcation diagram against the parameter ε

$$
\begin{aligned}
\dot{x}_{1a1} &= -x_{1a1} - \alpha x_{3a1} \\
\dot{x}_{2a1} &= -x_{2a1} - \alpha x_{1a1} \\
\dot{x}_{3a1} &= -\beta(\gamma + 1)x_{3a1} - \alpha\beta x_{2a1} - \beta\delta y_d \\
\dot{x}_{1b1} &= -x_{1b1} - \alpha x_{3b1} \\
\dot{x}_{2b1} &= -x_{2b1} - \alpha x_{1b1} \\
\dot{x}_{3b1} &= -(\varepsilon + 1)x_{3b1} - \alpha x_{2b1} + \delta y_d
\end{aligned}
\tag{1}
$$

where

$$
y_d = \begin{cases} x_{3a1} - x_{3b1} - 1 & for\ x_{3a1} - x_{3b1} > 1 \\ 0 & for\ |x_{3a1} - x_{3b1}| \leq 1 \\ x_{3a1} - x_{3b1} + 1 & for\ x_{3a1} - x_{3b1} < -1 \end{cases}
$$

The equations in (1) generate chaos for different sets of parameters. Bifurcation diagram against the parameter $\varepsilon = R/R_2$ is constructed. As shown in Fig. 1, the given system generates chaos for the parameter ε over a wide range ($1.81 < \varepsilon < 3.13$) which points out that the nonideal effect on the performance of the chaotic system is not critical. The chaotic attractor ($Horizontal : x_{3b1}$, $Vertical : x_{2b1}$) given in Fig. 2 is obtained from the numerical analysis of the system with $\alpha = 3.7$, $\beta = 0.1$, $\gamma = 1$, $\delta = 100$ and $\varepsilon = 2.5$.

Target random number generation mechanism is described in [11] where bit generation method is based on jittered oscillator sampling technique. As depicted in [11], the output of a fast oscillator is sampled on the rising edge of a jittered slower clock using a D flip-flop where the jittered slow clock is realized by a chaotic ring oscillator circuit.

In this design, if the fast and the slower clock frequencies are known as well as the starting phase difference ΔT, the output of the fast oscillator, sampled at the rising edge of the jittered slower clock, can be predicted. It can be shown that the output

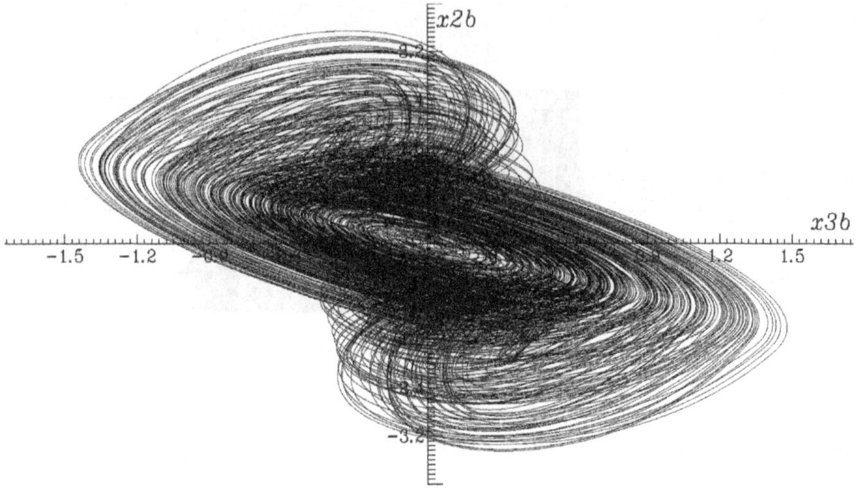

Fig. 2 Numerical analysis results of the chaotic system for $\alpha = 3.7, \beta = 0.1, \gamma = 1, \delta = 100$, and $\varepsilon = 2.5$

bit sequence $S_{(bit)i}$ is the inverse of least significant bit of the ratio between the total periods of the jittered slower clock and period of the fast clock:

$$S_{(bit)i} = \left(\frac{\lfloor \frac{(\sum_{j=1}^{i} T_{slow\,j}) - \Delta T}{T_{fast}/2} \rfloor mod2}{(2d_{fast})} \right)' \tag{2}$$

where $T_{fast} = \frac{1}{f_{fast}}$, f_{fast}, d_{fast} are the period, frequency and the duty cycle of the fast clock, respectively, and the periods of the jittered slower clock $T_{slow\,j}$ are obtained at times t satisfying:

$$s(t) = x_{3a1}(t) = Q\ with\ \frac{ds}{dt} > 0 \tag{3}$$

where $x_{3a1}(t)$ is the chaotic signal and Q is the logic threshold of the D flip-flop. We have numerically verified that, for high $\frac{f_{fast}}{f_{slow\,center}}$ ratios, the effect of ΔT becomes negligible and the mean value (m_{output}) of the output sequence S_{bit} approaches the fast clock duty cycle d_{fast} where frequency of the chaotic signal, corresponding to mean frequency of the jittered slower clock $f_{slow\,center}$, determines the throughput data rate (f_{rng}). It should be noted that anyone who knows the chaotic signal output can reproduce the same output bit sequence.

The authors of [11] have preferred to use NIST 800-22 [13, 18] statistical test suite in order to analyze output randomness of their RNG design. However, Big Crush [14] and Diehard [15] statistical test suites which are available at the publication date of target paper were not applied to output bit stream of the RNG. It should be noted that, the target random number generation system [11] does not satisfy the first secrecy criteria, which states that "RNG must pass all the statistical tests of randomness."

3 Attack System

After the seminal work on chaotic systems by Pecora and Carroll [16], synchronization of chaotic systems have been an increasingly active area of research [17]. In this paper, the convergence of attack and target systems is numerically demonstrated using master–slave synchronization scheme [17]. In order to provide cryptanalysis of the target random number generation system, an attack system is proposed which is given by Eq. 4:

$$\begin{aligned}
\dot{x_{1a2}} &= -x_{1a2} - \alpha x_{3a2} \\
\dot{x_{2a2}} &= -x_{2a2} - \alpha x_{1a2} + c(x_{2a1} - x_{2a2}) \\
\dot{x_{3a2}} &= -\beta(\gamma + 1)x_{3a2} - \alpha\beta x_{2a2} - \beta\delta y_d \\
\dot{x_{1b2}} &= -x_{1b2} - \alpha x_{3b2} \\
\dot{x_{2b2}} &= -x_{2b2} - \alpha x_{1b2} \\
\dot{x_{3b2}} &= -(\varepsilon + 1)x_{3b2} - \alpha x_{2b2} + \delta y_d
\end{aligned} \tag{4}$$

$$where$$

$$y_d = \begin{cases} x_{3a2} - x_{3b2} - 1 & for\ x_{3a2} - x_{3b2} > 1 \\ 0 & for\ |x_{3a2} - x_{3b2}| \leq 1 \\ x_{3a2} - x_{3b2} + 1 & for\ x_{3a2} - x_{3b2} < -1 \end{cases}$$

where c is the coupling strength between the target and attack systems. The only information available are the structure of the target random number generation system and a scalar time series observed from x_{2a1}.

In this paper, we construct the attack system expressed by Eq. 4 that synchronizes ($x_{2a2} \to x_{2a1}$ for $t \to \infty$) where t is the normalized time. We define the error signals as $e_{x1a} = x_{1a1} - x_{1a2}$, $e_{x2a} = x_{2a1} - x_{2a2}$, and $e_{x3a} = x_{3a1} - x_{3a2}$ where the aim of the attack is to design the coupling strength such that $|e(t)| \to 0$ as $t \to \infty$.

The master–slave synchronization of attack and target systems is verified by the conditional Lyapunov Exponents (CLEs), and as first reported in [16], is achievable if the largest CLE is negative. Largest CLEs are calculated for different values of coupling strength c while a scalar time series is observable from x_{2a1}. When c is greater than 0.42 then the largest CLE is negative and hence identical synchronization of target and attack systems starting with different initial conditions is achieved and stable [16]. (Largest conditional Lyapunov Exponent is -0.0108588 for $c = 0.5$). However for c is less than 0.42, largest CLE is positive and identical synchronization is unstable.

Log $|e_{x1a}(t)|$, Log $|e_{x2a}(t)|$, and Log $|e_{x3a}(t)|$ are shown in Fig. 3, Fig. 4, and Fig. 5 respectively, for $c = 3$, where the synchronization effect is better than that of $c = 0.5$. As shown in the given figures, the attack system converges to target system and master–slave synchronization is achieved in less than $145t$.

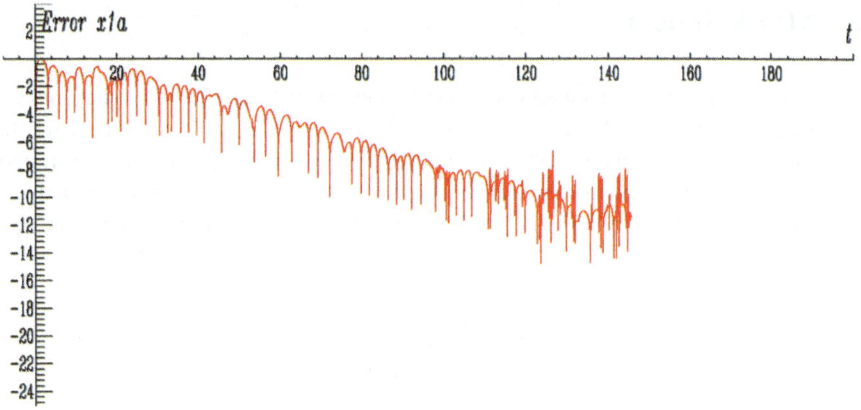

Fig. 3 Synchronization error Log $|e_{x1a}(t)|$

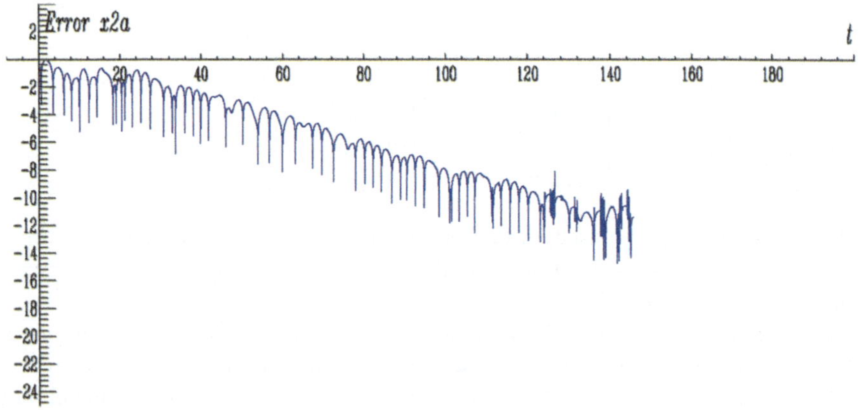

Fig. 4 Synchronization error Log $|e_{x2a}(t)|$

4 Numerical Results

We numerically demonstrate the proposed attack system using a fourth-order Runge–Kutta algorithm with fixed step size and its convergence is illustrated in Fig. 3, Fig. 4, and Fig. 5, respectively. Numerical results of $x_{1a1} - x_{1a2}$, $x_{2a1} - x_{2a2}$, and $x_{3a1} - x_{3a2}$ are also given in Fig. 6, Fig. 7, and Fig. 8, respectively illustrating the unsynchronized behavior and the synchronization of target and attack systems.

It is observed from the given figures that master–slave synchronization is achieved and stable. As shown by black lines in these figures, no synchronous phenomenon is observed before $145t$. In time, the proposed attack system converges to the target system and identical synchronization is achieved where colored lines depict synchronized behaviors of chaotic states in Fig. 6, Fig. 7, and Fig. 8, respectively.

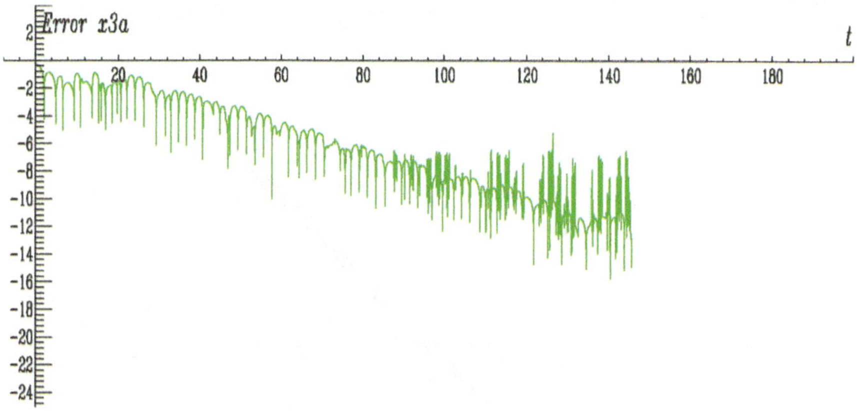

Fig. 5 Synchronization error Log $|e_{x3a}(t)|$

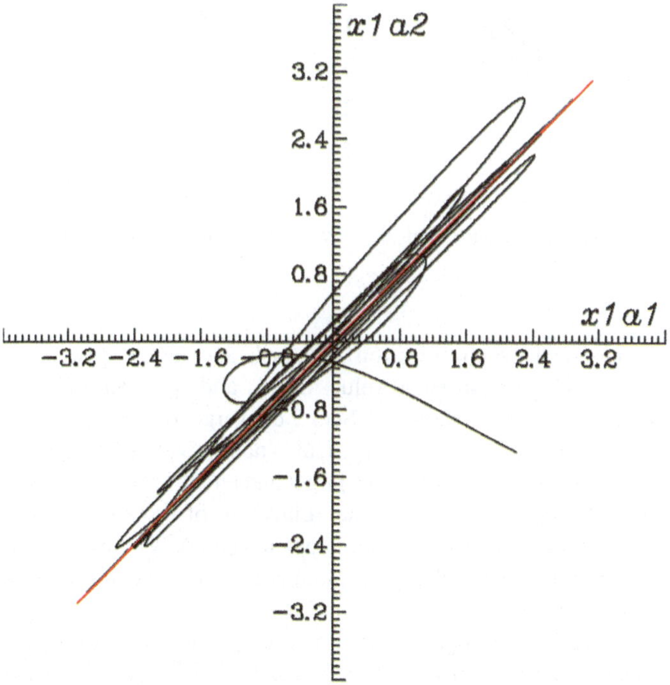

Fig. 6 Unsynchronized behavior and the synchronization of target and attack systems: $Horizontal : x_{1a1}$, $Vertical : x_{1a2}$

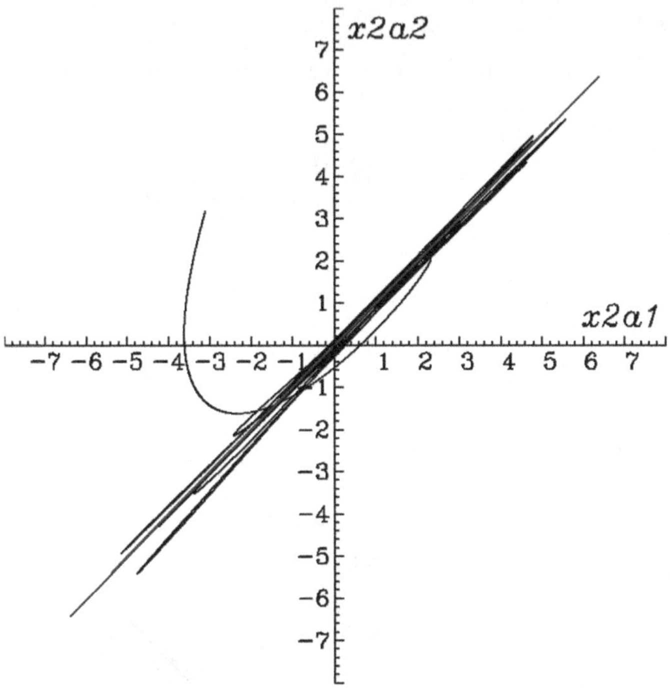

Fig. 7 Unsynchronized behavior and the synchronization of target and attack systems: $Horizontal : x_{2a1}$, $Vertical : x_{2a2}$

Since the identical synchronization of attack and target systems is achieved ($x_{2a2} \rightarrow x_{2a1}$) in $145t$, the estimated values of x_{3a1} and $S_{(bit)i}$ bit which is generated according to the procedure explained in Sect. 2 converge to their corresponding fixed values. As a result, it is obvious that identical synchronization of chaotic systems is achieved and hence output bit streams of target and attack systems are synchronized.

It is clearly shown master–slave synchronization of proposed attack system is achieved. Hence, output bit sequences of target and attack systems are synchronized. In conclusion, cryptanalysis of the target random number generation system not only predicts the previous and the next random bit but also demonstrates that the same output bit sequence of the target random number generation system can be reproduced. As a result, the target random number generation system [11] satisfies neither the second nor the third secrecy criterion that an RNG must satisfy. It should be noted that deterministic chaos itself cannot be pointed out as the source of randomness.

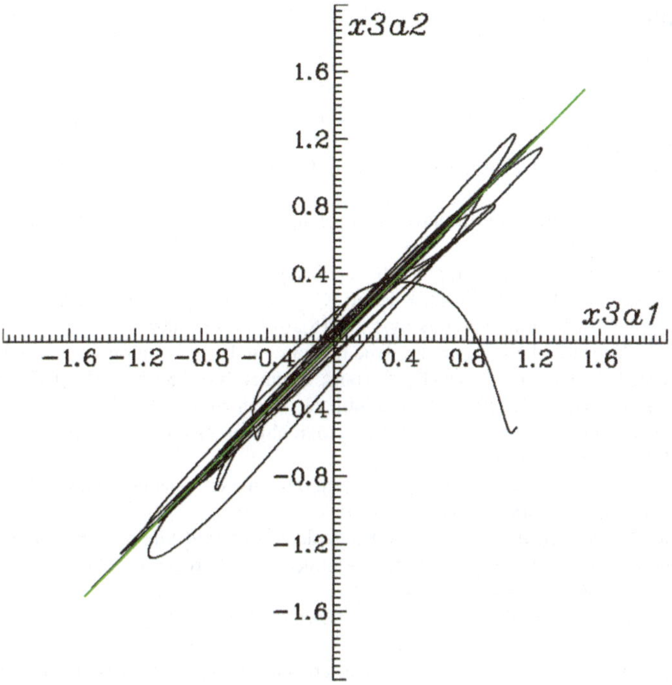

Fig. 8 Unsynchronized behavior and the synchronization of target and attack systems: *Horizontal* : x_{3a1}, *Vertical* : x_{3a2}

5 Conclusions

In this paper, we propose a cryptanalysis method for a chaos-based random number generator (RNG). An attack system is introduced to discover the security weaknesses of the chaos-based RNG and its convergence is proved using master–slave synchronization scheme. Although the only information available are the structure of the target RNG and a scalar time series observed from the target chaotic system, identical synchronization of target and attack systems is achieved and hence output bit streams are synchronized. The target RNG does not fulfill Big Crush and Diehard statistical test suites, the previous and the next bit can be predicted, while the same output bit sequence of the RNG can be reproduced. Simulation results presented in this paper not only verify the feasibility of the proposed method but also encourage its use for the cryptanalysis of the other chaos-based RNG designs. The proposed attack renders the generated bit streams predictable, thereby qualifying the target RNG to be used as a not true but pseudorandom source.

References

1. Jun, B., Kocher, P.: The Intel Random Number Generator, Cryptography Research, Inc., white paper prepared for Inter Corp. Apr 1999. http://www.cryptography.com/resources/whitepapers/IntelRNG.pdf
2. Menezes, A., Van Oorschot, P., Vanstone, S.: Handbook of Applied Cryptology. CRC Press (1996)
3. Schrift, A.W., Shamir, A.: On the universality of the next bit test. In: Proceeding of the CRYPTO, pp. 394–408 (1990)
4. Schneier, B.: Applied Cryptography 2th edn. Wiley (1996)
5. Petrie, C.S., Connelly, J.A.: A noise-based IC random number generator for applications in cryptography. IEEE Trans. Circuits Syst. I **47**(5), 615–621 (2000)
6. Göv, N.C., Mıhçak, M.K., Ergün, S.: True random number generation via sampling from flat band-limited Gaussian processes. IEEE Trans. Circuits Syst. I **58**(5), 1044–1051 (2011)
7. Bucci, M., Germani, L., Luzzi, R., Trifiletti, A., Varanonuovo, M.: A high speed oscillator-based truly random number source for cryptographic applications on a smartcard IC. IEEE Trans. Comput. **52**, 403–409 (2003)
8. Stojanovski, T., Kocarev, L.: Chaos-based random number generators-Part I: Analysis. IEEE Trans. Circuits Syst. I **48**(3), 281–288 (2001)
9. Callegari, S., Rovatti, R., Setti, G.: Embeddable ADC-based true random number generator for cryptographic applications exploiting nonlinear signal processing and chaos. IEEE Trans. Signal Process. **53**(2), 793–805 (2005)
10. Stojanovski, T., Pihl, J., Kocarev, L.: Chaos-based random number generators-Part II: Practical realization. IEEE Trans. Circuits Syst. I **48**(3), 382–385 (2001)
11. Çicek, İ., Dündar, G.: A hardware efficient chaotic ring oscillator based true random number generator. In: Proceedings of IEEE International Conference on Electronics, Circuits and Systems, pp. 430–433 (2011)
12. Ergün, S., Güler, Ü., Asada, K.: A high speed IC truly random number generator based on chaotic sampling of regular waveform. IEICE Trans. Fundam. Electron. Commun. Comput. Sci. **E94-A**(1) 180–190 (2011)
13. National Institute of Standard and Technology, FIPS PUB 140-2, Security Requirements for Cryptographic Modules, NIST, Gaithersburg, MD 20899 (2001)
14. L'Ecuyer, P.: Universit'e de Montr'eal., "Empirical Testing of Random Number Generators" (2002). http://www.iro.umontreal.ca/~lecuyer/
15. Marsalgia, G.: Diehard: a battery of tests of randomness (1997). http://stat.fsu.edu/~geo/diehard.htm
16. Pecora, L.M., Carroll, T.L.: Synchronization in chaotic systems. Phys. Rev. Lett. **64**(8), 821–824 (1990)
17. Hasler, M.: Synchronization principles and applications. In: Toumazou, C. (ed.) Tutorials IEEE International Symposium on Circuits and Systems (ISCAS '94), pp. 314–327. London, England (1994)
18. National Institute of Standard and Technology, "A Statistical Test Suite for Random and Pseudo Random Number Generators for Cryptographic Applications". http://csrc.nist.gov/groups/ST/toolkit/rng

Further Results on a Robust Multivariate Time Series Analysis in Nonlinear Models with Autoregressive and t-Distributed Errors

Hamza Alkhatib, Boris Kargoll and Jens-André Paffenholz

Abstract We investigate a time series model which can generally be explained as the additive combination of a multivariate, nonlinear regression model with multiple univariate, covariance stationary autoregressive (AR) processes whose white noise components obey independent scaled t-distributions. These distributions enable the stochastic modeling of heavy tails or outlier-afflicted observations and present the framework for a partially adaptive, robust maximum likelihood (ML) estimation of the deterministic model parameters, of the AR coefficients, of the scale parameters, and of the degrees of freedom of the underlying t-distributions. To carry out the ML estimation, we derive a generalized expectation maximization (GEM) algorithm, which takes the form of linearized, iteratively reweighted least squares. In order to derive a quality assessment of the resulting estimates, we extend this GEM algorithm by a Monte Carlo based bootstrap algorithm that enables the computation of the covariance matrix with respect to all estimated parameters. We apply the extended GEM algorithm to a multivariate global navigation satellite system (GNSS) time series, which is approximated by a three-dimensional circle while taking into account the colored measurement noise and partially heavy-tailed white noise components. The precision of the circle model fitted by the GEM algorithm is superior to that of the previous standard estimation approach.

H. Alkhatib (✉) · B. Kargoll · J.-A. Paffenholz
Geodätisches Institut, Leibniz Universität Hannover, Nienburger Str. 1, 30167 Hannover, Germany
e-mail: alkhatib@gih.uni-hannover.de
URL: https://www.gih.uni-hannover.de

B. Kargoll
e-mail: kargoll@gih.uni-hannover.de

J.-A. Paffenholz
e-mail: paffenholz@gih.uni-hannover.de

I. Rojas et al. (eds.), *Time Series Analysis and Forecasting*,
Contributions to Statistics, https://doi.org/10.1007/978-3-319-96944-2_3

Keywords Multivariate time series · Nonlinear regression model · AR process Scaled t-distribution · Partially adaptive estimation · Robust parameter estimation GEM algorithm · Bootstrapping · GNSS time series

1 Introduction

The estimation of the parameters of a measurement process is an important part of modeling and monitoring in many applications. Unfortunately, the measured data may contain multiple outliers. In such a case, the probability density function (pdf) of the random deviations is presumed to be heavy-tailed. Robust parameter estimation is then used to minimize the influence of outliers and to avoid the removal of measurements. Robust approaches utilize estimators that are less affected by outliers than the least-squares (LS) estimator. Maximum likelihood (ML) type estimators (M-estimators) have been developed to be robust to outliers, e.g., the Huber M-estimator [1]. These estimators basically take advantage of using reduced weights for outliers. In addition, (Student's) t regression models were introduced, which assume scaled t-distributed errors in robust ML estimation (cf. [2]). The computation of corresponding ML estimates can be easily realized by the iterative reweighted least-square (IRLS) approach, as shown in [3]. The variances of the random deviations are rescaled by means of the resulting weights conforming to their locations under the density function. This procedure, a so-called (partially) adaptive estimator, allows for estimating the unknown degree of freedom (d.o.f.) of the t-distribution jointly with the regression parameters and the scale parameter.

Multivariate multiple regression is a technique that estimates more than one (possibly nonlinear) regression model with more than one outcome variable (deterministic regression function) and with random deviations that are generally assumed to obey the normal distribution. Alternatively, [4] presumed the multivariate t-distribution with unknown scale matrix and unknown d.o.f., and demonstrated different expectation maximization (EM) algorithms to estimate the unknown model parameters jointly with the scale matrix and the d.o.f. efficiently. In previous contributions, [5, 6] demonstrated that the expectation *conditional* maximization (ECM) and the ECM *either* (ECME) algorithms accelerate the convergence of the EM algorithm noticeably. To deal with models that do not allow for closed-form solutions by EM, the optimization principle of *generalized* expectation maximization (GEM) was suggested by [3]. This approach approximates the maximum inside of every EM step instead of reaching the maximum completely. GEM algorithms applying Newton–Raphson steps have been used commonly (see e.g. [7]). To handle nonlinear functional models, [2, 8] presented an IRLS algorithm for GEM with Gauss–Newton steps.

Many multivariate time series are expected to be correlated over time, which makes (partially adaptive) parameter estimation more challenging. Autocorrelations appear in many types of sensor data. For instance, inertial sensor data, satellite gravity gradiometry data, and global navigation satellite system (GNSS) data have random deviations that show colored noise characteristics (see, e.g., [9–12]). Such sensors can also produce multiple outliers, so that robust parameter estimation is preferable over the classical LS approach. In order to take the autocorrelatedness of random deviations in time series into account, [13] extended the aforementioned partially adaptive estimator for linear regression models based on the scaled t-distribution. They assumed autoregressive (AR) random deviations in a univariate time series, with the white noise components of the AR process following independently and identically a scaled t-distribution.

We demonstrated in [14] the extension of the preceding univariate, linear model to a multivariate and nonlinear (differentiable) regression model. Regarding the setup of the AR model, we simplified the complex general case of stochastically dependent time series to the case where each time series component is associated with a univariate AR process of individual order, independently of the AR processes of the other components. Thus, cross-correlations between the different colored noise processes are currently not considered. In this paper, we add two important issues that have not been addressed in [14]. First, we replace the algorithm regarding the estimation of the d.o.f. of the underlying t-distribution by an interval-based search approach, which generally is more reliable than the previously used conventional search algorithm.

Second, we extend the previous GEM algorithm by a Monte Carlo (MC) bootstrap algorithm (in the sense of [15, 16]) for computing the covariance matrix of all unknown model parameters. In the context of geodetic regression models with autocorrelated measurements, [16] employed for this purpose MC strategies, as extensions of an iterative preconditioned conjugate gradients multiple adjustment (PCGMA) algorithm, to estimate an accurate covariance matrix for the model parameters (including optimal weighting and regularization parameters). Bootstrap models are also well suited to time series analysis with combined regression and AR models [17]. The method used in the current paper is based on an MC bootstrap approach to covariance matrix estimation in the context of EM algorithms for missing data models, as explained in [7].

The remainder of the paper is structured as follows. In Sect. 2, we describe the general time series model. In Sect. 3, we derive a GEM algorithm for parameter estimation and provide a bootstrap algorithm concerning the associated covariance matrix. In Sect. 4, we analyze three-dimensional (3D) time series of GNSS observations, where the regression model is given by a 3D circle.

2 The Observation Model

The time series model we consider in the following is

$$Y_{k,t} = h_{k,t}(\xi_1, \ldots, \xi_m) + E_{k,t}, \tag{1}$$

$$E_{k,t} = \alpha_{k,1} E_{k,t-1} + \cdots + \alpha_{k,p_k} E_{k,t-p_k} + U_{k,t}, \tag{2}$$

$$U_{k,t} \overset{\text{ind.}}{\sim} t_{\nu_k}(0, \sigma_k^2), \tag{3}$$

where $k \in \{1, \ldots, N\}$ is a sensor component and $t \in \{1, \ldots, n\}$ a time instance. In the first equation, $Y_{k,t}$ is an observable or measurand described by a possibly nonlinear function $h_{k,t}(\xi_1, \ldots, \xi_m)$ of unknown parameters $\boldsymbol{\xi} = [\xi_1 \ldots \xi_m]^T$ and a random deviation or measurement error $E_{k,t}$. Thus, the random vectors $\mathbf{Y}_t = [Y_{1,t} \ldots Y_{N,t}]^T$ form a multivariate time series. For each sensor component, the random deviations are modeled by a covariance stationary AR(p_k) process (2) with component-dependent, given order p_k and unknown coefficients $\boldsymbol{\alpha}_k = [\alpha_{k,1} \ldots \alpha_{k,p_k}]^T$. By employing different AR processes, we allow for different forms of colored measurement noise in the various components, while assuming cross-correlations between errors of different sensor components to be negligible. Furthermore, according to (3), the error variables $U_{k,1}, \ldots, U_{k,n}$ of each component independently and identically follow a t-distribution with component-dependent, unknown scale factor σ_k^2 and unknown d.o.f. ν_k. Thus, we allow these white noise components to have different (unknown) levels of variability and individual (unknown) tail or outlier characteristics.

For brevity of expressions, we treat the regression part as the vector-valued functions $\mathbf{h}_t(\boldsymbol{\xi}) = [h_{1,t}(\boldsymbol{\xi}) \ldots h_{N,t}(\boldsymbol{\xi})]^T$. Let us also stack all of the unknown model parameters $\boldsymbol{\xi}, \boldsymbol{\alpha}_1, \ldots, \boldsymbol{\alpha}_N, \sigma_1^2, \ldots, \sigma_N^2, \nu_1, \ldots, \nu_N$ in the single vector $\boldsymbol{\theta}$. In addition, we write the colored and white noise components as the multivariate time series $\mathbf{E}_t = [E_{1,t} \ldots E_{N,t}]^T$ and $\mathbf{U}_t = [U_{1,t} \ldots U_{N,t}]^T$, respectively. The lower case $\mathbf{y}_t$, $\mathbf{e}_t$, and $\mathbf{u}_t$ constitute real-valued vectors as certain numerical realizations of the random vectors $\mathbf{Y}_t$, $\mathbf{E}_t$ and $\mathbf{U}_t$, respectively.

To estimate the parameters $\boldsymbol{\theta}$, we construct a likelihood function given each white noise path $\mathbf{u}_k$. Using the definition of the scaled t-distribution (cf. [2]) and the independence assumption in (3), we can write the joint pdf of each white noise series in the factorized form

$$f(\mathbf{u}_k) = \prod_{t=1}^{n} \frac{\Gamma\left(\frac{\nu_k+1}{2}\right)}{\sqrt{\nu_k \pi \sigma_k^2}\,\Gamma\left(\frac{\nu_k}{2}\right)} \left[1 + \left(\frac{u_{k,t}}{\sigma_k}\right)^2 / \nu_k\right]^{-\frac{\nu_k+1}{2}}, \tag{4}$$

where Γ is the gamma function. Furthermore, due to the lack of cross-correlations, the product of these pdfs yields the joint pdf $f(\mathbf{u}) = f(\mathbf{u}_1) \ldots f(\mathbf{u}_N)$. The idea is now to introduce the functional model (1) and the AR processes (2) into that pdf. Assuming the AR processes to be invertible, we can write (2) also as

$$u_{k,t} = e_{k,t} - \alpha_{k,1} e_{k,t-1} - \cdots - \alpha_{k,p_k} e_{k,t-p_k}. \tag{5}$$

Introducing the lag operator notation $L^j e_t = e_{t-j}$ and the lag polynomial $\boldsymbol{\alpha}_k(L) = 1 - \alpha_{k,1} L - \cdots - \alpha_{k,p_k} L^{p_k}$, we can abbreviate the right-hand side of (5) by $\boldsymbol{\alpha}_k(L) e_{k,t}$. Here, we may view $\boldsymbol{\alpha}_k(L)$ as a *decorrelation filter* since it transforms the colored noise series $e_{k,1}, \ldots, e_{k,n}$ into the white noise series $u_{k,1}, \ldots, u_{k,n}$. Let us now "invert" also the functional model (1) and substitute the resulting error equations $E_{k,t} = Y_{k,t} - h_{k,t}(\xi_1, \ldots, \xi_m)$ into (5). Then, the natural logarithm of the (factorized) pdf $f(\mathbf{u}_1) \ldots f(\mathbf{u}_N)$ in (4) takes the form

$$\log f(\mathbf{u}) = \sum_{k=1}^{N} \left(n \log \left[\frac{\Gamma\left(\frac{\nu_k+1}{2}\right)}{\sqrt{\nu_k \pi \sigma_k^2}\,\Gamma\left(\frac{\nu_k}{2}\right)} \right] - \frac{\nu_k + 1}{2} \sum_{t=1}^{n} \log \left[1 + \left(\frac{\boldsymbol{\alpha}_k(L)(y_{k,t} - h_{k,t}(\boldsymbol{\xi}))}{\sigma_k} \right)^2 / \nu_k \right] \right), \tag{6}$$

which (as a function of all unknown model parameters and observations) we define to be the log-likelihood function $\log \mathcal{L}(\boldsymbol{\theta}; \mathbf{y})$. As the inverted AR process (5) involves time instances $t = 0, -1, \ldots$ at which no data are available, we make the assumption that the associated observations and noise realizations all take the value 0. When fixing the "initial conditions" is this standard manner, $\log \mathcal{L}$ is sometimes referred to as a *conditional log-likelihood function.*

This function, however, will only be used directly for the ML estimation of the degrees of freedom ν_k as the likelihood equations for all other parameters are too complicated and too cumbersome to solve. We can obtain closed-form expressions for the parameters $\boldsymbol{\xi}$, $\boldsymbol{\alpha}_k$, and σ_k^2 if we replace the distributional assumption (3) by the equivalent model (cf. [2])

$$U_{k,t} \overset{\text{ind.}}{\sim} N(0, \sigma_k^2 / w_{k,t}) \mid w_{k,t}, \tag{7}$$

$$W_{k,t} \overset{\text{ind.}}{\sim} \frac{\chi^2_{\nu_k}}{\nu_k}. \tag{8}$$

Here, the random variables $W_{k,t}$ are unobservable data in the form of *latent variables*, which later take the role of observation weights in an IRLS algorithm. Small weights are associated with errors in the tails (i.e., with outliers) and lead to an increase of the variance through its rescaling $\sigma_k^2 / w_{k,t}$. For this reason, the outlier model (7)–(8)

is sometimes called a *variance-inflation model*. The two independence assumptions in that model allow for factorization of the joint pdf of the white noise and latent variables, so that we obtain for the logarithm

$$\log f(\mathbf{u}, \mathbf{w}) = \text{const.} - \frac{n}{2}\sum_{k=1}^{N}\log(\sigma_k^2) + \frac{n}{2}\sum_{k=1}^{N}\nu_k \log\left(\frac{\nu_k}{2}\right) - n\sum_{k=1}^{N}\log\Gamma\left(\frac{\nu_k}{2}\right)$$
$$-\sum_{k=1}^{N}\sum_{t=1}^{n}\frac{1}{2}\left[\nu_k + \left(\frac{\boldsymbol{\alpha}_k(L)(y_{k,t} - h_k(\boldsymbol{\xi}))}{\sigma_k}\right)^2\right] w_{k,t} + \sum_{k=1}^{N}\sum_{t=1}^{n}\frac{1}{2}(\nu_k - 1)\log w_{k,t} \tag{9}$$

(see [14] for details). This function serves in the sequel as the proxy log-likelihood function $\log\mathcal{L}(\boldsymbol{\theta}; \mathbf{y}, \mathbf{w})$, instead of $\log\mathcal{L}(\boldsymbol{\theta}; \mathbf{y})$. The idea of replacing or imputing the "missing data" $w_{k,t}$ by conditional expectations based on the stochastic model for $W_{k,t}$ leads us to the following GEM algorithm.

3 Generalized EM Algorithm

EM Algorithm for Estimating Unknown Parameters

In [14], we developed a GEM algorithm, which consists of an E- and an M-step. The Q-function needed in the E-step is defined as the conditional expectation of the foregoing log-likelihood function, given measurement results $\mathbf{y}$ and parameter values in the ith iteration step $\boldsymbol{\theta}^{(i)}$, that is,

$$Q(\boldsymbol{\theta}|\boldsymbol{\theta}^{(i)}) = E_{\mathbf{W}|\mathbf{y};\boldsymbol{\theta}^{(i)}}\{\log\mathcal{L}(\boldsymbol{\theta}; \mathbf{y}, \mathbf{W})\}. \tag{10}$$

In order to carry out the E-step (required for the determination of the weights within IRLS), initial parameter values are needed. In addition, we choose within the first iteration step equal weights $w_{k,t}^{(0)} = 1$ for each component of the multivariate time series.

To carry out the M-step, we maximize the Q-function given in (10) by determining the first partial derivatives of the Q-function with regard to the unknown parameters $\boldsymbol{\xi}$, $\boldsymbol{\alpha}_k$, σ_k^2, and ν_k grouped in $\boldsymbol{\theta}$ and subsequently setting these derivatives equal to zero. The current iteration step is denoted by $(i+1)$ and leads to the parameter solution $\boldsymbol{\theta}^{(i+1)}$, which replaces the solution $\boldsymbol{\theta}^{(i)}$ of the previous iteration step. As the determination of the derivatives with respect to the parameters $\boldsymbol{\xi}$ involves the linearization of the functions $h_{k,t}$, the maximum is approximated by taking a Gauss–Newton step (of length γ). Thus, we have a GEM algorithm, which is summarized in Algorithm 1. Important steps are tagged with a circle around a letter–number combination. These steps are described in detail in the following lines:

(N.1) This update is added entirely or partially to the trial solution (in the sense of a Gauss–Newton step with step size $\gamma \in (0, 1])$. In case the Q-function is decreased by the current step length (which is initialized by $\gamma = 1$ in each iteration step), the step length is halved and the Q-function evaluated again at the current estimates. The step length is reduced until the Q-function increases and thus approaches the maximum, as required by GEM.

(N.2) Since we aim for covariance stationary and invertible AR processes, it is necessary to determine whether all roots of $\boldsymbol{\alpha}_k^{(i+1)}(z) = 0$ are located within the unit circle. In case this is not true, we stabilize the preceding polynomial by mirroring all roots with magnitude exceeding 1 into the unit circle (cf. [18]), using MATLAB's `polystab` routine.

(N.3) The estimates $\nu_1^{(i+1)}, \ldots, \nu_N^{(i+1)}$ constitute the zeros of these equations, which are to be found numerically. We use a reliable zero search based on the one-dimensional interval Newton method described in [19].

Bootstrapping Algorithm for Determining the Covariance Matrix

We have presented in section "EM Algorithm for Estimating Unknown Parameters" a GEM algorithm, which allows for the estimation of the deterministic model parameters, of the AR coefficients, of the scale parameters, and of the d.o.f.s of the underlying t-distributions. Oftentimes, not only the estimated parameter is of interest, but also their (variance-)covariance matrix. This covariance matrix, which we denote by $\boldsymbol{\Sigma}\{\widehat{\boldsymbol{\theta}}\}$, characterizes the quality of the estimated parameters. As the estimator $\widehat{\boldsymbol{\theta}}$ is a complex nonlinear function of random deviations, propagation of variance–covariance information cannot be carried out directly. Bootstrapping, in combination with MC techniques, can however be applied in this situation without any difficulty. The calculation process to derive the covariance matrix of the estimated parameters is summarized in Algorithm 2.

Algorithm 1: GEM Algorithm

Input : $y_{k,t}$, $h_{k,t}(\boldsymbol{\xi})$, p_k, itermax, ε, ε_ν $(k = 1, \ldots, N;\ t = 1, \ldots, n)$

Output: $\hat{\boldsymbol{\xi}}$, $\hat{\sigma}_k^2$, $\hat{\nu}_k$, $\hat{\mathbf{e}}_k$, $\hat{\mathbf{u}}_k$, $\widehat{\mathbf{W}}_k$, $\widehat{\boldsymbol{\alpha}}_k$

Initialization: $\boldsymbol{\xi}^{(0)}$; $\nu_k^{(0)}$; $\sigma_k^{(0)}$; $\boldsymbol{\alpha}_k^{(0)}$

for $i = 0 \ldots$ itermax **do**

E-Step:

$$w_{k,t}^{(i)} = \frac{\nu_k^{(i)}+1}{\nu_k^{(i)}+\left(\frac{\boldsymbol{\alpha}_k^{(i)}(L)(y_{k,t}-h_{k,t}(\boldsymbol{\xi}^{(i)}))}{\sigma_k^{(i)}}\right)^2} \rightsquigarrow \mathbf{W}_k^{(i)} = \begin{bmatrix} w_{k,1}^{(i)} & 0 \cdots & 0 \\ 0 & \vdots \ddots & \vdots \\ 0 & 0 \cdots & w_{k,n}^{(i)} \end{bmatrix}$$

M-Step:

$$\Delta y_{k,t}^{(i)} = y_{k,t} - h_{k,t}(\boldsymbol{\xi}^{(i)}),\ \mathbf{A}_{k,t}^{(i)} = \frac{\partial h_{k,t}(\boldsymbol{\xi}^{(i)})}{\partial \boldsymbol{\xi}}$$

$$\overline{\Delta y}_{k,t}^{(i)} = \boldsymbol{\alpha}_k^{(i)}(L)\Delta y_{k,t}^{(i)}, \quad \overline{\mathbf{A}}_{k,t}^{(i)} = \boldsymbol{\alpha}_k^{(i)}(L)\mathbf{A}_{k,t}^{(i)}$$

$$\Delta\boldsymbol{\xi}^{(i+1)} = \left(\sum_{k=1}^{N} \frac{1}{(\sigma_k^2)^{(i)}} \overline{\mathbf{A}}_k^{(i),T} \mathbf{W}_k^{(i)} \overline{\mathbf{A}}_k^{(i)}\right)^{-1} \sum_{k=1}^{N} \frac{1}{(\sigma_k^2)^{(i)}} \overline{\mathbf{A}}_k^{(i),T} \mathbf{W}_k^{(i)} \overline{\Delta \mathbf{y}}_k^{(i)}$$

$$\gamma = 1$$

$$\boldsymbol{\xi}^{(i+1)} = \boldsymbol{\xi}^{(i)} + \gamma\Delta\boldsymbol{\xi}^{(i+1)} \text{ (halve } \gamma \text{ if necessary)} \quad \textbf{(N.1)}$$

$$e_{k,t}^{(i+1)} = y_{k,t} - h_{k,t}(\boldsymbol{\xi}^{(i+1)}) \quad \rightsquigarrow \quad \mathbf{E}_k^{(i+1)} = \begin{bmatrix} e_{k,0}^{(i+1)} & \cdots & e_{k,1-p_k}^{(i+1)} \\ \vdots & & \vdots \\ e_{k,n-1}^{(i+1)} & \cdots & e_{k,n-p_k}^{(i+1)} \end{bmatrix}$$

$$\boldsymbol{\alpha}_k^{(i+1)} = \left(\mathbf{E}_k^{(i+1),T} \mathbf{W}_k^{(i)} \mathbf{E}_k^{(i+1)}\right)^{-1} \mathbf{E}_k^{(i+1),T} \mathbf{W}_k^{(i)} \mathbf{e}_k^{(i+1)}$$

(stabilize $\boldsymbol{\alpha}_k^{(i+1)}$ if necessary) **(N.2)**

$$u_{k,t}^{(i+1)} = e_{k,t}^{(i+1)} - \alpha_{k,1}^{(i+1)} e_{k,t-1}^{(i+1)} - \cdots - \alpha_{k,p_k}^{(i+1)} e_{k,t-p_k}^{(i+1)} = \boldsymbol{\alpha}_k^{(i+1)}(L) e_{k,t}^{(i+1)}$$

$$(\sigma_k^2)^{(i+1)} = \frac{1}{n} \sum_{t=1}^{n} w_{k,t}^{(i)} \left(u_{k,t}^{(i+1)}\right)^2 = \frac{\mathbf{u}_k^{(i+1),T} \mathbf{W}_k^{(i)} \mathbf{u}_k^{(i+1)}}{n}$$

$$0 = 1 + \log \nu_k^{(i+1)} - \psi\left(\frac{\nu_k^{(i+1)}}{2}\right) + \psi\left(\frac{\nu_k^{(i+1)}+1}{2}\right) - \log\left(\nu_k^{(i+1)} + 1\right)$$
$$+ \frac{1}{n} \sum_{t=1}^{n} \left(\log \frac{\nu_k^{(i+1)}+1}{\nu_k^{(i+1)}+\left(u_{k,t}^{(i+1)}/\sigma_k^{(i+1)}\right)^2} - \frac{\nu_k^{(i+1)}+1}{\nu_k^{(i+1)}+\left(u_{k,t}^{(i+1)}/\sigma_k^{(i+1)}\right)^2}\right) \quad \textbf{(N.3)}$$

if $\max_{j,k}(|\xi_j^{(i)} - \xi_j^{(i+1)}|, |(\sigma_k^2)^{(i)} - (\sigma_k^2)^{(i+1)}|) < \varepsilon$ *and*
$\max_k(|\nu_k^{(i)} - \nu_k^{(i+1)}|) < \varepsilon_\nu$ **then**
break

Algorithm 2: Bootstrapping Algorithm

Input : $\hat{\boldsymbol{\xi}}, \hat{\sigma}_k^2, \hat{\nu}_k, \hat{\mathbf{e}}_k, \hat{\mathbf{u}}_k, \widehat{\mathbf{W}}_k, \widehat{\boldsymbol{\alpha}}_k$

Output: $\boldsymbol{\Sigma}\{\widehat{\boldsymbol{\theta}}\}$

For every component $k = 1, \ldots, N$, every time instance $t = 1, \ldots, n$ and every $b = 1, \ldots, B$,

- Generate the white noise samples using the estimated t-distributions: $u_{k,1}^{(b)}, \ldots, u_{k,n}^{(b)} \overset{\text{ind.}}{\sim} t_{\hat{\nu}_k}(0, \hat{\sigma}_k^2)$.
- Compute the colored noise samples using the estimated AR models: $e_{k,t}^{(b)} = \widehat{\alpha}_{k,1} e_{k,t-1}^{(b)} + \cdots + \widehat{\alpha}_{k,p_k} e_{k,t-p_k}^{(b)} + u_{k,t}^{(b)}$.
- Compute the observation samples using the fitted deterministic functions: $\ell_{k,t}^{(b)} = h_{k,t}(\widehat{\boldsymbol{\xi}}) + e_{k,t}^{(b)}$.
- Use Algorithm 1 to compute the bootstrap solutions: $\widehat{\boldsymbol{\theta}}^{(1)}, \ldots, \widehat{\boldsymbol{\theta}}^{(b)}$.

  and compute the mean bootstrap solution: $\bar{\widehat{\boldsymbol{\theta}}} = \frac{1}{B} \sum_{b=1}^{B} \widehat{\boldsymbol{\theta}}^{(b)}$.
- Compute the bootstrap covariance matrix: $\boldsymbol{\Sigma}\{\widehat{\boldsymbol{\theta}}\} \approx \frac{1}{B} \sum_{b=1}^{B} \left(\widehat{\boldsymbol{\theta}}^{(b)} - \bar{\widehat{\boldsymbol{\theta}}}\right)\left(\widehat{\boldsymbol{\theta}}^{(b)} - \bar{\widehat{\boldsymbol{\theta}}}\right)^T$.

4 Field Experiment Setup and Its Results

The multi-sensor system (MSS) considered for geo-referencing in this contribution consists of a terrestrial laser scanner (TLS) and two GNSS antennas/receivers. As shown in Fig. 1, the TLS is the core sensor of the MSS, which rotates about its vertical axis with a constant angular velocity. The GNSS receivers are connected to two eccentric GNSS antennas, which are mounted such that the centroid of antenna reference points (ARPs) coincides with the TLS's rotating axis. In addition to these GNSS receivers, we assume to have a nearby reference GNSS station with known position. During the data acquisition, the MSS performs a complete 360° rotation about its vertical axis while collecting both TLS data (i.e., a 3D point cloud) and GNSS measurements, which are synchronized through GNSS receiver event marker (see [20] for details regarding the MSS).

The objective of the navigation system is to provide the position (the centroid of ARPs) and the pointing direction (heading) of the laser scanner. In [20], the standard real-time kinematic (RTK) positioning [21] was used to estimate individual rotating antenna positions. Then, a constrained nonlinear filtering method, in particular an extended Kalman filter, was used to obtain the above parameters. Nadarajah et al.

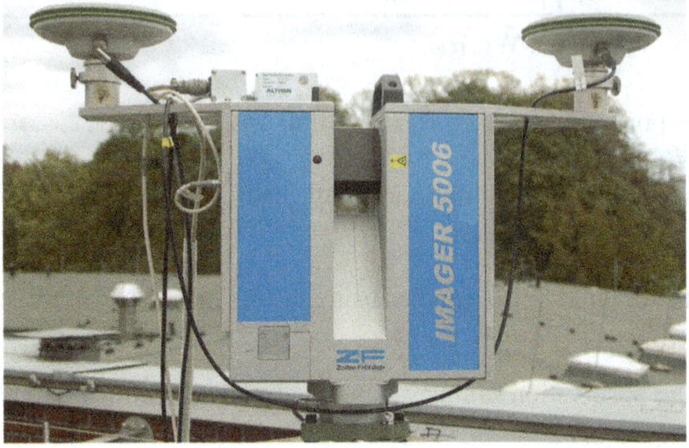

Fig. 1 Prototypical realization of the MSS formed by a laser scanner (blue) and two eccentrically mounted GNSS antennas (green). This MSS was used for the field experiment

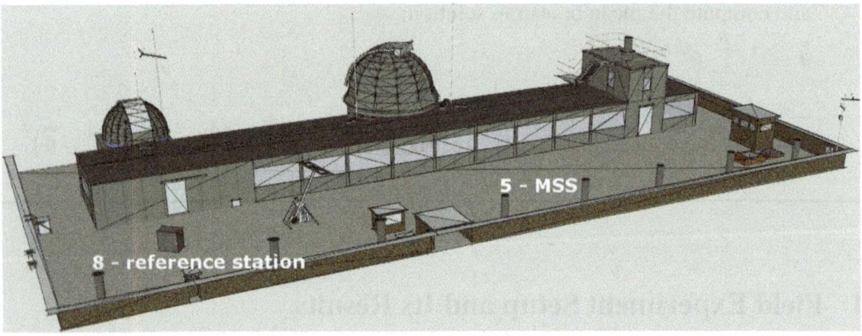

Fig. 2 Location of the practical experiment on the roof of the building of the Geodetic Institute (Messdach) at the Leibniz Universität Hannover, Germany. The MSS from Fig. 1 is mounted on pillar 5 while reference station is located at pillar 8

[22] utilized a constrained integer least-squares and array-aided positioning enabling improved ambiguity resolution and improved positioning accuracy.

For numerical analyses, we used the GNSS observations from a field experiment on the building of the Geodetic Institute (Messdach) at the Leibniz Universität Hannover, Germany. The MSS was mounted on pillar 5 (see Fig. 2) and equipped with a TLS *Z+F Imager 5006* as well as two individually and absolutely calibrated *LEIAX1202GG* GNSS antennas about 0.6 m apart. These antennas are connected to two dual frequency GNSS receivers *JAVAD TRE_G3TH DELTA*. The reference station (and thus the origin of the coordinates) is located at pillar 8 (about 20 m from the MSS) and equipped with a *JAVAD TRE_G3TH DELTA* GNSS receiver and *LEIAR25.R3 LEIT* antenna. A full 360° rotation consists of approximately 7609 points (acquired with a data rate of 10 Hz) with respect to one antenna.

We applied the GEM algorithm to approximate a measured and preprocessed 3D GNSS time series (see [20, 22]) by a 3D circle with the three components North (X_t), East (Y_t), and Up (Z_t):

$$\begin{pmatrix} X_t \\ Y_t \\ Z_t \end{pmatrix} = \begin{pmatrix} -r\cos(T_t)\sin(\Phi) + r\sin(T_t)\cos(\theta)\cos(\Phi) + C_x \\ r\cos(T_t)\cos(\Phi) + r\sin(T_t)\cos(\theta)\sin(\Phi) + C_y \\ -r\sin(T_t)\sin(\theta) + C_z \end{pmatrix} + \begin{pmatrix} E_{1,t} \\ E_{2,t} \\ E_{3,t} \end{pmatrix} \quad (11)$$

The multivariate (three-dimensional) nonlinear regression model in (11) contains six model parameters: two for the orientation (azimuth angle $\Phi \in [-\pi, \pi]$ and zenith angle $\theta \in [0, \pi]$) of its unit normal vector, one for the radius (r), and three for the circle center (C_x, C_y, C_z) (cf. [23]). In this application, $n = 7896$ time instances in (11) were registered. Concerning the random deviations $\mathbf{E}_t$, we determined three different AR models. To identify a correct and computationally inexpensive model for the given real data set, we applied the GEM algorithm (Algorithm 1) for different AR orders concerning the North-, East-, and Up-time series, beginning with small orders. We used the maximum cumulated periodogram test described in [13] for testing if each estimated noise series $\hat{u}_{k,1}, \ldots, \hat{u}_{k,n}$ differs significantly from theoretical white noise. In this way, we determined for the North- and East-component an AR(15) model each, and for the Up-component an AR(18) model (all of which passed the white noise tests). In Fig. 3, the estimated cumulated periodograms of the decorrelated residuals for the North- and East-component are depicted.

Figure 4 shows the adjusted circle and the observed 3D points. Having obtained an estimated d.o.f. of 120 and 88, respectively, for the North- and East-component, we conclude that the white noise components of these GNSS series are almost normally

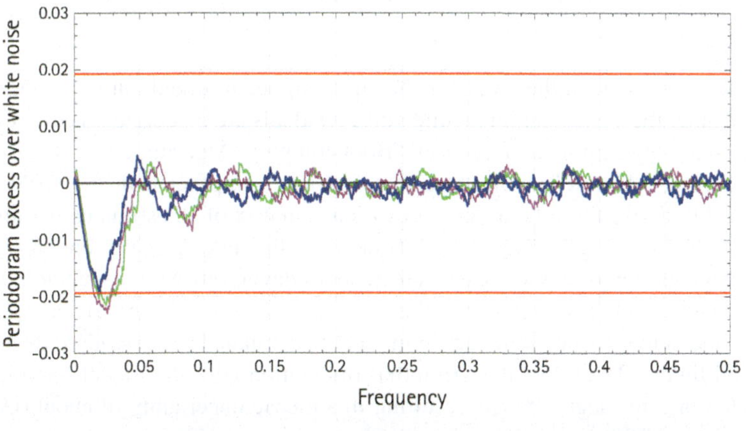

Fig. 3 Excess of the estimated periodogram of the decorrelated residuals for the North- and East-component AR(15) model (blue and green) and for the Up component AR(18) model (magenta) with respect to the theoretical white noise periodogram (black) and 99% significance bounds (red)

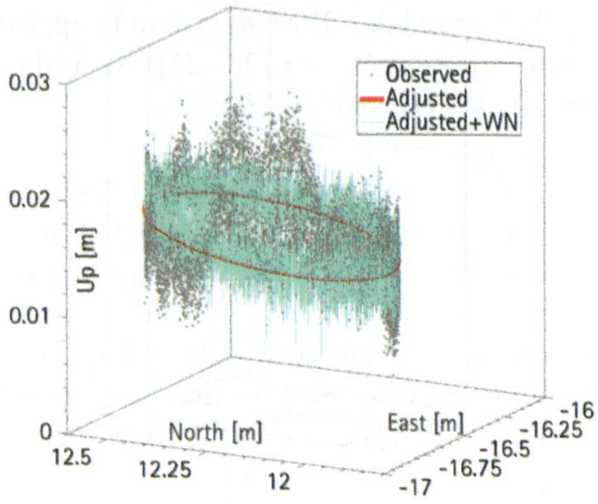

Fig. 4 3D view of observed (black points) and adjusted circle (red line) for $n = 7827$ real three-dimensional GNSS measurements taken from [22], displayed in a North East Up (NEU) coordinate system

Table 1 Results of the bootstrap algorithm using 500 replications. The rows 1–6 of this table indicate the means of estimated model parameters, their standard deviations, and the 95% confidence intervals. The rows 7–9 give the results for estimated scale factors for the North- (X), East- (Y), and Up- (Z) components

	Mean	Std	95% confidence interval
r (m)	0.2971	0.00022	[0.2968 0.2974]
Φ (rad)	−0.00078	0.00061	[−0.00183 0.0003]
θ (rad)	−0.00665	0.00283	[−0.0112 − 0.00217]
C_x (m)	12.2340	0.00013	[12.2338 12.2342]
C_y (m)	−16.6317	0.00038	[−16.6321 − 16.6312]
C_z (m)	0.01628	0.00138	[0.01458 0.01776]
σ_X^2	7.3×10^{-7}	1.2×10^{-8}	$\left[7.1 \times 10^{-7}\ 7.6 \times 10^{-7}\right]$
σ_Y^2	1.4×10^{-6}	2.2×10^{-8}	$\left[1.3 \times 10^{-6}\ 1.4 \times 10^{-6}\right]$
σ_Z^2	3.0×10^{-6}	4.0×10^{-8}	$\left[2.9 \times 10^{-6}\ 3.1 \times 10^{-6}\right]$

distributed. In contrast, the d.o.f. for the Up-component turned out to be 33, which indicates that the corresponding white noise residuals are moderately heavy-tailed.

We used Algorithm 2 of section "Bootstrapping Algorithm for Determining the Covariance Matrix" to illustrate the bootstrap sampling with the GEM algorithm and to derive the full variance–covariance matrix of all estimated parameters. For this purpose, we generated $B = 500$ samples. In Table 1, we present 95% confidence interval alongside the means and standard deviations with respect to both the six estimated circle parameters and the three scale factors. The metric components (radius and center point) of the circle model are estimated with standard deviations at submillimeter level. For the azimuthal orientation (Φ), the model parameter is estimated in centi-degree range, resulting in a metric uncertainty of about 0.018 m at a distance of 30 m. In comparison to the approach in [20], these results constitute an improvement of the estimated parameters of main interest within the direct geo-referencing of 3D point clouds. The estimated zenith angle (θ) is dominated by the Up-component of the GNSS observations. Since this Up-component is typically characterized by a higher measurement noise than the horizontal components

(North and East), we expected a higher standard deviation for the zenith angle. The same behavior can be seen for the estimated scale factors and corresponding confidence intervals, whereas the horizontal components perform better in contrast to the Up-component. In case of a carefully performed leveling process of the MSS, the variations of the zenith angle can be minimized and, therefore, would play a secondary role.

5 Conclusions and Outlook

In this paper, we considered multivariate regression time series with both AR colored noise and outlier-afflicted/heavy-tailed white noise components. The latter were modeled by scaled t-distributions with estimated d.o.f., in which distributions are heavy-tailed and thus frequently used as a way to achieve a data-adaptive robust ML estimator. We described the theory and implementation of a GEM algorithm, in which the deterministic model parameters, the AR coefficients, the scale factors, and the d.o.f. for the multiple time series can be estimated in the form of IRLS. In order to derive the covariance matrix of these parameters as a quality measure for the estimated parameters, we demonstrated an MC-based bootstrap algorithm, which allows also for the computation of confidence intervals concerning all parameters for a given error probability. The presented algorithm was also tested in a real data experiment using GNSS measurements. A model selection with respect to the order of the AR process was performed by employing a periodogram-based white noise test. Finally, the analysis of the estimated d.o.f. throughout the different multivariate time series showed that the white noise residuals of the Up-component of the GNSS time series deviate from a normal distribution. In the future, we intend to extend the GEM algorithm by modeling in addition cross-correlations between the individual time series.

References

1. Huber, P.J.: Robust estimation of a location parameter. Ann. Math. Stat. **35**, 73–101 (1964)
2. Lange, K.L., Little, R.J.A., Taylor, J.M.G.: Robust statistical modeling using the t-distribution. J. Am. Stat. Assoc. **84**, 881–896 (1989)
3. Dempster, A.P., Laird, N.M., Rubin, D.B.: Maximum likelihood from incomplete data via the EM algorithm. J. R. Stat. Soc. (Ser. B) **39**, 1–38 (1977)
4. Liu, C.H.: ML estimation of the multivariate t distribution and the EM algorithm. J. Multivar. Anal. **63**, 296–312 (1997)
5. Liu, C.H., Rubin, D.B.: The ECME algorithm: a simple extension of EM and ECM with faster monotone convergence. Biometrika **81**, 633–648 (1994)
6. Meng, X., Rubin, D.B.: Maximum likelihood estimation via the ECM algorithm: a general framework. Biometrika **80**, 267–278 (1993)
7. McLachlan, G.J., Krishnan, T.: The EM Algorithm and Extensions. Wiley, Hoboken, New Jersey (2008)

8. Phillips, R.F.: Least absolute deviations estimation via the EM algorithm. Stat. Comput. **12**, 281–285 (2002)
9. Wang, K., Xiong, S., Li, Y.: Modeling with noises for inertial sensors. In: Position Location and Navigation Symposium (PLANS) 2012, IEEE/ION, pp. 625-632 (2012)
10. Park, M., Gao, Y.: Error and performance analysis of MEMS-based inertial sensors with a low-cost GPS receiver. Sensors **8**, 2240–2261 (2008)
11. Schuh, W.D.: The processing of band-limited measurements; filtering techniques in the least squares context and in the presence of data gaps. Space Sci. Rev. **108**, 67–78 (2003)
12. Luo, X.: GPS Stochastic Modelling: Signal Quality Measures and ARMA Processes. Springer, Berlin, Heidelberg (2013)
13. Kargoll, B., Omidalizarandi, M., Loth, I., Paffenholz, J.A., Alkhatib, H.: An iteratively reweighted least-squares approach to adaptive robust adjustment of parameters in linear regression models with autoregressive and t-distributed deviations. J. Geodesy. (2017). https://doi.org/10.1007/s00190-017-1062-6
14. Alkhatib, H., Kargoll, B., Paffenholz, J.A.: Robust multivariate time series analysis in nonlinear models with autoregressive and t-distributed errors. In: Valenzuela, O., Rojas, F., Pomares, H., Rojas, I. (eds.) Proceedings ITISE 2017—International work-conference on Time Series, vol. 1, pp. 23–36 (2017). ISBN: 978-84-17293-01-7
15. Efron, B.: Bootstrap methods: another look at the jackknife. Ann. Stat. **7**(1), 1–26 (1997)
16. Alkhatib, H.: On Monte Carlo methods with applications to the current satellite gravity missions, Dissertation. Universitt Bonn, Bonn. Institute for Geodesy and Geoinformation (2008). http://hss.ulb.uni-bonn.de/2007/1078/1078.pdf
17. Li, H., Maddala, G.S.: Bootstrapping time series models. Economet. Rev. **15**, 115–158 (1996)
18. Porat, B.: Digital Processing of Random Signals. Dover, Mineola/New York (1994)
19. Hargreaves, G.I.: Interval Analysis in MATLAB. Numerical Analysis Report No. 416, Manchester Centre for Computational Mathematics, The University of Manchester (2002). ISSN 1360-1725
20. Paffenholz, J.A.: Direct Geo-Referencing of 3D Point Clouds with 3D Positioning Sensors. German Geodetic Commission, Series C. Ph.D. thesis, No. 689, Munich (2012). https://dgk.badw.de/fileadmin/user_upload/Files/DGK/docs/c-689.pdf
21. Wübbena, G., Bagge, A., Schmitz, M.: RTK networks based on Geo++ GNSMART–concepts, implementation, results
22. Nadarajah, N., Paffenholz, J.A., Teunissen, P.J.G.: Integrated GNSS attitude determination and positioning for direct geo-referencing. Sensors **14**(7), 12715–12734 (2014). https://doi.org/10.3390/s140712715
23. Moon, P., Spencer, D.E.: Field Theory Handbook—Including Coordinate Systems, Differential Equations, and Their Solutions, 3rd edn. Springer, New York (1988)

A New Estimation Technique for AR(1) Model with Long-Tailed Symmetric Innovations

Ayşen Dener Akkaya and Özlem Türker Bayrak

Abstract In recent years, it is seen in many time series applications that innovations are non-normal. In this situation, it is known that the least squares (LS) estimators are neither efficient nor robust and maximum likelihood (ML) estimators can only be obtained numerically which might be problematic. The estimation problem is considered newly through different distributions by the use of modified maximum likelihood (MML) estimation technique which assumes the shape parameter to be known. This becomes a drawback in machine data processing where the underlying distribution cannot be determined but assumed to be a member of a broad class of distributions. Therefore, in this study, the shape parameter is assumed to be unknown and the MML technique is combined with Huber's estimation procedure to estimate the model parameters of autoregressive (AR) models of order 1, named as adaptive modified maximum likelihood (AMML) estimation. After the derivation of the AMML estimators, their efficiency and robustness properties are discussed through simulation study and compared with both MML and LS estimators. Besides, two test statistics for significance of the model are suggested. Both criterion and efficiency robustness properties of the test statistics are discussed, and comparisons with the corresponding MML and LS test statistics are given. Finally, the estimation procedure is generalized to AR(q) models.

Keywords Adaptive modified maximum likelihood · Autoregressive models
Least squares estimators · Hypothesis testing · Modified maximum likelihood
Estimation · Efficiency · Robustness

A. D. Akkaya (✉)
Middle East Technical University, Ankara, Turkey
e-mail: akkay@metu.edu.tr

Ö. T. Bayrak
Cankaya University, Ankara, Turkey
e-mail: ozlemt@cankaya.edu.tr

I. Rojas et al. (eds.), *Time Series Analysis and Forecasting*,
Contributions to Statistics, https://doi.org/10.1007/978-3-319-96944-2_4

1 Introduction

AR models are very useful and popular in modeling time series data in many different areas such as economics, finance, engineering, and so on. The classical AR models assume that the innovations are normally distributed which might be invalid in applications. Under non-normality, the LS estimators are neither efficient nor robust, and maximum likelihood (ML) estimators are elusive due to the implicit nature of likelihood functions. Use of iterative approach is weary due to convergence problems and produces bias in estimators especially for small samples. Therefore, in recent studies, this assumption is relaxed by the use of MML method developed by Tiku [10] to estimate the unknown parameters in such situations [1–4, 14, 15]. The MML estimators capture all the good statistical properties of ML estimators and they are explicit functions of sample observations. Besides, they are (i) considerably more efficient (unbiased and smaller variance) than the LS estimators for all sample sizes, particularly for large n, (ii) asymptotically fully efficient under very general regularity conditions and almost fully efficient for small samples, and (iii) robust to plausible deviations from the assumed distribution and mild data anomalies (outliers, inliers, etc.). For the detailed information about MML procedure and its applications, one can refer to [11, 13]. On the other hand, MML method is based on the assumption of a particular distribution, i.e., the shape parameter is known. Many different ways are suggested in the literature for determining the shape parameter including the use of Q–Q plots and data exploratory techniques [3, 6, 11]. Due to the intrinsic robustness of MML estimators [11], the values obtained by any of these methods will yield essentially the same estimates and standard errors for plausible alternatives. However, when the data is huge and machine learning methods are applied, it is important to estimate this parameter also since one has no opportunity to investigate the nature of the underlying distribution in this case. It can only be assumed that it is a member of a broad class of distributions. Inserting a likelihood equation related to the shape parameter into the likelihood equation system makes it unsolvable analytically even if the MML estimation method is used. Thus, there is a need to extend the MML method so that the assumption on the shape parameter is relaxed.

In studies [7, 8], M-estimators which are efficient and robust under a broad class of long-tailed symmetric (LTS) distributions are developed. In this study, following [5, 12, 16], we use an adapted form of MML estimators which combine the logic of MML with M-estimators named as AMML estimation in memory of Moti Lal Tiku who actually initiated the idea and thought this name. The parameters are estimated under the assumption that the innovations in AR(1) model belong to LTS family. The efficiency and robustness properties of them are discussed via simulation as well as their comparison with LS and MML estimators. Then, two significance tests of the model are suggested. The criterion and efficiency robustness properties of the tests are discussed and compared with the corresponding MML and LS test procedures. Finally, the estimation procedure is expanded to AR(q) models.

2 Estimation of the Model Parameters

Consider the time series model

$$y_t = \mu + \phi y_{t-1} + \varepsilon_t, \quad (1 \le t \le n), \quad (-1 < \phi < 1) \tag{1}$$

where the innovations ε_t are independent and identically distributed (iid), and have one of the distributions in LTS family

$$f(\varepsilon) = \frac{\Gamma(p)}{\sigma\sqrt{k}\Gamma(1/2)\Gamma(p-1/2)}\left(1 + \frac{\varepsilon^2}{k\sigma^2}\right)^{-p}, -\infty < \varepsilon < \infty; \tag{2}$$

where $k = 2p - 3$ and $p \geq 2$. E(ε)=0 and V(ε) $= \sigma^2$. For $p = \infty$, Eq. (2) reduces to normal $N(0, 1)$. Kurtosis of the distribution assumes the values ∞, 9, 4.2, 3.4, and 3 for p =2.5, 3.5, 5, 10, and ∞, respectively. Note that the distribution of $t = \sqrt{\upsilon/k}(\varepsilon/\sigma)$ is student's t with $\upsilon = 2p - 1$ degrees of freedom. The likelihood function is

$$L \propto \sigma^{-n} \prod_{i=1}^{n} \left(1 + \frac{\varepsilon_i^2}{k\sigma^2}\right)^{-p} \tag{3}$$

In fact, Eq. (3) is the likelihood function conditional on $y_0 = \varepsilon_0/\sqrt{1-\phi^2}$ where ε_0 is an independent innovation that has the same distribution as $\varepsilon_i (1 \le i \le n)$. Actually, this is Model 2 of [17] which is more general than their Model 1.

Modified Maximum Likelihood Estimators

The likelihood equations for known p are obtained in terms of $z_i = (y_i - \phi y_{i-1} - \mu)/\sigma = \varepsilon_i/\sigma$, $(1 \le i \le n)$ as follows:

$$\begin{aligned}
\frac{\partial lnL}{\partial \mu} &= \frac{2p}{k\sigma}\sum_{i=1}^{n} g(z_i) = 0, \\
\frac{\partial lnL}{\partial \phi} &= \frac{2p}{k\sigma}\sum_{i=1}^{n} g(z_i) y_{i-1} = 0, \\
\frac{\partial lnL}{\partial \sigma} &= -\frac{n}{\sigma} + \frac{2p}{k\sigma}\sum_{i=1}^{n} z_i g(z_i) = 0,
\end{aligned} \tag{4}$$

where $g(z_i) = \frac{z_i}{\left(1+\frac{z_i^2}{k}\right)}$.

Since Eq. (4) includes nonlinear function, $g(z_i)$, they have no explicit solutions and iterative solutions are problematic. MML method is used to find estimators, which are known to be asymptotically equivalent to ML estimators [11]. Estimation procedure is carried out in three steps: (i) the maximum likelihood equations are expressed in terms of the order statistics of $z_{(i)} = \left(y_{[i]} - \phi y_{[i]-1} - \mu\right)/\sigma$ where $\left(y_{[i]}, y_{[i]-1}\right)$ are the concomitants of $z_{(i)}$, i.e., the pair $\left(y_j, y_{j-1}\right)(j = [i])$ associated with the ith ordered value, $z_{(i)}$ so that the ordering of the time series data is not lost; (ii) the nonlinear function $g\left(z_{(i)}\right)$ is replaced by linear approximations $g\left(z_{(i)}\right) \cong \alpha_i + \beta_i z_{(i)}$, $1 \le i \le n$ where the constant coefficients α_i and β_i are obtained from the first two terms of a Taylor series expansion of $g\left(z_{(i)}\right)$ around the ith population quantile, $t_{(i)} = E\left(z_{(i)}\right)$. Here, we use approximate values of $t_{(i)}$ calculated from

$$\frac{\Gamma(p)}{\sqrt{k}\Gamma(1/2)\Gamma(p-1/2)} \int_{-\infty}^{t_{(i)}} \left(1 + \frac{z^2}{k}\right)^{-p} dz = \frac{i}{n+1}(1 \le i \le n). \quad (5)$$

The resulting α_i and β_i are

$$\alpha_i = (2/k)t_{(i)}^3/\left\{1 + (1/k)t_{(i)}^2\right\}^2 \text{and } \beta_i = \left[1 - (1/k)t_{(i)}^2\right]/\left\{1 + (1/k)t_{(i)}^2\right\}^2, \quad (6)$$

and (iii) Eq. (6) is incorporated in Eq. (4) and by solving the modified (linearized) likelihood equations $\partial lnL^*/\partial\mu = 0$, $\partial lnL^*/\partial\phi = 0$, and $\partial lnL^*/\partial\sigma = 0$, the MML estimators are obtained as

$$\hat{\mu} = \sum_{i=1}^{n} \beta_i \left(y_{[i]} - \hat{\phi} y_{[i]-1}\right)/m,$$
$$\hat{\phi} = K + D\hat{\sigma}, \hat{\sigma} = \left(B + \sqrt{B^2 + 4nC}\right)/2\sqrt{n(n-1)} \quad (7)$$

where

$$m = \sum_{i=1}^{n} \beta_i, \; K = \frac{\sum_{i=1}^{n} \beta_i y_{[i]} y_{[i]-1} - \frac{1}{m} \sum_{i=1}^{n} \beta_i y_{[i]} \sum_{i=1}^{n} \beta_i y_{[i]-1}}{\sum_{i=1}^{n} \beta_i y_{[i]-1}^2 - \frac{1}{m}\left(\sum_{i=1}^{n} \beta_i y_{[i]-1}\right)^2},$$
$$D = \frac{\sum_{i=1}^{n} \alpha_i y_{[i]-1}}{\sum_{i=1}^{n} \beta_i y_{[i]-1}^2 - \frac{1}{m}\left(\sum_{i=1}^{n} \beta_i y_{[i]-1}\right)^2},$$
$$B = \frac{2p}{k} \sum_{i=1}^{n} \alpha_i \left(y_{[i]} - \bar{y}_{[.]} - K\left(y_{[i]-1} - \bar{y}_{[.]-1}\right)\right),$$
$$C = \frac{2p}{k} \sum_{i=1}^{n} \beta_i \left(y_{[i]} - \bar{y}_{[.]} - K\left(y_{[i]-1} - \bar{y}_{[.]-1}\right)\right)^2,$$
$$\bar{y}_{[.]} = \sum_{i=1}^{n} \beta_i y_{[i]}/m, \; \bar{y}_{[.]-1} = \sum_{i=1}^{n} \beta_i y_{[i]-1}/m.$$

Comment: The coefficients β_i $(1 \le i \le n)$ increase until the middle value and then decrease in a symmetric fashion. Therefore, if β_1 is positive, then all the β_i coefficients are positive and $\hat{\sigma}$ is real and positive. For small p and large n, however, β_1 (and a few other β_i coefficients) can be negative and needed to be rectified. Thus, if β_1 turns out to be negative, we replace α_i by $\alpha_i^* = (1/k)t_{(i)}^3/\left\{1 + (1/k)t_{(i)}^2\right\}^2$ and β_i by $\beta_i^* = 1/\left\{1 + (1/k)t_{(i)}^2\right\}^2 (1 \le i \le n)$. Realize that the extreme observations automatically receive small weights, which deplete their effect due to the umbrella ordering of β_i values and this is instrumental in achieving robustness to long-tailed symmetric distributions and outliers in a sample.

Computations: The estimates of the parameters require figuring out the concomitants found out by sorting the innovations. Therefore, there is a need to obtain the initial estimates for the parameters in the model. This is done by the use of LS estimators $\tilde{\mu}$ and $\tilde{\phi}$, given in Eq. (8) since LS estimators do not need any distributional assumptions. Then, the initial estimates of innovations $\tilde{\varepsilon}_i = y_i - \tilde{\phi} y_{i-1} - \tilde{\mu} (1 \le i \le n)$ are ordered to obtain the concomitants $\left(y_{[i]}, y_{[i]-1}\right)$ corresponding to the ith ordered value of the estimated residual. By the use of these concomitants, the MML estimators are calculated from Eq. (7). To eliminate the effects of the initial estimates, the LS estimators $\tilde{\mu}$ and $\tilde{\phi}$ are then replaced by $\hat{\mu}$ and $\hat{\phi}$, respectively, and the corresponding innovations $\hat{e}_i = y_i - \hat{\phi} y_{i-1} - \hat{\mu}$ are ordered to obtain the new concomitants. The revised MML estimators are computed from these new concomitants. The process is repeated one more time for the estimates to stabilize sufficiently.

Least Squares Estimators

Regardless of the underlying distribution, the LS estimators are

$$\tilde{\mu} = \frac{\sum_{i=1}^n y_i}{n} - \tilde{\phi}\frac{\sum_{i=1}^n y_{i-1}}{n}, \; \tilde{\phi} = \frac{\sum_{i=1}^n y_i y_{i-1} - \sum_{i=1}^n y_i \sum_{i=1}^n y_{i-1}/n}{\sum_{i=1}^n y_{i-1}^2 - \left(\sum_{i=1}^n y_{i-1}\right)^2/n},$$

$$\text{and } \tilde{\sigma} = \sqrt{\frac{\sum_{i=1}^n \left(y_i - \tilde{\phi} y_{i-1} - \tilde{\mu}\right)^2}{n-2}}. \tag{8}$$

Adaptive Modified Maximum Likelihood Estimators

Since the shape parameter p is unknown, the coefficients α_i and β_i have to be estimated from the sample data. The idea of Huber [8] is implemented for this purpose. Let T_0 and S_0 be the initial estimators of μ and σ, respectively, given as

$$T_0 = med\left\{y_i - \hat{\phi}_0 y_{i-1}\right\} \text{and } S_0 = 1.483\, med\left\{\left|y_i - \hat{\phi}_0 y_{i-1} - T_0\right|\right\}, (1 \le i \le n) \tag{9}$$

where $\hat{\phi}_0 = med\left\{\frac{y_2-y_1}{y_1-y_0}, \frac{y_3-y_2}{y_2-y_1}, \ldots, \frac{y_n-y_{n-1}}{y_{n-1}-y_{n-2}}\right\}, (i = 1, 2, \ldots, n-1)$.

Then, t_i values in Eq. (6) can be estimated by $\hat{t}_i = \frac{y_i - \hat{\phi}_0 y_{i-1} - T_0}{S_0}$ and the revised estimated values of coefficients α_i and β_i are obtained as follows:

$$\hat{\alpha}_i = (2/k)\hat{t}_i/\left\{1 + (1/k)\hat{t}_i^2\right\}^2 \text{and } \hat{\beta}_i = 1/\left\{1 + (1/k)\hat{t}_i^2\right\}^2. \tag{10}$$

Note that the MML estimators do not have bounded influence functions, so coefficients α_i and β_i are revised to make them bounded. Besides, they completely depend on the observations not the presumed values of the parameter ϕ and p.

The adaptive modified maximum likelihood estimators are obtained as in Eq. (7) by replacing α_i and β_i with the estimated coefficients $\hat{\alpha}_i$ and $\hat{\beta}_i$ given in Eq. (10) and the concomitants $\left(y_{[i]}, y_{[i]-1}\right)$ with the original observations (y_i, y_{i-1}). Since t_i values are not obtained from the quantiles of the distribution but estimated directly from the sample and complete sums are invariant to ordering, we do not need to use the concomitants anymore.

Realize that $\hat{\alpha}_i$ and $\hat{\beta}_i$ depend on the value k $(2p - 3)$. If k is chosen very large, then the coefficients $\hat{\beta}_i$ reduce to 1 and lose its effect to achieve robustness. On the other hand, if k is chosen as small, estimators become inefficient. Therefore, the choice $k = 30$ (i.e., $p = 16.5$) turns out to be a good compromise as suggested in [5, 12].

Computations: (i) First using sample observations calculate the initial values of $\hat{\phi}_0$, T_0, and S_0; (ii) use these initial values to calculate $\hat{\mu}$, $\hat{\phi}$ and $\hat{\sigma}$ from Eq. (7) by the use of coefficients $\hat{\alpha}_i$ and $\hat{\beta}_i$ given in Eq. (10); (iii) replace $\hat{\phi}_0$, T_0, and S_0 by $\hat{\phi}$, $\hat{\mu}$, and $\hat{\sigma}$,, respectively; and (iv) repeat the process one more time and calculate $\hat{\mu}$, $\hat{\phi}$ and $\hat{\sigma}$ which are the desired AMML estimators.

3 Efficiency and Robustness Comparisons of the Estimators

To evaluate the efficiency and robustness of the AMML, MML, and LS estimators, [100,000/n] (integer value) Monte Carlo runs (simulations) are used. The distribution of ε, known as population model, is taken as LTS with $p = 16.5$. As alternative sample models first the following (1) to (5) are used.

(1) Normal with mean 0 and variance σ^2, and the LTS family with
(2) $p = 5.0$;
(3) $p = 3.5$;
(4) $p = 2.5$;
(5) $p = 2.0$

The results for $n = 30$, 50, and 100, where $\phi = 0.5$, $\mu = 0.0$, and $\sigma = 1.0$, are given in Table 1.

It can be seen from Table 1 that all estimators are unbiased except AMML estimate of σ which is slightly less and its bias decreases as sample size increases as expected. However, in all cases, the variances of AMML estimates are smaller than that of the LS estimates and similar to MML estimates except Model (1) where they are close. Thus, although AMML slightly underestimates σ, its mean squared error is less than the others. All methods underestimate σ in Model (5). Therefore, we can conclude that AMML is efficient and robust to misspecification errors. Similar results are obtained for other presumed values of ϕ, μ, and σ, so they are not reported here for conciseness.

Then, the outlier models where $(n - r)X_i$ come from $N\left(0, \sigma^2\right)$ and r (we do not know which) come from

(6) $N\left(0, 4\sigma^2\right)$;
(7) $N\left(0, 16\sigma^2\right)$; $r = [0.5 + 0.1n]$ (integer value),

and the mixture models

(8) $0.90N\left(0, \sigma^2\right) + 0.10N\left(0, 4\sigma^2\right)$ and
(9) $0.90N\left(0, \sigma^2\right) + 0.10N\left(0, 16\sigma^2\right)$

are taken as alternative sample models. The innovations are scale corrected to make their variances equal to σ^2. The results for $n = 30$, 50, and 100, where $\phi = 0.5$, $\mu = 0.0$, and $\sigma = 1.0$ are given in Table 2.

It can be seen from Table 2 that the results are similar to misspecification ones given in Table 1. Again similar results are obtained for other presumed values of ϕ, μ, and σ, so they are not reported here for conciseness.

Finally, the extreme alternative sample models

(10) Student's t distribution with two degrees of freedom;
(11) Cauchy distribution; and
(12) Slash (Normal/Uniform) distribution

are taken as alternative sample models. It must be noted that Model (10) has finite mean but nonexistent variance, and Models (11) and (12) have nonexistent mean and variance. Since the differences between the AMML and the others become very striking due to exploding variances, only the results for AMML estimators are given in Table 3. However, it must be noted that Model (10) is still comparable due to negligible bias in μ and ϕ with unacceptable variances. Besides, σ is overestimated by other methods in this case again with unacceptable variances. Therefore, under such extreme alternatives, only AMML method is valid and robust.

Table 1 Simulated values of the mean and variance of LS, MML, and AMML estimators under alternative models (1)–(5)

n		μ	ϕ	σ	μ	ϕ	σ	μ	ϕ	σ
		LS			*MML*			*AMML*		
Model (1)										
30	Mean	0.00	0.42	0.97	0.01	0.42	1.04	0.00	0.42	0.92
	Var	1.44	0.83	0.50	1.47	0.87	0.63	1.48	0.84	0.52
50	Mean	0.00	0.45	0.98	0.00	0.45	1.06	0.00	0.45	0.94
	Var	1.36	0.79	0.51	1.30	0.86	0.63	1.40	0.80	0.53
100	Mean	0.00	0.48	0.99	0.00	0.48	1.06	0.00	0.48	0.96
	Var	1.10	0.71	0.49	1.23	0.78	0.58	1.12	0.72	0.52
Model (2)										
30	Mean	0.00	0.42	0.97	0.00	0.42	1.01	0.00	0.42	0.89
	Var	1.46	0.84	0.77	1.31	0.80	0.75	1.38	0.82	0.61
50	Mean	0.00	0.45	0.98	0.00	0.46	1.03	0.00	0.45	0.91
	Var	1.27	0.76	0.81	1.24	0.79	0.77	1.17	0.73	0.62
100	Mean	0.00	0.48	0.99	0.00	0.48	1.03	0.00	0.48	0.93
	Var	1.17	0.73	0.83	1.05	0.69	0.77	1.10	0.69	0.62
Model (3)										
30	Mean	0.00	0.41	0.97	0.00	0.42	0.97	0.00	0.42	0.86
	Var	1.39	0.79	1.00	1.20	0.75	0.75	1.24	0.75	0.64
50	Mean	0.00	0.45	0.98	0.00	0.46	1.02	0.00	0.45	0.88
	Var	1.28	0.78	1.17	1.09	0.72	0.95	1.09	0.72	0.62
100	Mean	−0.01	0.48	0.98	0.00	0.47	1.02	−0.01	0.48	0.89

(continued)

Table 1 (continued)

n		μ	ϕ	σ	μ	ϕ	σ	μ	ϕ	σ
		LS			*MML*			*AMML*		
	Var	1.06	0.78	1.01	0.97	0.66	0.81	0.90	0.71	0.78
Model (4)										
30	Mean	0.00	0.42	0.95	0.00	0.43	0.94	0.00	0.43	0.80
	Var	1.43	0.82	1.91	1.09	0.73	1.03	1.09	0.73	0.65
50	Mean	0.00	0.45	0.96	0.00	0.46	0.98	0.00	0.45	0.82
	Var	1.29	0.79	1.91	0.99	0.68	1.29	0.98	0.67	0.65
100	Mean	0.00	0.47	0.97	0.00	0.48	0.97	0.00	0.48	0.83
	Var	1.07	0.78	2.16	0.88	0.59	1.14	0.82	0.64	0.61
Model (5)										
30	Mean	0.00	0.42	0.92	0.00	0.43	0.86	0.00	0.43	0.70
	Var	1.41	0.77	4.02	0.81	0.67	1.95	0.80	0.66	0.56
50	Mean	0.00	0.45	0.94	0.00	0.46	0.90	0.00	0.46	0.71
	Var	1.26	0.72	5.80	0.77	0.62	1.90	0.71	0.56	0.57
100	Mean	0.00	0.47	0.96	0.01	0.48	0.91	0.00	0.48	0.73
	Var	1.05	0.68	6.29	0.68	0.51	2.82	0.57	0.50	0.55

Table 2 Simulated values of the mean and variance of LS, MML, and AMML estimators under alternative models (6)–(9)

n		μ	ϕ	σ	μ	ϕ	σ	μ	ϕ	σ
		LS			*MML*			*AMML*		
Model (6)										
30	Mean	0.00	0.41	1.10	−0.01	0.42	1.15	0.00	0.41	1.01
	Var	1.83	0.95	0.93	1.80	0.89	0.83	1.72	0.93	0.70
50	Mean	0.00	0.45	1.12	0.00	0.45	1.18	0.00	0.45	1.03
	Var	1.60	0.95	0.90	1.51	0.91	0.83	1.49	0.90	0.72
100	Mean	0.00	0.47	1.13	0.00	0.48	1.18	0.00	0.47	1.05
	Var	1.39	1.06	0.95	1.39	0.95	0.86	1.28	0.98	0.72
Model (7)										
30	Mean	0.00	0.40	1.05	0.00	0.40	0.98	0.00	0.40	0.79
	Var	1.63	1.55	2.08	1.05	1.17	0.95	1.05	1.18	0.54
50	Mean	0.01	0.43	1.05	0.00	0.45	1.03	0.00	0.43	0.81
	Var	1.40	1.86	2.17	0.95	1.36	1.32	0.94	1.29	0.54
100	Mean	0.00	0.46	1.09	−0.01	0.47	1.02	0.00	0.46	0.83
	Var	1.26	2.17	2.48	0.79	1.53	1.18	0.74	1.44	0.59

(continued)

Table 2 (continued)

n		μ	ϕ	σ	μ	ϕ	σ	μ	ϕ	σ
		LS			*MML*			*AMML*		
Model (8)										
30	Mean	0.00	0.42	0.97	0.00	0.42	1.01	0.00	0.42	0.88
	Var	1.48	0.81	0.76	1.29	0.80	0.75	1.39	0.80	0.58
50	Mean	0.00	0.45	0.98	0.00	0.46	1.03	0.00	0.45	0.91
	Var	1.38	0.80	0.80	1.13	0.78	0.74	1.26	0.77	0.57
100	Mean	0.00	0.47	0.99	0.00	0.48	1.04	0.00	0.47	0.92
	Var	1.17	0.78	0.89	1.04	0.75	0.72	1.10	0.74	0.63
Model (9)										
30	Mean	0.00	0.43	0.95	0.00	0.43	0.88	0.00	0.44	0.71
	Var	1.35	0.73	2.63	0.82	0.63	1.23	0.77	0.59	0.66
50	Mean	0.00	0.45	0.96	0.00	0.46	0.94	0.00	0.46	0.73
	Var	1.22	0.73	2.63	0.85	0.58	1.72	0.67	0.56	0.60
100	Mean	0.00	0.48	0.97	0.00	0.48	0.92	0.00	0.49	0.74
	Var	1.03	0.80	2.78	0.67	0.49	1.50	0.59	0.55	0.63

Table 3 Simulated values of the mean and variance of AMML estimators under alternative models (10)–(12)

n		Model (10)			Model (11)			Model (12)		
		μ	ϕ	σ	μ	ϕ	σ	μ	ϕ	σ
30	Mean	0.00	0.45	1.37	−0.01	0.47	1.94	0.01	0.47	2.68
	Var	2.95	0.54	2.96	6.32	0.28	11.73	11.08	0.27	19.08
50	Mean	0.00	0.47	1.40	0.01	0.48	1.93	−0.01	0.49	2.71
	Var	2.59	0.46	2.49	4.55	0.19	10.43	8.95	0.18	16.97
100	Mean	0.00	0.48	1.42	−0.01	0.49	1.92	0.01	0.49	2.72
	Var	2.32	0.30	2.80	4.29	0.09	9.54	7.56	0.10	16.96

4 Significance Test of the Model

Testing the null hypotheses H_0: $\phi = 0$ is of practical interest. If the null hypothesis is true, then the model given in Eq. (1) reduces to a random process. Therefore, this hypothesis actually tests the significance of the model.

Lemma 1 *If $\phi = 0$, then the conditional distribution (σ known) of $\hat{\phi}(\sigma)$ is asymptotically normal with mean ϕ and variance $k\sigma^2/\{2p\sum_{i=1}^{n}\beta_i(y_{i-1}-\bar{y}_{i-1})^2\}$.*

The result follows from the asymptotic equivalence of likelihood and modified likelihood equations [11, 15] and $\partial lnL^*/\partial\phi$ assumes the form [9]

$$\frac{\partial lnL^*}{\partial\phi} = \frac{2p}{k\sigma^2}\left\{\sum_{i=1}^{n}\beta_i(y_{i-1}-\bar{y}_{i-1})^2\right\}\left[\hat{\phi}(\sigma)-\phi\right]. \tag{11}$$

Since $\hat{\phi}$ and $\hat{\sigma}$ converge to ϕ and σ, respectively, as n becomes large, to test H_0 we propose a test statistic

$$T_1 = \sqrt{(1.1)\sum_{i=1}^{n}\beta_i(y_{i-1}-\bar{y}_{i-1})^2}\left(\hat{\phi}/\hat{\sigma}\right) \tag{12}$$

where $2p/k = 1.1$, $p = 16.5$. Large values of T_1 lead to the rejection of H_0: $\phi = 0$ in favor of H_1: $\phi > 0$.

The corresponding test statistic based on the MML procedure is obtained by replacing ϕ and σ by their MML estimators, (y_i, y_{i-1}) pair by their concomitants $\left(y_{[i]}, y_{[i]-1}\right)$ and related β_i coefficients in Eq. (12). The test statistic based on MML estimators is denoted by $T_1^{'}$.

The test statistic based on the LS estimators is

$$T_2 = \sqrt{\sum_{i=1}^{n}(y_{i-1}-\bar{y}_{i-1})^2}\left(\hat{\phi}/\hat{\sigma}\right). \tag{13}$$

It is important for a test statistic to have both criterion and efficiency robustness which are defined as having a Type I error not substantially higher than the presumed value for any plausible alternative and its power is high, respectively. Thus, the robustness properties of the power function of tests based on AMML, MML, and LS estimators are examined through [100,000/*n*] (integer value) Monte Carlo runs (simulations).

The distribution of ε, known as population model, is taken as LTS with $p = 16.5$. As alternative sample models, first Models (1)–(5) given in Sect. 3 are used. Type I errors and power values of the test statistics based on AMML, MML, and LS estimators for $n = 30$, 50, and 100, where $\mu = 0.0$ and $\sigma = 1.0$ are given in Table 4.

It can be seen from Table 4 that AMML, MML, and LS give similar results under Model (1). However, as the sample models deviate more from normality, AMML has

Table 4 Simulated power values of test statistics under alternative models (1)–(5)

	$n=30$			$n=50$			$n=100$		
	T_1	T_1'	T_2	T_1	T_1'	T_2	T_1	T_1'	T_2
ϕ	Model (1)								
0.00	0.037	0.032	0.030	0.043	0.042	0.042	0.045	0.042	0.037
0.05	0.06	0.05	0.05	0.08	0.07	0.07	0.11	0.10	0.10
0.10	0.10	0.10	0.10	0.13	0.12	0.12	0.24	0.23	0.23
0.20	0.19	0.19	0.19	0.34	0.34	0.34	0.59	0.58	0.59
0.30	0.37	0.39	0.38	0.59	0.60	0.60	0.89	0.88	0.88
0.40	0.57	0.60	0.60	0.83	0.84	0.84	0.99	0.99	0.99
0.50	0.76	0.78	0.78	0.94	0.95	0.95	1.0	1.0	1.0
ϕ	Model (2)								
0.00	0.035	0.033	0.034	0.034	0.031	0.032	0.044	0.044	0.041
0.05	0.06	0.06	0.06	0.08	0.07	0.07	0.12	0.11	0.11
0.10	0.09	0.09	0.09	0.13	0.12	0.12	0.25	0.23	0.23
0.20	0.22	0.22	0.21	0.34	0.34	0.34	0.63	0.60	0.60
0.30	0.41	0.41	0.39	0.61	0.61	0.61	0.90	0.90	0.90
0.40	0.59	0.59	0.59	0.83	0.85	0.84	0.99	0.99	0.99
0.50	0.78	0.79	0.78	0.96	0.96	0.96	1.00	1.00	1.00
ϕ	Model (3)								
0.00	0.028	0.023	0.024	0.038	0.035	0.036	0.035	0.032	0.034
0.05	0.06	0.06	0.05	0.07	0.06	0.06	0.10	0.09	0.10
0.10	0.09	0.09	0.08	0.13	0.11	0.10	0.24	0.22	0.20

(continued)

Table 4 (continued)

	$n=30$			$n=50$			$n=100$		
	T_1	T_1'	T_2	T_1	T_1'	T_2	T_1	T_1'	T_2
0.20	0.22	0.21	0.19	0.36	0.35	0.33	0.60	0.57	0.55
0.30	0.38	0.38	0.36	0.65	0.62	0.60	0.92	0.91	0.88
0.40	0.60	0.60	0.58	0.85	0.84	0.82	0.99	0.99	0.99
0.50	0.78	0.80	0.79	0.96	0.96	0.94	1.00	1.00	1.00
ϕ	Model (4)								
0.00	0.029	0.025	0.026	0.043	0.035	0.039	0.047	0.038	0.038
0.05	0.05	0.05	0.04	0.08	0.07	0.06	0.13	0.10	0.08
0.10	0.09	0.09	0.08	0.14	0.12	0.10	0.24	0.21	0.18
0.20	0.23	0.21	0.19	0.38	0.35	0.32	0.68	0.63	0.60
0.30	0.44	0.39	0.37	0.66	0.63	0.60	0.94	0.91	0.89
0.40	0.64	0.62	0.59	0.87	0.86	0.83	0.99	0.99	0.99
0.50	0.79	0.78	0.76	0.96	0.95	0.93	1.00	1.00	1.00
ϕ	Model (5)								
0.00	0.027	0.022	0.023	0.037	0.032	0.033	0.032	0.037	0.039
0.05	0.05	0.05	0.04	0.08	0.07	0.05	0.10	0.08	0.07
0.10	0.09	0.08	0.08	0.16	0.12	0.10	0.26	0.21	0.19
0.20	0.25	0.20	0.19	0.41	0.33	0.30	0.73	0.64	0.61
0.30	0.46	0.40	0.38	0.72	0.67	0.64	0.93	0.91	0.88
0.40	0.65	0.62	0.60	0.90	0.88	0.85	1.00	0.99	0.98
0.50	0.84	0.82	0.80	0.99	0.97	0.96	1.00	1.00	1.00

both criterion and efficiency robustness and converges to 1.00 faster as n becomes larger.

Then, the outlier Models (6) and (7) and mixture Models (8) and (9) are considered as alternative sample models. The innovations are scale corrected to make their variances equal to σ^2. The results for $n = 30$, 50, and 100, where $\mu = 0.0$ and $\sigma = 1.0$ are given in Table 5. It can be seen that the results are similar to misspecification ones given in Table 4.

Furthermore, following [3, 15 (Appendix A)]

$$\lim_{n\to\infty} \frac{1}{n}\beta_i(y_{i-1} - \bar{y}_{i-1})^2 \cong QE(y_{i-1} - \bar{y}_{i-1})^2$$
$$= QV(y_{i-1}) = \frac{(p-1/2)(p+1/2)}{p(p+1)}V(y_{i-1}) \tag{14}$$

where $V(y_{i-1}) = \sigma^2/(1-\phi^2)$, the following result is obtained.

Lemma 2 *If $\phi = 0$, then the conditional distribution (σ known) of $\hat{\phi}(\sigma)$ is asymptotically normal with mean ϕ and variance $k\sigma^2/\{2pnQV(y_{i-1})\}$.*

Based on Lemma 2, we propose an alternative test statistic as

$$T_3 = \sqrt{(1.1)n\frac{(p-1/2)(p+1/2)}{p(p+1)}V(y_{i-1})}\left(\hat{\phi}/\hat{\sigma}\right) \tag{15}$$

where $p = 16.5$ and $V(y_{i-1}) = \hat{\sigma}^2$ under H_0. Again, large values of T_3 lead to the rejection of H_0: $\phi = 0$ in favor of H_1: $\phi > 0$.

The corresponding test statistic based on the MML procedure is obtained by replacing ϕ and σ by their MML estimators and Q by $\frac{(p-1/2)}{(p+1)}$ if $\beta_1 > 0$, else keeping the same as the one used in T_3 (see comment in Sect. 2.1) in Eq. (12). The test statistic based on MML estimators is denoted by T_3'.

The test statistic based on the LS estimator is

$$T_4 = \sqrt{nV(y_{i-1})}\left(\tilde{\phi}/\tilde{\sigma}\right). \tag{16}$$

The robustness properties of these test statistics for the Models (1)–(9) are given in Tables 6 and 7.

We suggest using T_1 rather than T_3 since it has better robustness properties. This is due to the fact that the estimators are almost unbiased and T_1 considers the variance of the estimator while T_3 does not. Actually, this is the reason that the performances of the T_3 and corresponding MML and LS test statistics are similar.

Table 5 Simulated power values of test statistics under alternative models (6)–(9)

	$n=30$			$n=50$			$n=100$		
	T_1	T_1'	T_2	T_1	T_1'	T_2	T_1	T_1'	T_2
ϕ	Model (6)								
0.00	0.035	0.031	0.030	0.036	0.034	0.031	0.034	0.028	0.029
0.05	0.06	0.06	0.05	0.07	0.06	0.06	0.10	0.09	0.08
0.10	0.09	0.09	0.08	0.12	0.11	0.09	0.24	0.21	0.19
0.20	0.22	0.21	0.20	0.35	0.34	0.31	0.63	0.63	0.60
0.30	0.39	0.39	0.37	0.60	0.60	0.58	0.91	0.90	0.89
0.40	0.59	0.60	0.58	0.84	0.84	0.82	0.98	0.99	0.99
0.50	0.82	0.80	0.77	0.94	0.95	0.93	1.00	1.00	1.00
0.60	0.89	0.90	0.88	0.99	0.99	0.98			
0.70	0.96	0.95	0.94	1.00	1.00	1.00			
ϕ	Model (7)								
0.00	0.027	0.026	0.026	0.038	0.031	0.032	0.041	0.034	0.035
0.05	0.06	0.05	0.04	0.08	0.06	0.05	0.12	0.08	0.06
0.10	0.10	0.07	0.08	0.16	0.12	0.11	0.28	0.22	0.20
0.20	0.26	0.19	0.17	0.45	0.35	0.33	0.75	0.65	0.61
0.30	0.50	0.41	0.38	0.73	0.65	0.61	0.95	0.92	0.91
0.40	0.70	0.65	0.62	0.90	0.87	0.84	1.00	0.99	0.98
0.50	0.84	0.82	0.80	0.97	0.97	0.96	1.00	1.00	1.00
0.60	0.92	0.92	0.91	0.99	0.99	0.99			
0.70	0.97	0.98	0.97	1.00	1.00	1.00			

(continued)

Table 5 (continued)

	$n = 30$			$n = 50$			$n = 100$		
	T_1	T_1'	T_2	T_1	T_1'	T_2	T_1	T_1'	T_2
ϕ	Model (8)								
0.00	0.026	0.026	0.029	0.040	0.036	0.036	0.060	0.050	0.047
0.05	0.06	0.05	0.05	0.08	0.07	0.06	0.12	0.11	0.10
0.10	0.09	0.09	0.09	0.14	0.13	0.12	0.25	0.23	0.22
0.20	0.22	0.22	0.22	0.37	0.36	0.34	0.60	0.59	0.57
0.30	0.38	0.39	0.38	0.62	0.61	0.58	0.91	0.90	0.89
0.40	0.60	0.61	0.60	0.84	0.84	0.82	0.99	0.99	0.99
0.50	0.76	0.78	0.78	0.94	0.95	0.93	1.00	1.00	1.00
0.60	0.89	0.91	0.90	0.98	0.98	0.97			
0.70	0.95	0.96	0.96	1.00	1.00	1.00			
ϕ	Model (9)								
0.00	0.027	0.025	0.025	0.036	0.026	0.027	0.027	0.026	0.034
0.05	0.05	0.04	0.04	0.07	0.05	0.05	0.12	0.10	0.10
0.10	0.10	0.08	0.08	0.16	0.11	0.11	0.29	0.24	0.23
0.20	0.26	0.20	0.19	0.44	0.35	0.33	0.74	0.65	0.60
0.30	0.49	0.42	0.40	0.74	0.66	0.62	0.95	0.92	0.90
0.40	0.70	0.65	0.62	0.91	0.87	0.84	0.99	0.99	0.97
0.50	0.83	0.81	0.78	0.97	0.97	0.96	1.00	1.00	1.00
0.60	0.92	0.92	0.90	0.99	0.99	0.99			
0.70	0.96	0.97	0.97	1.00	1.00	1.00			

Table 6 Simulated power values of alternative test statistics under models (1)–(5)

	$n = 30$			$n = 50$			$n = 100$		
	T_3	T_3'	T_4	T_3	T_3'	T_4	T_3	T_3'	T_4
ϕ	Model (1)								
0.00	0.030	0.029	0.029	0.038	0.041	0.041	0.041	0.039	0.037
0.05	0.05	0.05	0.05	0.07	0.07	0.07	0.09	0.10	0.10
0.10	0.09	0.10	0.10	0.12	0.12	0.12	0.23	0.24	0.23
0.20	0.17	0.19	0.19	0.32	0.34	0.33	0.57	0.59	0.58
0.30	0.34	0.38	0.37	0.56	0.59	0.59	0.88	0.88	0.88
0.40	0.54	0.59	0.59	0.81	0.84	0.84	0.99	0.99	0.99
0.50	0.74	0.78	0.77	0.93	0.95	0.95	1.00	1.00	1.00
ϕ	Model (2)								
0.00	0.028	0.033	0.033	0.029	0.030	0.031	0.041	0.041	0.042
0.05	0.05	0.05	0.05	0.07	0.07	0.07	0.10	0.10	0.11
0.10	0.08	0.08	0.08	0.11	0.12	0.12	0.22	0.23	0.22
0.20	0.19	0.20	0.21	0.31	0.32	0.33	0.60	0.59	0.60
0.30	0.35	0.39	0.38	0.58	0.61	0.60	0.89	0.90	0.89
0.40	0.55	0.58	0.58	0.82	0.84	0.84	0.99	0.99	0.99
0.50	0.75	0.78	0.78	0.95	0.96	0.96	1.00	1.00	1.00
ϕ	Model (3)								
0.00	0.019	0.021	0.021	0.031	0.034	0.034	0.029	0.028	0.034
0.05	0.05	0.05	0.05	0.06	0.06	0.06	0.08	0.09	0.09
0.10	0.07	0.08	0.09	0.10	0.11	0.11	0.20	0.22	0.22

(continued)

Table 6 (continued)

	$n=30$			$n=50$			$n=100$		
	T_3	T_3'	T_4	T_3	T_3'	T_4	T_3	T_3'	T_4
0.20	0.18	0.20	0.20	0.32	0.34	0.34	0.55	0.56	0.55
0.30	0.33	0.37	0.37	0.59	0.61	0.61	0.90	0.90	0.90
0.40	0.55	0.59	0.59	0.82	0.83	0.83	0.99	0.99	0.99
0.50	0.75	0.78	0.78	0.95	0.95	0.95	1.00	1.00	1.00
ϕ	Model (4)								
0.00	0.019	0.023	0.025	0.025	0.032	0.038	0.033	0.033	0.038
0.05	0.04	0.05	0.05	0.05	0.06	0.06	0.09	0.10	0.11
0.10	0.07	0.08	0.08	0.10	0.11	0.12	0.18	0.19	0.20
0.20	0.17	0.19	0.20	0.31	0.33	0.33	0.61	0.62	0.60
0.30	0.35	0.37	0.37	0.60	0.61	0.60	0.91	0.89	0.89
0.40	0.57	0.60	0.60	0.85	0.85	0.84	0.99	0.99	0.99
0.50	0.74	0.77	0.77	0.95	0.95	0.95	1.00	1.00	1.00
ϕ	Model (5)								
0.00	0.014	0.020	0.021	0.022	0.029	0.032	0.020	0.031	0.041
0.05	0.03	0.04	0.05	0.05	0.06	0.07	0.05	0.07	0.08
0.10	0.06	0.07	0.07	0.09	0.11	0.12	0.17	0.18	0.20
0.20	0.16	0.18	0.19	0.29	0.31	0.31	0.62	0.62	0.61
0.30	0.35	0.37	0.37	0.63	0.64	0.63	0.90	0.90	0.88
0.40	0.58	0.61	0.59	0.87	0.86	0.85	0.99	0.99	0.98
0.50	0.78	0.80	0.79	0.97	0.97	0.96	1.00	1.00	1.00

Table 7 Simulated power values of alternative test statistics under models (6)–(9)

	$n=30$			$n=50$			$n=100$		
	T_3	T_3'	T_4	T_3	T_3'	T_4	T_3	T_3'	T_4
ϕ	Model (6)								
0.00	0.026	0.027	0.028	0.032	0.031	0.031	0.028	0.028	0.028
0.05	0.05	0.05	0.05	0.06	0.06	0.07	0.08	0.08	0.09
0.10	0.08	0.08	0.08	0.10	0.10	0.11	0.21	0.20	0.20
0.20	0.19	0.20	0.20	0.31	0.32	0.33	0.61	0.62	0.62
0.30	0.35	0.37	0.37	0.58	0.59	0.59	0.89	0.90	0.89
0.40	0.55	0.59	0.59	0.82	0.83	0.82	0.98	0.99	0.99
0.50	0.75	0.79	0.79	0.94	0.95	0.95	1.00	1.00	1.00
0.60	0.86	0.89	0.89	0.98	0.99	0.99			
0.70	0.93	0.95	0.95	1.00	1.00	1.00			
ϕ	Model (7)								
0.00	0.016	0.021	0.023	0.018	0.027	0.032	0.023	0.031	0.035
0.05	0.04	0.04	0.04	0.04	0.05	0.06	0.06	0.07	0.08
0.10	0.05	0.07	0.07	0.08	0.10	0.11	0.16	0.18	0.20
0.20	0.16	0.17	0.17	0.31	0.32	0.32	0.62	0.62	0.60
0.30	0.36	0.37	0.37	0.63	0.61	0.60	0.93	0.92	0.91
0.40	0.62	0.62	0.60	0.86	0.85	0.84	0.99	0.99	0.98
0.50	0.80	0.80	0.79	0.96	0.97	0.96	1.00	1.00	1.00
0.60	0.90	0.91	0.91	0.99	0.99	0.99			
0.70	0.96	0.97	0.97	1.00	1.00	1.00			

(continued)

Table 7 (continued)

	$n=30$			$n=50$			$n=100$		
	T_3	T_3'	T_4	T_3	T_3'	T_4	T_3	T_3'	T_4
ϕ	Model (8)								
0.00	0.020	0.025	0.025	0.032	0.032	0.034	0.048	0.047	0.047
0.05	0.05	0.05	0.05	0.06	0.07	0.07	0.10	0.11	0.11
0.10	0.08	0.08	0.09	0.12	0.12	0.13	0.22	0.23	0.23
0.20	0.18	0.20	0.21	0.34	0.35	0.35	0.56	0.58	0.57
0.30	0.34	0.37	0.37	0.58	0.60	0.60	0.89	0.89	0.89
0.40	0.56	0.60	0.59	0.81	0.83	0.83	0.99	0.99	0.99
0.50	0.74	0.78	0.77	0.93	0.95	0.94	1.00	1.00	1.00
0.60	0.88	0.90	0.90	0.98	0.98	0.98			
0.70	0.95	0.96	0.96	0.99	1.00	1.00			
ϕ	Model (9)								
0.00	0.015	0.021	0.024	0.014	0.021	0.025	0.014	0.023	0.032
0.05	0.03	0.04	0.04	0.04	0.05	0.05	0.06	0.09	0.10
0.10	0.05	0.07	0.07	0.08	0.10	0.11	0.18	0.21	0.23
0.20	0.15	0.18	0.18	0.30	0.32	0.32	0.63	0.62	0.60
0.30	0.35	0.39	0.39	0.63	0.63	0.62	0.93	0.92	0.90
0.40	0.61	0.62	0.61	0.86	0.86	0.84	0.99	0.99	0.98
0.50	0.78	0.80	0.79	0.97	0.97	0.96	1.00	1.00	1.00
0.60	0.91	0.92	0.91	0.99	0.99	0.99			
0.70	0.96	0.97	0.97	1.00	1.00	1.00			

5 Generalization to AR(q) Model

Consider the general stationary autoregressive model of order p

$$y_t = \mu + \sum_{j=1}^{q} \phi_j y_{t-j} + \varepsilon_t, \quad (1 \le t \le n) \tag{17}$$

where the innovations ε_t are independent and identically distributed (iid), and have one of the distributions in LTS family given in Eq. (2).

Take initially that $\phi_1 = \phi_2 = \cdots = \phi_q$ and let T_0 and S_0 be the initial estimators of μ and σ, respectively, given as

$$\begin{aligned} T_0 &= med\left\{y_i - \hat{\phi}_0\left(y_{i-1} + y_{i-2} + \cdots + y_{i-q}\right)\right\}\text{and} \\ S_0 &= 1.483\, med\left\{\left|y_i - \hat{\phi}_0\left(y_{i-1} + y_{i-2} + \cdots + y_{i-q}\right) - T_0\right|\right\}, (1 \le i \le n) \end{aligned} \tag{18}$$

where $\hat{\phi}_0 = med\left\{\frac{y_{q+1}-y_q}{y_q-y_0}, \frac{y_{q+2}-y_{q+1}}{y_{q+1}-y_1}, \ldots, \frac{y_n-y_{n-1}}{y_{n-1}-y_{n-q-1}}\right\}, (i = 1, 2, \ldots, n-1)$.

The AMML estimators are

$$\begin{aligned} \hat{\mu} &= \sum_{i=1}^{n} \beta_i \left(y_i - \hat{\phi}_1 y_{i-1} - \hat{\phi}_2 y_{i-2} - \cdots - \hat{\phi}_q y_{i-q}\right)/m \\ \hat{\phi} &= C^{-1}(K + D\hat{\sigma}), \hat{\sigma} = \left(B + \sqrt{B^2 + 4nC}\right)/2n \end{aligned} \tag{19}$$

where

$$\hat{\phi} = \begin{bmatrix} \hat{\phi}_1 \\ \hat{\phi}_2 \\ \vdots \\ \hat{\phi}_q \end{bmatrix}, \quad D = \begin{bmatrix} \sum_{i=1}^{n} \alpha_i y_{i-1} \\ \sum_{i=1}^{n} \alpha_i y_{i-2} \\ \vdots \\ \sum_{i=1}^{n} \alpha_i y_{i-q} \end{bmatrix}, \quad K = \begin{bmatrix} \sum_{i=1}^{n} \beta_i y_{i-1}(y_i - \bar{y}_0) \\ \sum_{i=1}^{n} \beta_i y_{i-2}(y_i - \bar{y}_0) \\ \vdots \\ \sum_{i=1}^{n} \beta_i y_{i-q}(y_i - \bar{y}_0) \end{bmatrix},$$

$$C = \begin{bmatrix} C_{11} & \cdots & C_{1q} \\ \vdots & \ddots & \vdots \\ C_{q1} & \cdots & C_{qq} \end{bmatrix}, C_{jk} = \sum_{i=1}^{n} \beta_i y_{i-k}(y_{i-j} - \bar{y}_j), \bar{y}_j = \sum_{i=1}^{n} \beta_i y_{i-j}/m$$

$$m = \sum_{i=1}^{n} \beta_i, B = \frac{2p}{k}\left(\sum_{i=1}^{n} \alpha_i y_i - K'C^{-1}D\right), C = \frac{2p}{k}\left(\sum_{i=1}^{n} \beta_i (y_i - \bar{y}_0) - K'C^{-1}K\right).$$

$\hat{\alpha}_i$ and $\hat{\beta}_i$ are as given in Eq. (10) where $\hat{t}_i = \frac{y_i - \hat{\phi}_0(y_{i-1}+y_{i-2}+\cdots+y_{i-q})-T_0}{S_0}$ again estimated from the sample.

Computation is similar to the AR(1) case. First, use the initial values to calculate the estimates of the model parameters. Then, revise them with the AMML estimates and repeat the process one more time to calculate the desired final AMML estimate values of the model parameters.

6 Conclusion

In this study, for AR(1) models, MML technique is adapted to machine data processing, where the distribution family is known rather than the exact distribution. For this purpose, the idea of Huber M-estimation is inserted to MML technique. Then, the efficiency and robustness properties of the most widely used LS estimators, MML and AMML estimators are examined through simulations and observed that MML and AMML estimators are more efficient than LS estimators as expected. However, AMML underestimates σ in all cases but have much smaller variances than the others yielding less mean squared errors. Therefore, if there is an opportunity to examine the distribution, one should prefer the use of MML rather than AMML. Otherwise, like in machine data processing, one can safely use AMML estimators having in mind the bias in σ which cannot be corrected since the exact distribution is not known. Then, the two different test statistics for the significance test of the model are proposed. Their criterion and efficiency robustness properties are examined via simulation and compared with the corresponding ones using MML and LS estimators in the test statistics. It is observed that under normality they all perform similar. However, as the innovations deviate from normality, more AMML- and MML-based test statistics become more robust especially for detecting small deviations. Again, the performances of AMML and MML are almost similar though MML can be preferred in cases where p is known due to its slightly better performance. Finally, the estimation procedure is expanded to AR(q) models.

References

1. Akkaya, A.D., Tiku, M.L.: Estimating parameters in autoregressive models in non-normal situations: asymmetric innovations. Commun. Stat. Theory M **30**, 517–536 (2001). https://doi.org/10.1081/STA-100002095
2. Akkaya, A.D., Tiku, M.L.: Time series AR(1) model for short-tailed distributions. Statistics **39**, 117–132 (2005). https://doi.org/10.1080/02331880512331344036
3. Akkaya, A.D., Tiku, M.L.: Autoregressive models in short-tailed symmetric distributions. Statistics **42**(3), 207–221 (2008). https://doi.org/10.1080/02331880701736663
4. Bayrak, O.T., Akkaya, A.D.: Estimating parameters of a multiple autoregressive model by the modified maximum likelihood method. J. Comput. Appl. Math. **233**, 1762–1772 (2010). https://doi.org/10.1016/j.cam.2009.09.013

5. Dönmez, A.: Adaptive estimation and hypothesis testing methods. PhD thesis, Middle East Technical University (2010)
6. Hamilton, L.C.: Regression with graphics. Brooks/Cole Publishing Company, California (1992)
7. Hampel, F.R., Ronchetti, E.M., Rousseeuw, P.J., Stahel, W.A.: Robust Statistics: The Approach Based on Influence Functions. Wiley, New York (1986)
8. Huber, P.J.: Robust Statistics. Wiley, New York (1981)
9. Kendall, M.G., Stuart, A.: The Advanced Theory of Statistics, vol. 2. Charles Griffin, London (1979)
10. Tiku, M.L.: Estimating the mean and standard deviation from a censored normal sample. Biometrika **54**, 155–165 (1964). https://doi.org/10.2307/2333859
11. Tiku, M.L., Akkaya, A.D.: Robust estimation and hypothesis testing. New Age International (P) Ltd., New Delhi, India (2004)
12. Tiku, M.L., Sürücü, B.: MMLEs are as good as M-estimators or better. Stat. Prob. Lett. **79**(7), 984–989 (2009). https://doi.org/10.1016/j.spl.2008.12.001
13. Tiku, M.L., Tan, W.Y., Balakrishnan, N.: Robust Inference. Marcel Dekker, New York (1986)
14. Tiku, M.L., Wong, W.K., Bian, G.: Estimating parameters in autoregressive models in non-normal situations: symmetric innovations. Commun. Stat. Theory M **28**, 315–341 (1999). https://doi.org/10.1080/03610929908832300
15. Tiku, M.L., Wong, W.K., Vaughan, D.C., Bian, G.: Time series models in non-normal situations. J. Time Ser. Anal. **21**, 571–596 (2000). https://doi.org/10.1111/1467-9892.00199
16. Ülgen, B.E.: Robust estimation and hypothesis testing in microarray analysis. PhD thesis, Middle East Technical University (2010)
17. Vinod, H.D., Shenton, L.R.: Exact moments for autoregressive and random walk models for a zero or stationary initial value. Economet. Theory **12**, 481–499 (1996). https://doi.org/10.1017/S0266466600006824

Prediction of High-Dimensional Time Series with Exogenous Variables Using Generalized Koopman Operator Framework in Reproducing Kernel Hilbert Space

Jia-Chen Hua, Farzad Noorian, Philip H. W. Leong, Gemunu Gunaratne and Jorge Gonçalves

Abstract We propose a novel methodology to predict high-dimensional time series with exogenous variables using Koopman operator framework, by assuming that the time series are generated by some underlying unknown dynamical system with input as exogenous variables. In order to do that, we first generalize the definition of the original Koopman operator to allow for input to the underlying dynamical system. We then obtain a formulation of the generalized Koopman operator in reproducing kernel Hilbert space (RKHS) and a new derivation of its numerical approximation methods, namely, Extended Dynamic Mode Decomposition (EDMD) and its kernel-based version. We also obtain a statistical interpretation of kernel-based EDMD developed for deterministic Koopman operator by utilizing the connection between RKHS and Gaussian processes regression, and relate it to the stochastic Koopman and Perron–Frobenius operator. In applications, we found that the prediction performance of this methodology is promising in forecasting real-world high-dimensional time series with exogenous variables, including financial markets data. We believe that this methodology will be of interest to the community of scientists and engineers working on quantitative finance, econometrics, system biology, neurosciences, meteorology, oceanography, system identification and control, data mining, machine learning, computational intelligence, and many other fields involving high-dimensional time series and spatiotemporal data.

J.-C. Hua (✉) · J. Gonçalves
Luxembourg Centre for Systems Biomedicine, University of Luxembourg, 4367 Belvaux, Luxembourg
e-mail: jia.chen.hua@ntnu.no

F. Noorian · P. H. W. Leong
School of Electrical and Information Engineering, University of Sydney, Sydney, NSW 2006, Australia

G. Gunaratne
Department of Physics, University of Houston, Houston, TX 77204, USA

I. Rojas et al. (eds.), *Time Series Analysis and Forecasting*, Contributions to Statistics, https://doi.org/10.1007/978-3-319-96944-2_5

Keywords High-dimensional time series · Spatiotemporal dynamics · Complex system · Koopman operator · Perron–Frobenius operator · Dynamical system · Reproducing kernel Hilbert space · Gaussian processes · Machine learning Data mining · Econophysics · Financial markets modeling · Energy forecasting Collective behavior

1 Introduction

In many application fields, a high-dimensional time series $\{\boldsymbol{x}_t\}$ may be considered as being generated by or sampled from some underlying dynamical system $(\mathbb{R}^N, t, \boldsymbol{F}^t)$ that is often nonlinear and stochastic, where $\boldsymbol{x}_t \mapsto \boldsymbol{F}^\tau(\boldsymbol{x}_t) = \boldsymbol{x}_{t+\tau} \in \mathbb{R}^N$ is the N-dimensional state vector evolved by the flow $\boldsymbol{F}$. However, these high-dimensional state vectors are usually extrinsic measurements or outputs of the underlying lower dimensional true state's dynamics $z_{t+\tau} = \hat{\boldsymbol{F}}^\tau(z_t)$. When the task is to predict each component of the output $\boldsymbol{x}_t$, using some machine learning techniques to learn its superficial dynamics $\boldsymbol{F}^t$ without identifying the true state dynamics $z_{t+\tau} = \hat{\boldsymbol{F}}^\tau(z_t)$ may be computationally heavy and not optimal. Nevertheless, if there is no need for identification other than prediction of outputs, it is often favorable to find some intrinsic feature maps $\{\varphi_i(\boldsymbol{x}_t)\}_{i=1}^M$ (where $M < N$ and the M-dimensional feature vector $\boldsymbol{\varphi}(\boldsymbol{x}_t)$ is not necessarily the same as the underlying true state z_t) to embed the high-dimensional output to lower dimensional intrinsic manifold, or in other words, to learn both the geometry *and* dynamics for *simultaneous* dimensionality reduction and prediction, as shown in the schematic Fig. 1. Therefore, the question is how to find such $\{\varphi_i(\boldsymbol{x}_t)\}_{i=1}^M$ that given the latest output $\boldsymbol{x}_t$, one can predict feature maps' future values $\boldsymbol{\varphi}(\boldsymbol{x}_{t+\tau})$ and transform back to the pre-image $\boldsymbol{x}_{t+\tau}$, and how is $\boldsymbol{\varphi}(\boldsymbol{x}_{t+\tau})$ related to $\boldsymbol{\varphi}(\boldsymbol{x}_t)$ on the intrinsic manifold. The key to these questions is the Koopman operator of dynamical systems [1–3], whose eigenfunctions can serve as the desired intrinsic feature maps $\{\varphi_i(\boldsymbol{x}_t)\}$. The Koopman operator is a linear operator that enables investigation of a nonlinear dynamical system using linear theories and techniques, and since it has been developed as a data-driven framework [4, 5], most of its applications up to now are dealing with high-dimensional time series. There have been several major numerical methods developed to extract the spectral properties

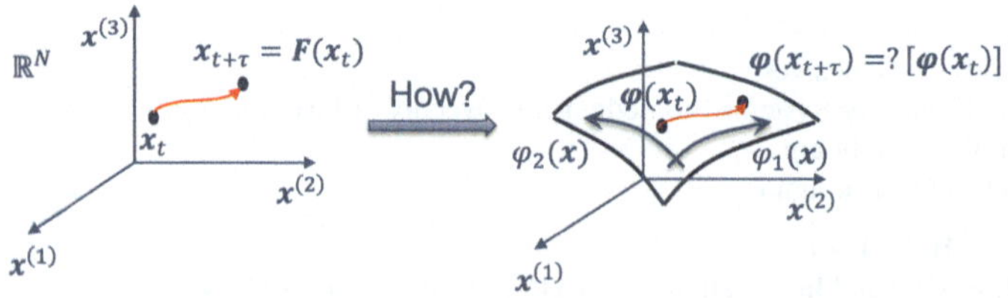

Fig. 1 Manifold learning that is capable of simultaneous dimensionality reduction and prediction

of Koopman operator from time series data, and utilizing these properties for time series prediction has several major advantages [6].

In this paper, we generalize the Koopman operator framework to systems with inputs as exogenous variables. By taking the simplest generalization approach [7], we found that the techniques and methods that we developed for Kernel KMR [6] methodology can be utilized almost directly with minimal modification. Hence, we can generalize Kernel KMR to Kernel GKMRX (Kernel-based Generalized Koopman Mode Regression with eXogenous variables) to predict high-dimensional time series with exogenous variables. In the theory part of this paper, we formulate the Koopman operator in reproducing kernel Hilbert space (RKHS), which is the most important function space in modern machine learning, and we obtain a new derivation of the Extended Dynamic Mode Decomposition (EDMD) algorithm [5] and its kernel-based extension [8] by exploiting the Dirac bra–ket notation [9]. Moreover, we obtain a statistical interpretation of these numerical methods developed for deterministic Koopman operator by exploiting the connection between RKHS and Gaussian processes regression, and relate it to the stochastic Koopman and Perron–Frobenius operator. In the application part, we test our new prediction methodology for various types of data from different fields and obtain promising initial results.

2 Theory

Koopman Operator of Dynamical System and Its Generalization to Systems with Input

Consider a high-dimensional time series $\{\boldsymbol{x}_n\}$ sampled from an underlying dynamical system $(\mathcal{M}, n, \boldsymbol{F})$, where $n \in \mathbb{Z}$ is discrete time, $\mathcal{M} \subset \mathbb{R}^N$ is the N-dimensional state space containing the $\{\boldsymbol{x}_n\}$, and $\boldsymbol{x}_i \mapsto \boldsymbol{F}(\boldsymbol{x}_i) = \boldsymbol{x}_{i+1}$ defines the evolution law. For continuous-time dynamical system $(\mathcal{M}, t, \boldsymbol{F}^t)$, the flow $\boldsymbol{F}^t$ evolves the system state as $\boldsymbol{x}_0 \mapsto \boldsymbol{F}^t(\boldsymbol{x}_0) = \boldsymbol{x}_t$. Since time series data are often sampled with a fixed time gap τ, the adjacent two snapshots of the system are related by $\boldsymbol{F}^\tau(\boldsymbol{x}_t) = \boldsymbol{x}_{t+\tau}$. When the context is clear, we will drop the τ in $\boldsymbol{F}^\tau$ to denote either the discrete time map or continuous-time flow of a fixed time gap τ. Here, we restrict to stationary time series, or at least locally stationary time series, which can be considered as being sampled from autonomous dynamical systems. We will generalize the Koopman operator to systems with input later.

The (deterministic) Koopman operator $\mathcal{K} : \mathcal{F} \to \mathcal{F}$ is defined as $(\mathcal{K}\phi)(\boldsymbol{x}) = (\phi \circ \boldsymbol{F})(\boldsymbol{x}) = \phi(\boldsymbol{F}(\boldsymbol{x}))$, where $\circ$ denotes the composition of ϕ with $\boldsymbol{F}$, and $\mathcal{F}$ is the "feature space" consisting of scalar observables or functions of state space $\phi : \mathcal{M} \to \mathbb{C}$. Since $\mathcal{K}\phi$ is another element in $\mathcal{F}$, the Koopman operator defines a new dynamical system $(\mathcal{F}, n, \mathcal{K})$ where $\mathcal{K}$ evolves an observable or feature $\phi \in \mathcal{F}$ to a new function $\mathcal{K}\phi$ that gives the value of ϕ at "one step in the future." Unlike $\boldsymbol{F}$ which is finite dimensional, $\mathcal{K}$ is infinite dimensional because it acts on func-

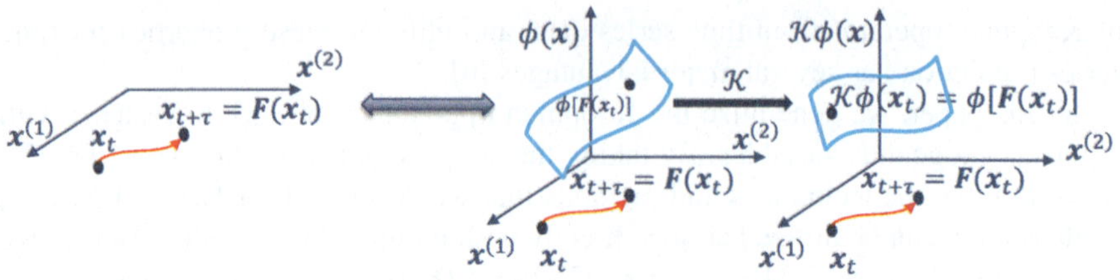

Fig. 2 Koopman operator is the "pullback" of $\boldsymbol{F}$: when the system evolves from $\boldsymbol{x}_t$ to $\boldsymbol{F}(\boldsymbol{x}_t)$ in state space, a function ϕ defined on the state space is evolved by $\mathcal{K}$ such that the new function $\mathcal{K}\phi$ evaluated at old state $\boldsymbol{x}_t$ is taking the value of old function ϕ evaluated at new state $\boldsymbol{F}(\boldsymbol{x}_t)$

tion space $\mathcal{F}$. However, it is also linear even when $\boldsymbol{F}$ is nonlinear, and hence one can investigate its spectral properties, i.e., eigenvalues and eigenfunctions, which we refer to as Koopman eigenvalues $\{\mu_k\}$ and eigenfunctions $\{\varphi_k\}$.

Technical results showed that $\mathcal{K}$ fully characterizes $\boldsymbol{F}$ under very general conditions [2, 3], so dynamical systems $(\mathcal{M}, n, \boldsymbol{F})$ and $(\mathcal{F}, n, \mathcal{K})$ are two different representations of the same evolution, as shown in the schematic Fig. 2. The link between these two representations is the "full state observable" $\boldsymbol{g}(\boldsymbol{x}) = \boldsymbol{x}$, where $\boldsymbol{x} \mapsto \boldsymbol{F}(\boldsymbol{x})$, and $g_i \mapsto (\mathcal{K}g_i) = g_i \circ \boldsymbol{F}$ where $g_i \in \mathcal{F}$ is the ith component of the *vector-valued observable* $\boldsymbol{g} : \mathcal{M} \to \mathbb{R}^N$. Assuming g_i is in the span of a set of K Koopman eigenfunctions $\{\varphi_k\}_{k=1}^K$, where K could (and often will) be infinite, then it can be projected as $g_i = \sum_{k=1}^K \xi_{ik}\varphi_k$ with $\xi_{ik} \in \mathbb{C}$. Hence, $\boldsymbol{g}$ can be obtained by "stacking" these weights into vectors (i.e., $\boldsymbol{\xi}_j = [\xi_{1j}, \xi_{2j}, \ldots, \xi_{Nj}]^T$). As a result,

$$\boldsymbol{x} = \boldsymbol{g}(\boldsymbol{x}) = \sum_{k=1}^{K} \boldsymbol{\xi}_k \varphi_k(\boldsymbol{x}), \tag{1}$$

where $\boldsymbol{\xi}_k$ is the kth *Koopman mode* corresponding to the eigenfunction φ_k. To make prediction or arrive at the system state of "one step in the future," one can either evolve $\boldsymbol{x}$ through $\boldsymbol{F}$ directly, or evolve the full state observable $\boldsymbol{g}(\boldsymbol{x})$ through the Koopman operator $\mathcal{K}$ as $\boldsymbol{g}(\boldsymbol{F}(\boldsymbol{x})) = (\mathcal{K}\boldsymbol{g})(\boldsymbol{x}) = \sum_{k=1}^K \boldsymbol{\xi}_k(\mathcal{K}\varphi_k)(\boldsymbol{x}) = \sum_{k=1}^K \mu_k \boldsymbol{\xi}_k \varphi_k(\boldsymbol{x})$. Similarly, for continuous-time case, we have $\boldsymbol{x}_{t+\tau} = \boldsymbol{F}^{\tau}(\boldsymbol{x}_t) = \boldsymbol{g}(\boldsymbol{F}^{\tau}(\boldsymbol{x}_t)) = (\mathcal{K}_\tau \boldsymbol{g})(\boldsymbol{x}_t)$ $= \sum_{k=1}^K \mathrm{e}^{\lambda_k \tau} \boldsymbol{\xi}_k \varphi_k(\boldsymbol{x}_t)$, where λ_k and φ_k are the kth eigenvalue and eigenfunction of the infinitesimal generator $\hat{\mathcal{K}} \triangleq \frac{\mathrm{d}}{\mathrm{d}t}$ of the semi-group of Koopman operators $\{\mathcal{K}_t\}_{t \in \mathbb{R}^+}$, and $\mu_k = \mathrm{e}^{\lambda_k \tau}$ is the kth eigenvalue of finite-time Koopman operator $\mathcal{K}_\tau = \mathrm{e}^{\tau \hat{\mathcal{K}}}$.

In order to compute $\{(\mu_k, \varphi_k, \boldsymbol{\xi}_k)\}_{k=1}^K$ of Koopman eigenvalues, eigenfunctions, and modes from data, one has to find a matrix representation of $\mathcal{K}$ by projecting it into some subspace of $\mathcal{F}$ spanned by a basis $\{\psi_k(\boldsymbol{x})\}_{k=1}^K$. For computational feasibility and convenience, we usually require $\psi_k(\cdot) \in L^2(\mathcal{M})$, such that we can compute inner products using training data $\{(\boldsymbol{x}_1, \boldsymbol{y}_1), \ldots, (\boldsymbol{x}_M, \boldsymbol{y}_M)\}$ where $\boldsymbol{y}_i = \boldsymbol{F}(\boldsymbol{x}_i)$, in order to require $\{\psi_k(\boldsymbol{x})\}_{k=1}^K$ to be orthonormal by computing the Moore–Penrose pseudoinverse of the data matrix $\boldsymbol{\Psi}_x^+$, where $[\boldsymbol{\Psi}_x]_{ij} = \psi_j(\boldsymbol{x}_i)$.

Exploiting Dirac's bra–ket notation [9] to write functions, functionals, inner products, and linear operators in a compact way, we denote the ith row of $\boldsymbol{\Psi}_x^+$ as $\langle\psi_i|$ such that the inner product $\langle\psi_i|\psi_j\rangle_{L^2} = \delta_{ij}$, where δ_{ij} is the Kronecker delta. Hence, in this "feature space" $\mathcal{F}_K \triangleq \mathrm{span}\{\psi_k(\cdot)\}_{k=1}^K$, the identity operator can be written as $\mathbb{1}_{\mathcal{F}_K} = \sum_{k=1}^K |\psi_k\rangle\langle\psi_k|$, and $\mathcal{K}$ projected to $\mathcal{F}_K$ can be written as $\mathcal{K} = \mathcal{K}\mathbb{1}_{\mathcal{F}_K} = \mathcal{K}\sum_{k=1}^K |\psi_k\rangle\langle\psi_k| = \sum_{k=1}^K |\psi_k \circ \boldsymbol{F}\rangle\langle\psi_k|$. Therefore, the elements of matrix representation $\boldsymbol{K}$ of $\mathcal{K}$ are $\boldsymbol{K}_{ij} = \langle\psi_i|\mathcal{K}|\psi_j\rangle_{L^2} = \langle\psi_i|\psi_j \circ \boldsymbol{F}\rangle_{L^2}$, and $\boldsymbol{K} = \boldsymbol{\Psi}_x^+\boldsymbol{\Psi}_y$, where the jth column of $\boldsymbol{\Psi}_y$ is $|\psi_j \circ \boldsymbol{F}\rangle$ and $[\boldsymbol{\Psi}_y]_{ij} = \psi_j \circ \boldsymbol{F}(\boldsymbol{x}_i) = \psi_j(\boldsymbol{y}_i)$. Eigenvalue problem $\mathcal{K}|\varphi_k\rangle = \mu_k|\varphi_k\rangle$ becomes eigenvalue equation of $\boldsymbol{K}$ as $\boldsymbol{K}\boldsymbol{v}_k = \mu_k\boldsymbol{v}_k$, where the ith component of $\boldsymbol{v}_k$ is $(\boldsymbol{v}_k)_i = \langle\psi_i|\varphi_k\rangle_{L^2}$, so the eigenfunction $|\varphi_k\rangle = \sum_{i=1}^K |\psi_i\rangle(\boldsymbol{v}_k)_i$, or $\boldsymbol{\Phi}_x = \boldsymbol{\Psi}_x\boldsymbol{V}$ in matrix notation, where $[\boldsymbol{\Phi}_x]_{ij} = \varphi_j(\boldsymbol{x}_i)$ and columns of $\boldsymbol{V}$ are $\{\boldsymbol{v}_k\}$. The *continuous-time eigenvalue* can be computed as $\lambda_k \triangleq \log(\mu_k)/\tau$, and according to Eq. (1), Koopman modes $\{\boldsymbol{\xi}_k\}$ can be computed by projecting $\boldsymbol{g}(\boldsymbol{x}) = \boldsymbol{x}$ onto $\{\varphi_k(\boldsymbol{x})\}$ as $\boldsymbol{\Xi} = \boldsymbol{\Phi}_x^+\boldsymbol{X}$, where the ith rows of $\boldsymbol{\Xi}$ and $\boldsymbol{X}$ are $\boldsymbol{\xi}_i^T$ and $\boldsymbol{x}_i^T$, respectively. This procedure is called Extended Dynamic Mode Decomposition (EDMD) [5], and it has become one of the most widely adopted numerical methods for data-driven Koopman spectral analysis, even outside the fluid dynamics community where the Koopman operator's spectral properties was thoroughly investigated for the first time [2].

Furthermore, there are several ways to generalize Koopman operator to systems with input [7]. One of the simplest ways is to augment the system state $\boldsymbol{x}_t$ with the current input $\boldsymbol{u}_t \in \mathbb{R}^{N'}$, such that the dimension of the extended system state $\tilde{\boldsymbol{x}}$ will be $N + N'$. The time evolution of the system will be extended as $\tilde{\boldsymbol{x}}_{t+\tau} = \tilde{\boldsymbol{F}}^\tau(\tilde{\boldsymbol{x}}_t) = \tilde{\boldsymbol{F}}^\tau(\boldsymbol{x}_t, \boldsymbol{u}_t)$, where the first N components of $\tilde{\boldsymbol{F}}$ and $\tilde{\boldsymbol{x}}$ are $\boldsymbol{x}_{t+\tau} = \boldsymbol{F}^\tau(\boldsymbol{x}_t, \boldsymbol{u}_t)$, and we assume that there is a purely formal map or flow that "shifts" the input as $\boldsymbol{u}_{t+\tau} = \boldsymbol{S}^\tau(\boldsymbol{x}_t, \boldsymbol{u}_t)$, since there is not necessarily any "dynamics" of the input. The generalized Koopman operator can be defined on this extended system as before $\mathcal{K}\phi(\tilde{\boldsymbol{x}}_t) = \phi \circ \tilde{\boldsymbol{F}}^\tau(\tilde{\boldsymbol{x}}_t)$. For prediction purposes, we are only interested in the original system state $\boldsymbol{x}$, so there is no need to project N'-dimensional full state observable of input $\boldsymbol{g}_u(\boldsymbol{x}_t, \boldsymbol{u}_t) = \boldsymbol{u}_t$ on Koopman eigenfunctions in order to compute the corresponding Koopman modes for input. Except for this trivial difference, all the available numerical procedures for Koopman spectral analysis and prediction can be applied with very little modification. Notice that this augmentation trick can be also applied to previous state and input, such that one can investigate a system with finite amount of memory in the same way as investigating a system without memory. For simplicity, we only consider memoryless system in this paper, and this topic will be left for future investigation.

Reproducing Kernel Hilbert Space and Gaussian Processes Regression

In this subsection, we briefly summarize the basic theory of reproducing kernel Hilbert space (RKHS) and its relation to Gaussian processes regression. For a more complete exposition of this topic with technical details, we refer the readers to Refs. [10, 11].

The RKHS is a Hilbert space of functions equipped with inner product $\langle\cdot|\cdot\rangle_{\mathcal{H}_k}$ satisfying: (1) $\forall \boldsymbol{x}$ fixed, $k(\boldsymbol{x}, \boldsymbol{y}) = k_{\boldsymbol{x}}(\cdot) \in \mathcal{H}_k$ is a function of $\boldsymbol{y}$; (2) $k(\cdot, \cdot)$ has the "reproducing" property: $\forall f \in \mathcal{H}_k$, $\langle f(\cdot)|k_{\boldsymbol{x}}(\cdot)\rangle_{\mathcal{H}_k} = f(\boldsymbol{x})$. It follows from (2) that $\langle k_{\boldsymbol{y}}(\cdot)|k_{\boldsymbol{x}}(\cdot)\rangle_{\mathcal{H}_k} = k_{\boldsymbol{y}}(\boldsymbol{x}) = k_{\boldsymbol{x}}(\boldsymbol{y}) = k(\boldsymbol{y}, \boldsymbol{x})$. Each RKHS has a unique k, and according to Moore–Aronszajn theorem, given any symmetric positive definite function $k(\boldsymbol{y}, \boldsymbol{x})$, there is a unique RKHS such that $k(\boldsymbol{y}, \boldsymbol{x})$ is the reproducing kernel. In fact, this theorem showed that this unique RKHS $\{f \in \mathcal{H}_k | f(\cdot) = \sum_{i=1}^{M\to\infty} \alpha_i k(\cdot, \boldsymbol{x}_i)\}$ can be built from defining the inner product $\langle f|g\rangle_{\mathcal{H}_k} = \sum_{j=1}^{M'\to\infty} \sum_{i=1}^{M\to\infty} \alpha_i \beta_j k(\boldsymbol{y}_j, \boldsymbol{x}_i)$, where $g(\cdot) = \sum_{j=1}^{M'\to\infty} \beta_j k(\cdot, \boldsymbol{y}_j)$. It satisfies the reproducing property $\langle f(\cdot)|k_{\boldsymbol{x}}(\cdot)\rangle_{\mathcal{H}_k}$ $= \sum_{i=1}^{M\to\infty} \langle \alpha_i k(\cdot, \boldsymbol{x}_i)|k_{\boldsymbol{x}}(\cdot)\rangle_{\mathcal{H}_k} = \sum_{i=1}^{M\to\infty} \alpha_i k(\boldsymbol{x}, \boldsymbol{x}_i) = f(\boldsymbol{x})$. The reproducing kernels can be considered as a basis of this RKHS, and they are also called "point evaluation functionals."

Another representation of RKHS is from Mercer's theorem, which states that a positive (semi-)definite function can be eigen-decomposed as $k(\boldsymbol{x}, \boldsymbol{x}') = \sum_{i=1}^{\infty} \sigma_i q_i(\boldsymbol{x}) q_i(\boldsymbol{x}')$, where $\{q_i(\cdot)\}$ are orthonormal in L^2, and $\{\sigma_i\}_{i=1}^{M\to\infty}$ is a nonincreasing sequence of eigenvalues with $\sigma_M \to 0$ when $M \to \infty$. It follows from this theorem that the unique RKHS associated to this $k(\boldsymbol{x}, \boldsymbol{x}')$ is $\{f \in L^2 | \sum_{i=1}^{\infty} \frac{\langle q_i|f\rangle_{L^2}^2}{\sigma_i} < \infty\}$, and the inner product is given by $\langle f|g\rangle_{\mathcal{H}_k} = \sum_{i=1}^{\infty} \langle f|q_i\rangle_{L^2} \frac{1}{\sigma_i} \langle q_i|g\rangle_{L^2}$. One consequence of this inner product is that the induced norm is $\|f\|_{\mathcal{H}_k}^2 = \langle f|f\rangle_{\mathcal{H}_k} = \sum_{i=1}^{\infty} \frac{\langle q_i|f\rangle_{L^2}^2}{\sigma_i}$, and in order to be bounded, the components $f_i = \langle q_i|f\rangle_{L^2}$ must decay quickly when i increases, which effectively imposes a smoothness requirement on L^2 in order for it to become a RKHS. Another consequence of this inner product is that one can define $\{p_i(\cdot) = \sqrt{\sigma_i} q_i(\cdot)\}$ such that it is an orthonormal basis of this unique RKHS, and as an analog to the Dirac delta which can be represented by $\delta_{\boldsymbol{x}}(\cdot) = \sum_{i=1}^{\infty} q_i(\boldsymbol{x}) q_i(\cdot)$, the reproducing kernel functions can be written as $k_{\boldsymbol{x}}(\cdot) = \sum_{i=1}^{\infty} p_i(\boldsymbol{x}) p_i(\cdot)$.

For a regularized optimization problem $J[f] = \frac{1}{2\lambda_M^2} \sum_{i=1}^{M} (y_i - f(\boldsymbol{x}_i))^2 + \frac{1}{2}\|f\|_{\mathcal{H}_k}^2$ given some training data or observations $\{(\boldsymbol{x}_1, y_1), (\boldsymbol{x}_2, y_2), \ldots, (\boldsymbol{x}_M, y_M)\}$, where $\boldsymbol{x}_i \in \mathbb{R}^N$ and $y_i \in \mathbb{R}$, the representer theorem [12] asserts that the minimizer $\hat{f}(\cdot) = \sum_{i=1}^{M} \alpha_i k(\cdot, \boldsymbol{x}_i)$, such that one can effectively minimize $J[\alpha_i]$ by setting the derivatives with respect to α_i equal to zeros, and then the α_i's can be solved as a column vector $\boldsymbol{\alpha} = (\boldsymbol{G} + \lambda_M^2 \boldsymbol{I})^{-1} \boldsymbol{y}$, where $\boldsymbol{y} = [y_1, \ldots, y_M]^T$ are the training outputs, $\boldsymbol{I}$ is the identity matrix, and $\boldsymbol{G}$ is the kernel Gramian matrix where $\boldsymbol{G}_{ij} = k(\boldsymbol{x}_i, \boldsymbol{x}_j)$. Given a new test data $\boldsymbol{x}_*$, the predicted function output is $\hat{f}(\boldsymbol{x}_*) = \boldsymbol{k}(\boldsymbol{x}_*)^T (\boldsymbol{G} + \lambda_M^2 \boldsymbol{I})^{-1} \boldsymbol{y}$, where $\boldsymbol{k}(\boldsymbol{x}_*)^T = [k(\boldsymbol{x}_*, \boldsymbol{x}_1), \ldots, k(\boldsymbol{x}_*, \boldsymbol{x}_M)]$. This is the same as the posterior mean of Gaussian processes regression with i.i.d. noise variance λ_M^2.

A more heuristic view of predicting the function output given a new test data is from the point evaluation at this new data. As an analog to the point evaluation in L^2 using Dirac delta $f(\boldsymbol{x}_*) = \langle \delta_{\boldsymbol{x}_*} | f \rangle_{L^2} = \int f(\boldsymbol{x})\delta(\boldsymbol{x} - \boldsymbol{x}_*)d\boldsymbol{x}$ (which is computationally infeasible using training data), one can work in the RKHS using the reproducing kernel function $k(\boldsymbol{x}, \boldsymbol{x}')$ as $f(\boldsymbol{x}_*) = \langle k_{\boldsymbol{x}_*} | f \rangle_{\mathcal{H}_k} = \sum_{i=1}^{M} \langle k_{\boldsymbol{x}_*} | q_i \rangle_{L^2} \frac{1}{\sigma_i} \langle q_i | f \rangle_{L^2}$, where the inner products in L^2 can be approximated by summation using training data as $\langle g | f \rangle_{L^2} = \int g(\boldsymbol{x}) f(\boldsymbol{x}) d\boldsymbol{x} \approx \sum_{i=1}^{M} g(\boldsymbol{x}_i) f(\boldsymbol{x}_i) = \sum_{i=1}^{M} \langle g | k_{\boldsymbol{x}_i} \rangle_{\mathcal{H}_k} \langle k_{\boldsymbol{x}_i} | f \rangle_{\mathcal{H}_k}$. Hence, one can obtain $\sum_{i=1}^{M} \langle k_{\boldsymbol{x}_*} | q_i \rangle_{L^2} \frac{1}{\sigma_i} \langle q_i | f \rangle_{L^2} \approx \sum_{ijl} \langle k_{\boldsymbol{x}_*} | k_{\boldsymbol{x}_j} \rangle_{\mathcal{H}_k} \langle k_{\boldsymbol{x}_j} | q_i \rangle_{\mathcal{H}_k} \frac{1}{\sigma_i} \langle q_i | k_{\boldsymbol{x}_l} \rangle_{\mathcal{H}_k} \langle k_{\boldsymbol{x}_l} | f \rangle_{\mathcal{H}_k}$. Notice that the kernel Gramian matrix has eigen decomposition $\boldsymbol{G} = \boldsymbol{Q} \boldsymbol{\Sigma}^2 \boldsymbol{Q}^T$, where $Q_{ij} = q_j(\boldsymbol{x}_i) = \langle k_{\boldsymbol{x}_i} | q_j \rangle_{\mathcal{H}_k}$ and $\boldsymbol{\Sigma}$ is diagonal with $\boldsymbol{\Sigma}_{ii} = \sqrt{\sigma_i}$. Hence, $\boldsymbol{G}^{-1} = \boldsymbol{Q} \boldsymbol{\Sigma}^{-2} \boldsymbol{Q}^T$ and $(\boldsymbol{G}^{-1})_{ij} = \sum_{l=1}^{M} \langle k_{\boldsymbol{x}_i} | q_l \rangle_{\mathcal{H}_k} \frac{1}{\sigma_l} \langle q_l | k_{\boldsymbol{x}_j} \rangle_{\mathcal{H}_k}$. Finally, one arrives at

$$f(\boldsymbol{x}_*) = \langle k_{\boldsymbol{x}_*} | f \rangle_{\mathcal{H}_k} = \boldsymbol{k}(\boldsymbol{x}_*)^T \boldsymbol{G}^{-1} [f(\boldsymbol{x}_1), \ldots, f(\boldsymbol{x}_M)]^T, \tag{2}$$

which is the same as the posterior mean in noiseless Gaussian processes regression. Replacing $\boldsymbol{G}^{-1}$ by the Moore–Penrose pseudoinverse $\boldsymbol{G}^+$ will be equivalent to regularization, or adding noise in Gaussian processes regression. A typical way of regularization using $\boldsymbol{G}^+$ is to truncate out some small eigenvalues σ_i's and the corresponding eigenvectors $q_i(\boldsymbol{x})$'s, although a more sophisticated way to perform this truncation is using a smooth cutoff developed in Ref. [6]. A useful result following the above derivation is that the inner product in RKHS can be approximated using training data as $\langle g | f \rangle_{\mathcal{H}_k} \approx \sum_{ij} \langle g | k_{\boldsymbol{x}_i} \rangle_{\mathcal{H}_k} [\boldsymbol{G}^{-1}]_{ij} \langle k_{\boldsymbol{x}_j} | f \rangle_{\mathcal{H}_k}$, which means that the projection operator into this RKHS can be approximated by training data as $\mathbb{1}_{\mathcal{H}_k} = \sum_{i=1}^{M} |p_i\rangle\langle p_i| \approx \sum_{ij} |k_{\boldsymbol{x}_i}\rangle_{\mathcal{H}_k} [\boldsymbol{G}^{-1}]_{ij} \, {}_{\mathcal{H}_k}\langle k_{\boldsymbol{x}_j}|$.

In summary, deterministic approximation of a function in RKHS or point evaluation of a function on new data can have a statistical interpretation via Gaussian processes regression. Moreover, since $\boldsymbol{k}(\boldsymbol{x}_*)^T \boldsymbol{G}^{-1}$ is a row vector of weights on the training outputs $[f(\boldsymbol{x}_1), \ldots, f(\boldsymbol{x}_M)]^T$, and if it sums up to 1 and if the amount of training data is sufficiently large, it may be considered as a density estimation for the posterior distribution of Gaussian processes, which will induce a density on the training data $[\boldsymbol{x}_1, \ldots, \boldsymbol{x}_M]^T$. A special case is the point evaluation on training data $f(\boldsymbol{x}_i) = \boldsymbol{k}(\boldsymbol{x}_i)^T \boldsymbol{G}^{-1} [f(\boldsymbol{x}_1), \ldots, f(\boldsymbol{x}_M)]^T$, where $\boldsymbol{k}(\boldsymbol{x}_i)^T \boldsymbol{G}^{-1}$ will become a row vector with every element equal to zero except for the ith one equal to 1, which is a probability mass function concentrated on $\boldsymbol{x}_i$ that approximates the Dirac delta distribution $\delta_{\boldsymbol{x}_i}(\cdot)$. Again, replacing $\boldsymbol{G}^{-1}$ by the Moore–Penrose pseudoinverse $\boldsymbol{G}^+$ effectively corresponds to Gaussian processes with additive noise such that the Dirac delta will become a narrow Gaussian centered at the training data.

Koopman Operator in Reproducing Kernel Hilbert Space

To interpret Koopman operator in RKHS, first notice that it can be also defined as an integral operator [13–16], which enables a better and uniform formulation of both deterministic and stochastic Koopman operator, and its Hermitian adjoint, namely, the Perron–Frobenius operator $\mathcal{L} = \mathcal{K}^{\dagger}$, where the $\dagger$ denotes Hermitian adjoint. Again, consider the dynamical system $(\mathcal{M}, t, \boldsymbol{F}^t)$. When $\boldsymbol{F}^t$ is highly nonlinear and/or stochastic, starting from an initial point on $\mathcal{M}$ and keeping track of its single trajectory along the time evolution will become meaningless, as any finite initial difference will blow up exponentially. Instead, a better strategy is to investigate the statistical behavior of a swarm of points' time evolution, which leads to the investigation of (probability) measure/density on $\mathcal{M}$ and its time evolution induced by $\boldsymbol{F}^t$. Consider a probability density function ρ defined on $\mathcal{M}$, and for computational convenience, we require $\rho \in \mathcal{F} \subseteq L^2(\mathcal{M})$. When $\boldsymbol{F}$ evolves an arbitrary swarm of points of system states on $\mathcal{M}$, i.e., evolves the pre-image $\boldsymbol{F}^{-1}(\mathbb{A})$ of any measurable domain $\mathbb{A} \subseteq \mathcal{M}$ to $\mathbb{A}$ at time τ later, the density ρ on $\boldsymbol{F}^{-1}(\mathbb{A})$ will be evolved by a linear operator to a new density on $\mathbb{A}$ as $\int_{\mathbb{A}} (\mathcal{L}_\tau \rho)(\boldsymbol{y}) d\boldsymbol{y} = \int_{\boldsymbol{F}^{-1}(\mathbb{A})} \rho(\boldsymbol{x}) d\boldsymbol{x}$, such that the probability measure in conserved, where the $\mathcal{L}_\tau$ is the Perron–Frobenius operator that evolves probability densities. If $\boldsymbol{F}$ is stochastic, which means that $\boldsymbol{F}(\boldsymbol{x})$ follows a *transition probability density* $p_\tau(\boldsymbol{y}|\boldsymbol{x})$, the Perron–Frobenius operator can be also defined as

$$(\mathcal{L}_\tau \rho)(\boldsymbol{y}) = \int_{\boldsymbol{F}^{-1}(\mathbb{A})} \rho(\boldsymbol{x}) p_\tau(\boldsymbol{y}|\boldsymbol{x}) d\boldsymbol{x}. \tag{3}$$

A special case is the deterministic system, where $p_\tau(\boldsymbol{y}|\boldsymbol{x})$ will become a Dirac delta distribution $\delta_{\boldsymbol{F}(\boldsymbol{x})}(\boldsymbol{y}) = \delta(\boldsymbol{y} - \boldsymbol{F}(\boldsymbol{x}))$, such that the center of an initial Dirac delta distribution $\delta_{\boldsymbol{x}}$ will be moved in consistence with the dynamics as $\mathcal{L}_\tau \delta_{\boldsymbol{x}}(\boldsymbol{y}) = \int_{\boldsymbol{F}^{-1}(\mathbb{A})} \delta(\boldsymbol{x} - \boldsymbol{x}')\delta(\boldsymbol{y} - \boldsymbol{F}(\boldsymbol{x}')) d\boldsymbol{x}' = \delta_{\boldsymbol{F}(\boldsymbol{x})}(\boldsymbol{y})$. Analogous to this, notice that Koopman operator for deterministic system is defined as $(\mathcal{K}_\tau h)(\boldsymbol{x}) = (h \circ \boldsymbol{F})(\boldsymbol{x}) = h(\boldsymbol{F}(\boldsymbol{x}))$, it can be also written as $(\mathcal{K}_\tau h)(\boldsymbol{x}) = \int_{\mathbb{A}} h(\boldsymbol{y}) \delta(\boldsymbol{y} - \boldsymbol{F}(\boldsymbol{x})) d\boldsymbol{y}$, and following this idea, the Koopman operator for stochastic system should be defined as

$$(\mathcal{K}_\tau h)(\boldsymbol{x}) = \int_{\mathbb{A}} h(\boldsymbol{y}) p_\tau(\boldsymbol{y}|\boldsymbol{x}) d\boldsymbol{y} = \mathbb{E}[h(\boldsymbol{F}(\boldsymbol{x}))|\boldsymbol{x}], \tag{4}$$

which is the conditional expectation of observable h's value at time τ later. Using these definitions, one can check that the Koopman operator and Perron–Frobenius operator are adjoint to each other for both deterministic and stochastic systems, by considering how the expectation value of an observable over some region evolves in time:

$$\begin{aligned}\mathbb{E}[h(y)] &= \int_{\mathbb{A}} (\mathcal{L}_\tau \rho)(y) h(y) dy = \langle \mathcal{L}_\tau \rho | h \rangle_{L^2} = \int_{\mathbb{A}} \int_{\boldsymbol{F}^{-1}(\mathbb{A})} \rho(\boldsymbol{x}) p_\tau(\boldsymbol{y}|\boldsymbol{x}) d\boldsymbol{x} h(\boldsymbol{y}) d\boldsymbol{y} \\ &= \int_{\boldsymbol{F}^{-1}(\mathbb{A})} \mathbb{E}[h(\boldsymbol{F}(\boldsymbol{x}))|\boldsymbol{x}] \rho(\boldsymbol{x}) d\boldsymbol{x} = \int_{\boldsymbol{F}^{-1}(\mathbb{A})} (\mathcal{K}_\tau h)(\boldsymbol{x}) \rho(\boldsymbol{x}) d\boldsymbol{x} = \langle \rho | \mathcal{K}_\tau h \rangle_{L^2},\end{aligned} \tag{5}$$

where $\mathcal{K}_\tau$ acting to the left on $\langle \rho|$ is $\langle \rho|\mathcal{K}_\tau|h\rangle_{L^2} = \langle \mathcal{K}_\tau^\dagger \rho|h\rangle_{L^2} = \langle \mathcal{L}_\tau \rho|h\rangle_{L^2}$. This formulation enables us to predict the expectation of a function's value at a later time when tracking and predicting a single trajectory is not meaningful due to high nonlinearity and/or stochasticity of $\boldsymbol{F}$, and we can relate this formulation to Koopman and Perron–Frobenius operators in RKHS as follows.

Recall from Eq. (2) that point evaluation in RKHS is the same as computing some expectation value such as the posterior mean of Gaussian processes, for example, $\langle k_{\boldsymbol{x}_i}|f\rangle_{\mathcal{H}_k} = \boldsymbol{k}(\boldsymbol{x}_i)^T \boldsymbol{G}^{-1}[f(\boldsymbol{x}_1), \ldots, f(\boldsymbol{x}_M)]^T$, where $\boldsymbol{k}(\boldsymbol{x}_i)^T \boldsymbol{G}^{-1}$ is a row vector with all zero elements except for the ith equal to 1, which may be considered as discrete approximation to Dirac delta distribution $\delta_{\boldsymbol{x}_i}$. Replacing $\boldsymbol{G}^{-1}$ by pseudoinverse $\boldsymbol{G}^+$ will be equivalent to regularization or adding noise to the Gaussian processes, such that $\boldsymbol{k}(\boldsymbol{x}_i)^T \boldsymbol{G}^+$ can approximate some narrow Gaussian centered at $\boldsymbol{x}_i$. Similarly, consider the projection of Koopman operator in RKHS by point evaluation of a function $|h\rangle$ evolved by $\mathcal{K}$ at a new state $\boldsymbol{x}_*$ as $\langle k_{\boldsymbol{x}_*}|\mathcal{K}|h\rangle_{\mathcal{H}_k} = \boldsymbol{k}(\boldsymbol{x}_*)^T \boldsymbol{G}^{-1}[\mathcal{K}h(\boldsymbol{x}_1), \ldots, \mathcal{K}h(\boldsymbol{x}_M)]^T$, where the $\boldsymbol{k}(\boldsymbol{x}_*)^T \boldsymbol{G}^{-1}$ is expected to approximate the initial density $\rho(\boldsymbol{x})$ before time evolution in Eq. (3), in the limit of infinite amount of training data, i.e., $M \to \infty$.

On the other hand, recall that the identity operator in RKHS $\mathbb{1}_{\mathcal{H}_k} = \sum_i |p_i\rangle\langle p_i| \approx \sum_{ij} |k_{\boldsymbol{x}_i}\rangle_{\mathcal{H}_k} \boldsymbol{G}^{-1}\ {}_{\mathcal{H}_k}\langle k_{\boldsymbol{x}_j}|$, and inner product can also be approximated as $\langle g|\mathbb{1}_{\mathcal{H}_k}|f\rangle_{\mathcal{H}_k} \approx \sum_{ij} \langle g|k_{\boldsymbol{x}_i}\rangle_{\mathcal{H}_k} [\boldsymbol{Q}\boldsymbol{\Sigma}^{-2}\boldsymbol{Q}_y^T]_{ij} \langle k_{\boldsymbol{y}_j}|f\rangle_{\mathcal{H}_k}$, where $[\boldsymbol{Q}_y^T]_{ij} = \langle q_i|k_{\boldsymbol{y}_j}\rangle_{\mathcal{H}_k} = \sum_l \langle q_i|q_l\rangle_{L^2} \frac{1}{\sigma_l} \langle q_l|k_{\boldsymbol{y}_j}\rangle_{L^2} \approx \sum_l \frac{1}{\sigma_l} \langle q_i|k_{\boldsymbol{x}_l}\rangle_{\mathcal{H}_k} \langle k_{\boldsymbol{x}_l}|k_{\boldsymbol{y}_j}\rangle_{\mathcal{H}_k} = [\boldsymbol{\Sigma}^{-2}\boldsymbol{Q}^T\boldsymbol{K}^T]_{ij}$, and $\boldsymbol{K}_{ij} = \langle k_{\boldsymbol{x}_i}|\mathcal{K}|k_{\boldsymbol{x}_j}\rangle_{\mathcal{H}_k} = k_{\boldsymbol{x}_j}(\boldsymbol{F}(\boldsymbol{x}_i)) = k(\boldsymbol{y}_i, \boldsymbol{x}_j) = \langle k_{\boldsymbol{y}_i}|k_{\boldsymbol{x}_j}\rangle_{\mathcal{H}_k}$. It follows that $\boldsymbol{Q}\boldsymbol{\Sigma}^{-2}\boldsymbol{Q}_y^T = \boldsymbol{Q}\boldsymbol{\Sigma}^{-2}\boldsymbol{Q}^T\boldsymbol{Q}\boldsymbol{\Sigma}^{-2}\boldsymbol{Q}^T\boldsymbol{K}^T = \boldsymbol{G}^{-2}\boldsymbol{K}^T$, and hence $\mathbb{1}_{\mathcal{H}_k}$ can also be approximated by $\mathbb{1}_{\mathcal{H}_k} \approx \sum_{ij} |k_{\boldsymbol{x}_i}\rangle_{\mathcal{H}_k} [\boldsymbol{G}^{-2}\boldsymbol{K}^T]_{ij}\ {}_{\mathcal{H}_k}\langle k_{\boldsymbol{y}_j}|$. After substituting the $\mathbb{1}_{\mathcal{H}_k}$'s in $\langle k_{\boldsymbol{x}_*}|\mathbb{1}_{\mathcal{H}_k} \mathcal{K}\mathbb{1}_{\mathcal{H}_k}|h\rangle_{\mathcal{H}_k}$ with appropriate approximations, one can obtain

$$\langle k_{\boldsymbol{x}_*}|\mathcal{K}|h\rangle_{\mathcal{H}_k} \approx \boldsymbol{k}(\boldsymbol{x}_*)^T \boldsymbol{G}^{-1}\boldsymbol{K}\boldsymbol{G}^{-2}\boldsymbol{K}^T[h(\boldsymbol{y}_1), \ldots, h(\boldsymbol{y}_M)]^T. \tag{6}$$

When the number of training snapshots pairs $M \to \infty$, we would expect that $\boldsymbol{k}(\boldsymbol{x}_*)^T \boldsymbol{G}^{-1}$ approximates $\rho(\boldsymbol{x})$, and $\boldsymbol{K}\boldsymbol{G}^{-2}\boldsymbol{K}^T$ approximates the transition density $p_\tau(\boldsymbol{y}|\boldsymbol{x})$, such that the matrix multiplication between $\boldsymbol{G}^{-1}$ and $\boldsymbol{K}$ in Eq. (6) approximates the integral over $\boldsymbol{x}$ in Eq. (3), and the matrix multiplication between $\boldsymbol{K}^T$ and $[h(\boldsymbol{y}_1), \ldots, h(\boldsymbol{y}_M)]^T$ in Eq. (6) approximates the integral over $\boldsymbol{y}$ in Eq. (4). Finally, we can consider Eq. (6) as an appropriate discrete approximation of Eq. (5) using training data, and the point evaluation of a function h evolved by Koopman operator in RKHS at a new data point $\langle k_{\boldsymbol{x}_*}|\mathcal{K}|h\rangle_{\mathcal{H}_k}$ is equivalent to predicting its expectation value $\mathbb{E}[h(\boldsymbol{y})]$ over training data at a later time. Notice that during the derivation of Eq. (6), *we did not use the definition of stochastic Koopman operator*, but using training data, we can indeed approximate $(\mathcal{K}_\tau h)(\boldsymbol{x}) = \int_{\mathbb{A}} h(\boldsymbol{y}) p_\tau(\boldsymbol{y}|\boldsymbol{x}) d\boldsymbol{y}$

$= \mathbb{E}[h(\boldsymbol{F}(\boldsymbol{x}))|\boldsymbol{x}]$ by the rows of $\boldsymbol{K}\boldsymbol{G}^{-2}\boldsymbol{K}^T[h(\boldsymbol{y}_1), \ldots, h(\boldsymbol{y}_M)]^T$, and approximate $(\mathcal{L}_\tau \rho)(\boldsymbol{y}) = \int_{\boldsymbol{F}^{-1}(\mathbb{A})} \rho(\boldsymbol{x}) p_\tau(\boldsymbol{y}|\boldsymbol{x}) d\boldsymbol{x}$ by the columns of $\boldsymbol{k}(\boldsymbol{x}_*)^T \boldsymbol{G}^{-1}\boldsymbol{K}\boldsymbol{G}^{-2}\boldsymbol{K}^T$. These nice relations are induced by the connection between deterministic approximation of a function in RKHS and Gaussian processes regression, and replacing $\boldsymbol{G}^{-1}$ by $\boldsymbol{G}^+$ will turn these almost singular densities to narrow Gaussians, which have even better statistical interpretation and correspond to regularized optimization in RKHS and noisy Gaussian processes regression that usually have better prediction accuracy.

In order to predict the future state of the system using the spectral properties of Koopman operator in RKHS, we first need to obtain a matrix representation of $\mathcal{K}$ projected in this space. Following the derivation of EDMD procedure in previous section, one can write $\mathbb{1}_{\mathcal{H}_k} = \sum_i |p_i\rangle\langle p_i| = \sum_i |q_i\rangle_{L^2} \frac{1}{\sigma_i}\,{}_{L^2}\langle q_i| = \sum_{ij} |q_j\rangle_{L^2} \frac{1}{\sigma_j} \langle q_j|p_i\rangle_{L^2} \langle p_i|$ $\approx \sum_{il} |k_{\boldsymbol{x}_l}\rangle_{\mathcal{H}_k} \langle k_{\boldsymbol{x}_l}|q_i\rangle_{\mathcal{H}_k} \frac{1}{\sqrt{\sigma_i}} \langle p_i| = \sum_{il} |k_{\boldsymbol{x}_l}\rangle_{\mathcal{H}_k} [\boldsymbol{Q}\boldsymbol{\Sigma}^+]_{li} \langle p_i| = \sum_{il} |p_i\rangle_{\mathcal{H}_k} [\boldsymbol{\Sigma}^+ \boldsymbol{Q}^T]_{il}\,{}_{\mathcal{H}_k}\langle k_{\boldsymbol{x}_l}|$, where $|p_i\rangle = \sqrt{\sigma_i}|q_i\rangle$ (in some literature they are called canonical features or Mercer's features due to Mercer's theorem). Then, $\mathcal{K}$ can be written as $\mathcal{K}\mathbb{1}_{\mathcal{H}_k} = \sum_k |p_k \circ \boldsymbol{F}\rangle\langle p_k|$, and its matrix representation is $\hat{\boldsymbol{K}}_{ij} = \langle p_i|\mathcal{K}|p_j\rangle_{\mathcal{H}_k}$ $= \langle p_i|\mathbb{1}_{\mathcal{H}_k}\mathcal{K}\mathbb{1}_{\mathcal{H}_k}|p_j\rangle_{\mathcal{H}_k} = [\boldsymbol{\Sigma}^+\boldsymbol{Q}^T\boldsymbol{K}\boldsymbol{Q}\boldsymbol{\Sigma}^+]_{ij}$, where we plugged in the last two expressions of $\mathbb{1}_{\mathcal{H}_k}$ above, and $\boldsymbol{K}_{ij} = \langle k_{\boldsymbol{x}_i}|\mathcal{K}|k_{\boldsymbol{x}_j}\rangle_{\mathcal{H}_k} = k_{\boldsymbol{x}_j}(\boldsymbol{F}(\boldsymbol{x}_i)) = k(\boldsymbol{y}_i, \boldsymbol{x}_j) =$ $\langle k_{\boldsymbol{y}_i}|k_{\boldsymbol{x}_j}\rangle_{\mathcal{H}_k}$ can be computed directly on training data. Similarly, the eigenvalue problem can be solved by computing eigenvalues and eigenvectors of $\hat{\boldsymbol{K}}$, where the ith component of eigenvector $\boldsymbol{v}_j$ is $(\boldsymbol{v}_j)_i = \langle p_i|\varphi_j\rangle_{\mathcal{H}_k}$, so the eigenfunction $|\varphi_j\rangle =$ $\sum_i |p_i\rangle(\boldsymbol{v}_j)_i$. The point evaluation of an eigenfunction on training data is $\langle k_{\boldsymbol{x}_i}|\varphi_j\rangle_{\mathcal{H}_k}$ $= \varphi_j(\boldsymbol{x}_i) = \sum_l \langle k_{\boldsymbol{x}_i}|p_l\rangle_{\mathcal{H}_k}(\boldsymbol{v}_j)_l = \sum_{nl} \langle k_{\boldsymbol{x}_i}|k_{\boldsymbol{x}_n}\rangle_{\mathcal{H}_k} \langle k_{\boldsymbol{x}_n}|q_l\rangle_{\mathcal{H}_k} \frac{1}{\sqrt{\sigma_l}}(\boldsymbol{v}_j)_l = [\boldsymbol{G}\boldsymbol{Q}\boldsymbol{\Sigma}^+ \boldsymbol{V}]_{ij}$, where columns of $\boldsymbol{V}$ are $\{\boldsymbol{v}_j\}$. By defining $[\boldsymbol{\Phi}_x]_{ij} = \langle k_{\boldsymbol{x}_i}|\varphi_j\rangle_{\mathcal{H}_k}$ and $[\boldsymbol{\Phi}_y]_{ij} =$ $\langle k_{\boldsymbol{y}_i}|\varphi_j\rangle_{\mathcal{H}_k}$, we can write the matrix of eigenfunctions evaluated on training data in a compact form as $\boldsymbol{\Phi}_x = \boldsymbol{G}\boldsymbol{Q}\boldsymbol{\Sigma}^+\boldsymbol{V}$ and $\boldsymbol{\Phi}_y = \boldsymbol{K}\boldsymbol{Q}\boldsymbol{\Sigma}^+\boldsymbol{V}$. Following the same convention and notation in derivation of EDMD, the matrix of Koopman modes can be solved as $\Xi = \boldsymbol{\Phi}_x^+\boldsymbol{X} = \boldsymbol{\Phi}_y^+\boldsymbol{Y} = [\text{diag}(e^{\lambda\tau})]^+\boldsymbol{\Phi}_x^+\boldsymbol{Y}$, where rows in $\boldsymbol{Y}$ are $\{\boldsymbol{y}^T\}$ and $[\text{diag}(e^{\lambda\tau})]$ is the diagonal matrix containing the finite-time eigenvalues $\mu_i = e^{\lambda_i\tau}$. This procedure is called kernel-based Koopman spectral analysis [8] and it is currently being adopted as a better approach for other applications [6]. Finally, given a new system state $\boldsymbol{x}_*$, the prediction of the lth component of system state $F_l(\boldsymbol{x}_*)$ will be a point evaluation of the $\mathcal{K}$-evolved observable g_l at $\boldsymbol{x}_*$ as

$$\begin{aligned}\langle k_{\boldsymbol{x}_*}|F_l\rangle_{\mathcal{H}_k} = \langle k_{\boldsymbol{x}_*}|\mathcal{K}|g_l\rangle_{\mathcal{H}_k} &= \sum_{i=1}^{M} \langle k_{\boldsymbol{x}_*}|\mathcal{K}|\varphi_i\rangle_{\mathcal{H}_k}\Xi_{il} = \sum_{i=1}^{M} \langle k_{\boldsymbol{x}_*}|\varphi_i\rangle_{\mathcal{H}_k} e^{\lambda_i\tau}\Xi_{il} \\ &= \sum_{i=1}^{M} k(\boldsymbol{x}_*, \boldsymbol{x}_i)[\boldsymbol{Q}\boldsymbol{\Sigma}^+\boldsymbol{V}[\text{diag}(e^{\lambda\tau})]\Xi]_{il},\end{aligned} \tag{7}$$

where Ξ_{il} is the Koopman mode associated with the ith eigenfunction when projecting $g_l(\boldsymbol{x})$ on $\boldsymbol{\Phi}_x$.

Another benefit of working in RKHS is that when properly choosing and/or designing the kernel functions (e.g., Gaussian RBF kernel), the unique associated

RKHS is dense in the space of continuous bounded functions, which means that these kernel functions are universal approximators to any function in this very large and general function space, and hence they should achieve better approximation and prediction in most cases, especially in computing Koopman eigenfunctions via point evaluation $\varphi_j(\boldsymbol{x}_i) = \langle k_{\boldsymbol{x}_i} | \varphi_j \rangle_{\mathcal{H}_k}$.

3 Numerical Algorithm

Recall Eq. (7), if one needs to predict all state variables at a future time, one can simply compute

$$\boldsymbol{F}(\boldsymbol{x}_*) = \boldsymbol{k}(\boldsymbol{x}_*)^T \boldsymbol{Q}\boldsymbol{\Sigma}^+ \boldsymbol{V}[\text{diag}(\text{e}^{\lambda\tau})]\boldsymbol{\Xi}, \tag{8}$$

where $\boldsymbol{k}(\boldsymbol{x}_*)^T = [k(\boldsymbol{x}_*, \boldsymbol{x}_1), \ldots, k(\boldsymbol{x}_*, \boldsymbol{x}_M)]$. Notice that for system with input, all the $\boldsymbol{x}_*$, $\boldsymbol{x}_i$, and $\boldsymbol{y}_i$ are extended states with input, but the Koopman modes $\boldsymbol{\Xi}$ will only contain N columns corresponding to the first N components of the extended state, which eliminates meaningless prediction on input. Another observation is that if we substitute $\boldsymbol{\Xi}$ in Eq. (8) with $\boldsymbol{\Xi} = [\text{diag}(\text{e}^{\lambda\tau})]^+ \boldsymbol{\Phi}_x^+ \boldsymbol{Y}$, after some simplification, we will get $\boldsymbol{k}(\boldsymbol{x}_*)^T \boldsymbol{G}^+ \boldsymbol{Y}$, which is exactly the regularized optimization in RKHS or Gaussian processes regression on each state variable one-by-one. As we elaborated in Ref. [6], one of the major advantages of utilizing the spectral properties of Koopman operator is to linearly decompose the system dynamics as a summation over individual modes, such that it is possible to regularize, sort, perform more "physical" cross-validation, and optimize these modes in order to generate an ensemble of predictors to achieve better prediction. When investigating time series with exogenous variables as a dynamical system with input, since the only major change on the numerical procedure is to neglect the Koopman modes associated with input, one can simply work with the remaining Koopman modes and all techniques and methods developed for Kernel-based Koopman modes regression (Kernel KMR) [6] can be employed almost unchanged. Hence, we achieved a simple yet useful extension of Kernel KMR, which we refer to as *Kernel-based Generalized Koopman Mode Regression with eXogenous variables* (Kernel GKMRX). For more details on the techniques and methods constituting the Kernel KMR, we suggest referring to Ref. [6].

4 Numerical Examples and Applications

We tested this new methodology by predicting high-dimensional stock prices' log returns while considering trading volumes of these stocks as exogenous variables. Due to the page limit rule, we refer the readers to Ref. [6] for detailed description of the stock markets data that we used. Compared to Kernel KMR, the Kernel GKMRX can achieve about 0.1% improvement in both root-mean-squared error (RMSE) and mean absolute error (MAE). This insignificant improvement is due to the fact that stock returns time series are very close to random walk, and trading volume as extra information will not change this fact to improve the prediction significantly. However, when applying to electricity generation and consumption time series with weather condition as exogenous variables, we expect some major improvement over Kernel KMR as reported in Ref. [6], and we are currently testing and summarizing these results.

5 Conclusion and Outlook

In this paper, we generalized our previously developed Kernel KMR methodology to Kernel GKMRX (Kernel-based Generalized Koopman Mode Regression with eXogenous variables) for prediction of high-dimensional time series with exogenous variables, by utilizing a simple yet useful generalization of Koopman operator to dynamical systems with input that generates the time series. We found that the techniques and methods that we developed for Kernel KMR can be employed in Kernel GKMRX with minimal modification. By formulating Koopman operator in reproducing kernel Hilbert space, we obtained a new derivation of the kernel-based EDMD and the original EDMD algorithms using Dirac bra–ket notation. Moreover, we obtained a statistical interpretation of these numerical methods developed for deterministic Koopman operator by exploiting the connection between RKHS and Gaussian processes regression, and relate them to the stochastic Koopman and Perron–Frobenius operators. This connection and the statistical interpretation are crucial to justify the application of existing data-driven deterministic Koopman spectral analysis to nondeterministic dynamical systems, and account for the advantage of kernel-based EDMD over original EDMD which relies on explicit choice of basis functions spanning the space where the Koopman operator is projected and approximated. In applications, we found that the prediction performance of this methodology is promising in forecasting real-world high-dimensional time series with exogenous variables, e.g., stock returns time series with trading volume as exogenous variables.

This generalization of Koopman operator to systems with input is not unique, and we are keen to investigate other generalization for prediction purposes. Moreover, even the very simple trick in this generalization that we used in this paper can be developed further to investigate system with memory in the same way as for memoryless systems. These will be left for future work. Another possible improvement, which is still an open question, is the design of kernel functions. When utilizing Gaussian RBF kernels, it should be possible to optimize the kernel widths as hyperparameters by some other more sophisticated techniques in machine learning. This, again, will be left for future investigation.

Acknowledgements The corresponding author would like to thank Dr. Alexandre Mauroy for insightful discussions on generalizing Koopman operator to systems with input.

References

1. Koopman, B.O.: PNAS **17**(5), 315 (1931)
2. Mezić, I.: Nonlinear Dyn. **41**(1–3), 309 (2005). https://doi.org/10.1007/s11071-005-2824-x
3. Budišić, M., Mohr, R.M., Mezić, I.: Chaos: an Interdisciplinary. J. Nonlinear Sci. **22**(4), 047510 (2012). https://doi.org/10.1063/1.4772195
4. Rowley, C.W., Mezić, I., Bagheri, S., Schlatter, P., Henningson, D.S.: J. Fluid Mech. **641**, 115 (2009). https://doi.org/10.1017/S0022112009992059
5. Williams, M.O., Kevrekidis, I.G., Rowley, C.W.: J. Nonlinear Sci. 1–40 (2015). https://doi.org/10.1007/s00332-015-9258-5
6. Hua, J.C., Noorian, F., Moss, D., Leong, P.H.W., Gunaratne, G.H.: Nonlinear Dyn. **90**(3), 1785 (2017). https://doi.org/10.1007/s11071-017-3764-y
7. Proctor, J.L., Brunton, S.L., Kutz, J.N.: arXiv:1602.07647 [math] (2016)
8. Williams, M.O., Rowley, C.W., Kevrekidis, I.G.: arXiv:1411.2260 [math] (2014)
9. Dirac, P.A.M.: Math. Proc. Camb. Philos. Soc. **35**(03), 416 (1939). https://doi.org/10.1017/S0305004100021162
10. Hofmann, T., Schölkopf, B., Smola, A.J.: Ann. Stat. **36**(3), 1171 (2008). https://doi.org/10.1214/009053607000000677
11. Rasmussen, C.E., Williams, C.K.I.: Adaptive computation and machine learning. In: Gaussian Processes for Machine Learning. MIT Press, Cambridge, MA (2006)
12. Schölkopf, B., Herbrich, R., Smola, A.J.: Computational Learning Theory. Lecture Notes in Computer Science, pp. 416–426. Springer, Berlin, Heidelberg (2001). https://doi.org/10.1007/3-540-44581-1_27
13. Lasota, A., Mackey, M.C.: Chaos, Fractals, and Noise. Applied Mathematical Sciences, vol. 97. Springer, New York, NY (1994). https://doi.org/10.1007/978-1-4612-4286-4
14. Cvitanović, P., Artuso, R., Mainieri, R., Tanner, G., Vattay, G.: Chaos: Classical and Quantum. Niels Bohr Inst., Copenhagen (2016)
15. Klus, S., Koltai, P., Schütte, C.: J. Comput. Dyn. **3**(1), 1 (2016). https://doi.org/10.3934/jcd.2016003
16. Klus, S., Nüske, F., Koltai, P., Wu, H., Kevrekidis, I., Schütte, C., Noé, F.: arXiv:1703.10112 [math] (2017)

Eigenvalues Distribution Limit of Covariance Matrices with AR Processes Entries

Zahira Khettab and Tahar Mourid

Abstract We consider a class of random matrices $B_N = X_N T_N X_N^t$, where X_N is a matrix $(N \times n(N))$ whose rows are independent, the entries X_{ij} in each row satisfy an autoregressive relation AR(1), and T_N is a diagonal matrix independent of X_N. Under some conditions, we show that if the empirical distribution function of eigenvalues of T_N converges almost surely to a proper probability distribution as $N \longrightarrow \infty$ and $\frac{n(N)}{N} \longrightarrow c > 0$, then the empirical distribution function of eigenvalues of B_N converges almost surely to a non-random limit function given by Marcenko and Pastur. Numerical simulations illustrate the behavior of kernel density estimators and density estimators of Stieltjes transform around the true density and we give a numerical comparison on the base of L_1 error varying different parameters.

Keywords Large dimensional random matrix · Empirical distribution function of eigenvalues · Covariance matrix · Autoregressive processes · Stieltjes transform · Kernel density estimators

1 Introduction

Theoretical studies on covariance matrices have a long history and appear in many domains in the real world and having links with practical problems (see [1] and [9]). For example, in multivariate statistics, spectral asymptotic results are used in solving the detection problem in signal process [9].

Consider the following random matrix:

$$B_N = X_N T_N X_N^t \tag{1}$$

Z. Khettab (✉) · T. Mourid
Laboratoire de Statistiques et Modélisations Aléatoires, University of Tlemcen, 13000 Tlemcen, Algeria
e-mail: khettab.zahira@yahoo.fr

T. Mourid
e-mail: t_mourid@mail.univ-tlemcen.dz

I. Rojas et al. (eds.), *Time Series Analysis and Forecasting*, Contributions to Statistics, https://doi.org/10.1007/978-3-319-96944-2_6

where $X_N = (\frac{1}{\sqrt{N}} X_{ij})$, $(i = 1, \ldots, N;\ j = 1, \ldots, n(N))$ is a matrix $(N \times n(N))$ with independent rows with the entries X_{ij} of each row satisfy an autoregressive relation AR(1) and T_N is a diagonal matrix $(n \times n)$ with real entries independent of X_N (X_N^t is the transpose matrix of X_N). More precisely, for each $i \geq 1$ we have

$$X_{ij+1} = \rho X_{ij} + \varepsilon_{ij+1}, \qquad j \geq 1 \tag{2}$$

where $\left(\varepsilon_{ij}, i, j \geq 1\right)$ are i.i.d. rv's (random values) with mean 0 and variance $\sigma^2 > 0$, such that ε_{ij} admits a continuous density function with respect to Lebesgue measure. The parameter ρ is such that $|\rho| < 1$ assuring a strictly stationary process. The diagonal matrix $T_N = diag\,(\tau_1, \ldots, \tau_n)$ is independent of X_N and the rv's τ_i are real.

The empirical distribution function ($e.d.f.$) of the eigenvalues (λ_i) of the symmetric matrix B_N is defined by

$$F^{B_N}(x) = \frac{1}{N} \sum_{i=1}^{N} 1_{(\lambda_i \leq x)},$$

where 1_A denoting the indicator function of the set A.

A large number of papers have dealt with the problem to identify the limit of the $e.d.f.$ of eigenvalues of random matrices B_N as $N \longrightarrow \infty$ and $\frac{n(N)}{N} \longrightarrow c > 0$. Marcenko and Pastur [7] originally studied this problem for more general forms of random matrices. They establish, under some conditions on moments, that if the $e.d.f.$ F^{T_N} converges to a proper distribution function H, then F^{B_N} converges in probability to a proper distribution function. Their method involves the Stieltjes transform, where they shown that the Stieltjes transform of the limiting distribution function satisfies a first-order partial differential equation, then via the characteristics they shown that this function is a solution of an algebraic equation identifying hence the limit.

Afterward, several authors [4, 5, 8, 11, 12] extended this result giving the almost sure convergence of the $e.d.f.$ of eigenvalues under mild conditions on the entries X_{ij}. Most of the previous papers employ the same transform as [7] and the entries X_{ij} of the matrices are independent random variables, except the paper [3], where dependent entries are considered.

Our goal in this paper is to study, under some assumptions, the limit of the $e.d.f.$ F^{B_N} of the random matrix $B_N = X_N T_N X_N^t$, where the entries X_{ij} of the matrix X_N satisfy an autoregressive relation AR(1) for each i. We follow the approach given in [8] where the authors apply Marcenko and Pastur method to study the limit of Stieltjes transform of the $e.d.f.$ F^{B_N} and then we identify the limit law. We illustrate by numerical simulations the behavior of kernel density estimators and density estimators of Stieltjes transform to identify the true density and give L_1 errors varying different parameters.

The paper is organized as follows. Section 2 provides the main result. Section 3 presents numerical simulations. The proof of the main result will be postponed in Sect. 4.

2 Main Result

First, we introduce some random variables and random matrices. We truncate and centralize the entries X_{ij} of the random matrix X_N to obtain new corresponding random matrices as follows: for $i = 1, \ldots, N$; $j = 1, \ldots, n(N)$, let

$$\hat{X}_{ij} = X_{ij} 1_{(|X_{ij}|<\sqrt{N})}, \hat{X}_N = \left(\frac{1}{\sqrt{N}} \hat{X}_{ij} \right), \hat{B}_N = \hat{X}_N T_N \hat{X}_N^t, \tag{3}$$

$$\tilde{X}_{ij} = \hat{X}_{ij} - E\left(\hat{X}_{ij}\right), \tilde{X}_N = \left(\frac{1}{\sqrt{N}} \tilde{X}_{ij} \right), \tilde{B}_N = \tilde{X}_N T_N \tilde{X}_N^t, \tag{4}$$

and

$$\begin{cases} \bar{X}_{ij} = \tilde{X}_{ij} 1_{(|X_{ij}| \le \ln N)} - E\tilde{X}_{ij} 1_{(|X_{ij}| \le \ln N)}, \\ \bar{X}_N = \left(\frac{1}{\sqrt{N}} \bar{X}_{ij} \right), \bar{B}_N = \bar{X}_N T_N \bar{X}_N^t. \end{cases} \tag{5}$$

We pointed out that the problem described above has been often handled by the method of Stieltjes transform. Let $\mathcal{M}(\mathbb{R})$ be the set of distribution functions on $\mathbb{R}$. Recall that the Stieltjes transform of a distribution function $F \in \mathcal{M}(\mathbb{R})$ is defined by

$$m_F(z) = \int \frac{1}{\lambda - z} dF(\lambda), \quad z \in \mathbb{C}^+ \equiv \{z \in \mathbb{C} : \Im\text{m} z > 0\},$$

where $\Im\text{m}$ is the imaginary part. The inversion formula is given by

$$F([a,b]) = \frac{1}{\pi} \lim_{\epsilon \to 0^+} \int_a^b \Im\text{m} m_F(x + i\epsilon)\, dx,$$

where a and b are continuity points of F. Also, the weak convergence of probability distribution functions is equivalent to the convergence of Stieltjes transforms (Theorem B.9, [1]). From the inversion formula, it follows that for any countable set $S \subset \mathbb{C}^+$ such that $\mathbb{R} \subset \bar{S}$ the closure of S, and a sequence $(F_N) \in \mathcal{M}(\mathbb{R})$, $F \in \mathcal{M}(\mathbb{R})$, we have the following equivalence:

$$\lim_{N \to \infty} m_{F_N}(z) = m_F(z), \forall z \in S \Longleftrightarrow F_N \to F \ \ as \ \ N \to \infty, \tag{6}$$

where $F_N \to F$ is the vague convergence of distributions functions.

Furthermore, we consider the following random matrices. For $j, l = 1, 2, \ldots, n(N)$, denote by $\bar{q}_j$ the jth column of $\bar{X}_N$ defined by (5) that is

$$\bar{q}_j = \frac{1}{\sqrt{N}} \left(\bar{X}_{1j}, \ldots, \bar{X}_{Nj} \right)^t := \frac{1}{\sqrt{N}} V_j, \tag{7}$$

and by

$$\bar{B}_{(j)} = \bar{B}_{(j)}^N := \bar{B}_N - \tau_j \bar{q}_j \bar{q}_j^t, \tag{8}$$

where τ_j are the elements of T_N, and define

$$x = x_N := \frac{1}{N} \sum_{j=1}^{n} \frac{\tau_j}{1 + \tau_j m_{F^{\bar{B}_N}}(z)}, \; x_{(j)} = x_{(j)}^N := \frac{1}{N} \sum_{l=1}^{n} \frac{\tau_l}{1 + \tau_l m_{F^{\bar{B}_{(j)}}}(z)}, \tag{9}$$

where $m_{F^{\bar{B}_N}}$ and $m_{F^{\bar{B}_{(j)}}}$ are Stieltjes transform of the matrices $\bar{B}_N$ and $\bar{B}_{(j)}$, respectively. Finally, set

$$C_{(j)}^1 := \left(\bar{B}_{(j)} - zI \right)^{-1}, \quad C_{(j)}^2 := \left(x_{(j)} - z \right)^{-1} \left(\bar{B}_{(j)} - zI \right)^{-1} \tag{10}$$

where I is the identity matrix.

Now, we state the main result of this paper giving the almost sure limit of the $e.d.f.$ of the eigenvalues of the random matrix B_N (tr is the trace of the matrix).

Theorem 1 *Assume*

(a) For $N = 1, 2, \ldots$, let $X_N = \left(\frac{1}{\sqrt{N}} X_{ij} \right)$ be a matrix $(N \times n(N))$ with independent rows and an AR(1) autoregressive relation (2) in each row. The entries $X_{ij}, i, j \geq 1$, have all their moments finite and $\frac{n(N)}{N} \to c > 0$ as $N \to \infty$.

(b) $T_N = diag\left(\tau_1, \ldots, \tau_n\right)$, $\tau_i \in \mathbb{R}$, and the e.d.f. of T_N converges almost surely to a distribution function H as $N \to \infty$.

(c) The matrices X_N and T_N are independent.

(d) For $k = 1, 2$ and $j = 1, 2, \ldots, n(N)$, the matrices $C_{(j)}^k$ defined in (10) satisfy

$E \left| V_j^t C_{(j)}^k V_j - tr C_{(j)}^k \right|^6 \leq K N^3$, *where V_j given by (7) and $K > 0$.*

Then, the e.d.f. F^{B_N} of the random matrix $B_N = X_N T_N X_N^t$ converges vaguely almost surely to a distribution function F, as $N \longrightarrow \infty$, whose Stieltjes transform $m_F(z)$ satisfies the following functional relation:

$$m_F(z) = -\left(z - c \int \frac{\tau dH(\tau)}{1 + \tau m_F(z)} \right)^{-1}; z \in \mathbb{C}^+. \tag{11}$$

Remark Assumption (a) is fulfilled in part if the white noise (ε_{ij}) has all moments (Gaussian white noise). Assumptions (b), (c) are standard and analogous of that

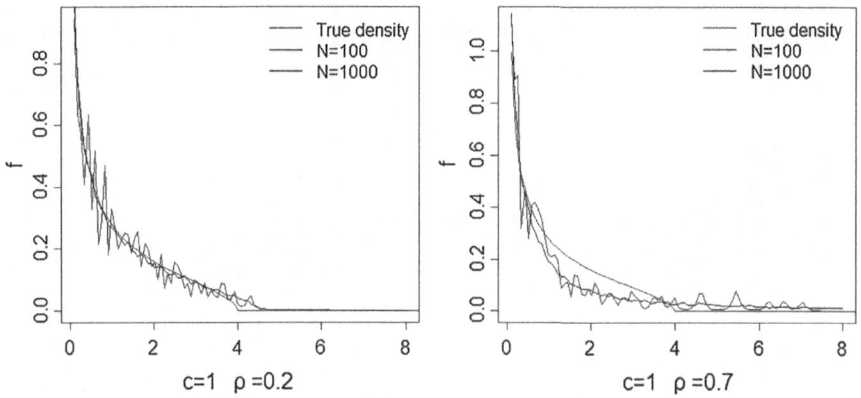

Fig. 1 Densities of the limit law and STE with T_N the identity matrix

existing in [8]. Assumption (d) requires a control of sixth moment of quadratic form of a matrix and its trace by a third power of size N. In the case of *i.i.d.* entries $X_{ij}, i, j \geq 1$, assumption (d) is fulfilled (cf. Lemma 3.1 in [8]).

3 Numerical Simulations

As a practical impact of the main result, we illustrated in [6] the behavior of the empirical density estimator of *e.d.f.* of eigenvalues ($\lambda_i, i = 1, \ldots, N$) of large random matrices B_N, and identify the density function of the limit law by numerical simulations. First, we recall the formulas giving density of limit law and the empirical Stieltjes transform estimator. From [10], we have for all $x \in \mathbb{R} - \{0\}$, and $z = x + iy$, $y > 0$, the distribution function F (limit of the *e.d.f.* F^{B_N}) has a continuous derivative f defined by $f(x) = (1/\pi)\,\Im m m_0(x)$, where $m_0(x)$ is given by Stieltjes transform $m_F(z)$ as $\lim_{z \to x} \; m_F(z) := m_0(x)$ (Figs. 1, 2 and Table 1).

The Stieltjes Transform Estimator (STE) is defined by

$$f_N(x) = (1/\pi)\,\Im m m_{F^{B_N}}(z),$$

where $m_{F^{B_N}}(z) = \frac{1}{N} tr\,(B_N - zI)^{-1} = \frac{1}{N} \sum_{i=1}^{N} (\lambda_i - z)^{-1}$.

Now, we apply Gaussian Kernel Estimators (GKE) defined by

$$\hat{f}_N(\lambda) = \frac{1}{N h_N} \sum_{i=1}^{N} K(\frac{\lambda - \lambda_i}{h_N}); \lambda \in \mathbb{R}$$

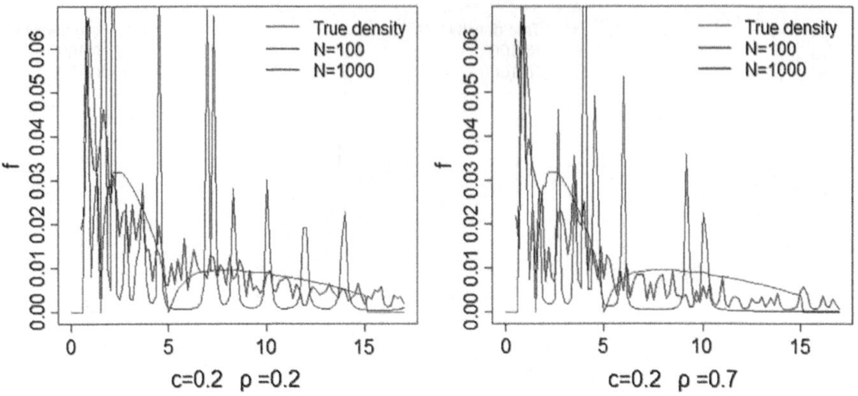

Fig. 2 Densities of the limit law and STE with T_N having three eigenvalues: 1, 3, 10

Table 1 L_1-errors of STE in Case 1: T_N identity matrix with $c = 1$ and Case 2: T_N diagonal matrix having three eigenvalues: 1, 3, 10 with $c = 0.2$. For weak and strong dependence and different sample size values N

	$\rho = 0.2$			$\rho = 0.7$		
N	100	500	1000	100	500	1000
Case 1	0.0297	0.0245	0.0200	0.0390	0.0327	0.0298
Case 2	0.0206	0.0020	0.0009	0.0151	0.0020	0.0009

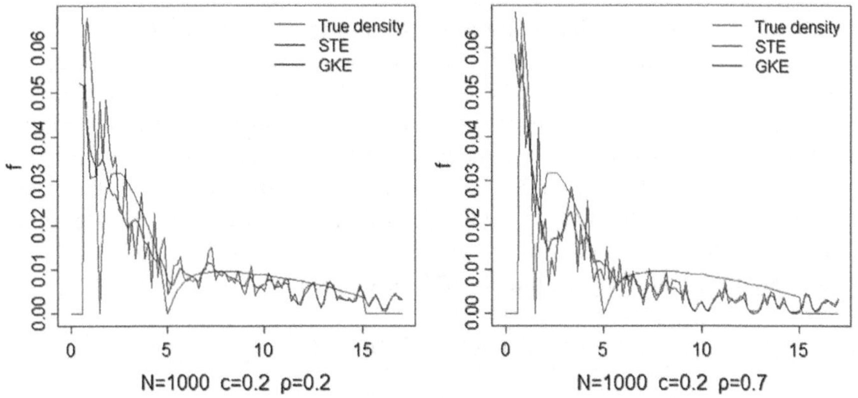

Fig. 3 Behavior of STE and GKE with matrix T_N having three eigenvalues: 1, 3, 10

where h_N is the bandwidth converging to 0 and $Nh_N \to \infty$, and K is a Gaussian kernel : $K(u) = \frac{1}{\sqrt{2\pi}} \exp(-\frac{1}{2}u^2)$.

We compare the performance of Stieltjes Transform Estimators (STE) and Gaussian Kernel Estimators (GKE) on the base of L_1-errors (Fig. 3 and Table 2).

Table 2 L_1-errors of the STE and GKE, with T_N having three eigenvalues and $c = 0.2$ for different sample size values N

	$\rho = 0.2$			$\rho = 0.7$		
N	100	500	1000	100	500	1000
STE	0.0160	0.0102	0.0092	0.0126	0.0102	0.0093
GKE	0.0109	0.0094	0.0088	0.0106	0.0093	0.0091

Conclusion
From more numerical simulations, we may observe that the performance of estimators strongly depends on the choices of the dimension c, AR parameter ρ, and sample size N. The variability of parameters has a direct impact on the stabilization and convergence rate of the estimators. Particular choices of parameters confirm a good performance of the estimators and lead to indicate optimal values for these parameters. We also observe an effect of the dimension c on density estimator convergence rate. For $c > 1$, for both weak and strong dependencies ($\rho = 0.2$, $\rho = 0.7$), the estimators perform well from on $N = 100$. However, for small values of c ($c < 0.2$), there is an influence of parameter values on the convergence rate. For weak dependence ($\rho = 0.2$) STE perform quite well for moderate value $N = 100$, whereas for strong dependence ($\rho = 0.7$) the estimator accurate enough well only for large N (when $N > 1000$). The number of eigenvalues of T_N has an effect on the behavior of the estimators as well as on their performance. Both estimators perform well and give a good representation of the true density with a small advantage of GKE.

4 Proof of the Main Result

Recall these well-known facts. For each i, the process $(X_{ij},\, j \in \mathbb{Z})$ satisfying relation (2) is a stationary AR(1) process, then it satisfies the geometric strong mixing property (G.S.M) with strong mixing coefficient $\alpha_k = \alpha_k(\mathcal{F}_0^m, \mathcal{F}_{m+k}^\infty) = O\left(\rho^k\right)$, where $0 < \rho < 1$ and $\mathcal{F}_a^b = \mathcal{F}_{a,i}^b = \sigma(X_{ij}, a \le j \le b)$, whenever ε_{ij} has a strictly positive continuous density (see [2] p. 58).

The covariance between two real-valued rv's is bounded as follows: if $\eta \in L^p$ and $\xi \in L^q$ are $\mathcal{F}_0^m$ and $\mathcal{F}_{m+k}^\infty$-measurable, respectively, then we have

$$|E(\eta\xi) - E(\eta)E(\xi)| \le 12 \|\eta\|_p \|\xi\|_q \alpha_k^{\frac{1}{r}} \tag{12}$$

for all $1 \le p, q, r \le \infty$ with $\frac{1}{p} + \frac{1}{q} + \frac{1}{r} = 1$, and the norm $\|.\|_p = E^{\frac{1}{p}} |.|^p$.

On the other hand, there exists a distance $D(.,.)$ on the space $\mathcal{M}(\mathbb{R})$, such that for two sequences $(F_N), (G_N) \in \mathcal{M}(\mathbb{R})$, we have (see [8]).

$$\lim_{N\to\infty} \|F_N - G_N\| = 0 \implies \lim_{N\to\infty} D(F_N, G_N) = 0, \tag{13}$$

where $\|.\|$ denotes the sup-norm of bounded functions from $\mathbb{R}$ to $\mathbb{R}$.

To lighten the writing, the dependency of most of variables on N will occasionally be dropped form the notation. Now, we replace T by a suitable matrix for further analysis: for $\theta \geq 0$ define, $T_\theta = diag\left(\tau_1 1_{(|\tau_1|\leq\theta)}, \ldots, \tau_n 1_{(|\tau_n|\leq\theta)}\right)$, and let Q be any $(N \times n)$ matrix. If θ and $(-\theta)$ are continuity points of H, then by Lemma 2.5 of [8] and assumption (b) of the Theorem 1, as $N \to \infty$ and $\frac{n}{N} \to c > 0$, we have

$$\left\| F^{QTQ^t} - F^{QT_\theta Q^t} \right\| \leq \frac{1}{N} rg\,(T - T_\theta) = \frac{1}{N} \sum_{j=1}^{n} 1_{(|\tau_j|>\theta)} \to cH\left\{[-\theta, \theta]^c\right\} \;\; a.s.$$

It follows that if $\theta = \theta_N \to \infty$, then

$$\left\| F^{QTQ^t} - F^{QT_\theta Q^t} \right\| \to 0 \;\; a.s. \tag{14}$$

Choose θ such that

$$\theta^4 \left[E^{\frac{2}{3}} \left|X_{11}\right|^2 1_{(|X_{11}|\geq \ln N)} + \frac{1}{N} \right] \to 0, \tag{15}$$

and

$$\sum_{N=1}^{\infty} \theta^8 \left[\frac{1}{N^{7/6}} E^{1/6} \left|X_{11}\right|^4 1_{(\ln N \leq |X_{11}| < \sqrt{N})} + \frac{1}{N^2} \right] < \infty. \tag{16}$$

For continue, we need the following result.

Lemma 1 *Let the* $(N \times n)$ *matrices* $X = \left(\frac{1}{\sqrt{N}} X_{ij}\right)$ *verifying assumption (a) of Theorem 1, and* $\hat{X} = \left(\frac{1}{\sqrt{N}} \hat{X}_{ij}\right)$ *where* $\hat{X}_{ij} = X_{ij} 1_{(|X_{ij}|<\sqrt{N})}$. *For* $\theta \geq 0$ *set* $T_\theta = diag\left(\tau_1 1_{(|\tau_1|\leq\theta)}, \ldots, \tau_n 1_{(|\tau_n|\leq\theta)}\right)$, $\tau_i \in \mathbb{R}$. *We have*

$$D\left(F^{XT_\theta X^t},\; F^{\hat{X}T_\theta \hat{X}^t}\right) \to 0 \;\; a.s.$$

Proof From Corollary A.42 of [1], we find

$$D^2\left(F^{XT_\theta X^t}, F^{\hat{X}T_\theta \hat{X}^t}\right) \leq \left[\frac{2}{N} tr\left(XX^t - \hat{X}\hat{X}^t\right) + \frac{4}{N} tr \hat{X}\hat{X}^t\right]\left[\frac{\theta^2}{N} tr\left(XX^t - \hat{X}\hat{X}^t\right)\right].$$

In order that this distance tends almost surely to 0, we can show by Borel–Cantelli lemma that $\left[\frac{\theta^2}{N} tr\left(XX^t - \hat{X}\hat{X}^t\right)\right]$ tends to 0 and $\left[\frac{4}{N} tr \hat{X}\hat{X}^t\right]$ is bounded almost surely. So the result.

Proof of the Theorem 1

For $(N \times n)$ matrix, $X = \left(\frac{1}{\sqrt{N}} X_{ij}\right)$ verifying assumption (a) of Theorem 1. With help to inequality (12) and the fact that (X_{ij}) satisfies the G.S.M. property, we obtain

$$M_1 \leq E\left(tr\left(XX^t\right)^2\right) \leq M_2 \tag{17}$$

where

$$M_1 = \frac{K}{N}\{nE\,|X_{11}|^4 + n\,(N+n-2)\,E^2\,|X_{11}|^2 - E^{\frac{2}{3}}\,|X_{11}|^6\},$$

$$M_2 = \frac{K}{N}\{nE\,|X_{11}|^4 + n\,(N+n-2)\,E^2\,|X_{11}|^2 + E^{\frac{2}{3}}\,|X_{11}|^6 + (N-1)\,E^{\frac{4}{3}}\,|X_{11}|^3\}.$$

With the same arguments, we may deduce a bound of the variance

$$var\left(tr\left(XX^t\right)^2\right) \leq \frac{K}{N^4}\{N^4 E\,|X_{11}|^4\,E^2\,|X_{11}|^2 \tag{18}$$

$$+N^3[E^{\frac{4}{3}}\,|X_{11}|^6 + E^{\frac{1}{2}}\,|X_{11}|^4\,E^{\frac{1}{3}}\,|X_{11}|^6\,E^{\frac{1}{3}}\,|X_{11}|^{12} + E\,|X_{11}|^2\,E^{\frac{2}{3}}\,|X_{11}|^9]$$
$$+N^2[E^{\frac{2}{3}}\,|X_{11}|^6\,E^{\frac{1}{6}}\,|X_{11}|^{24} + E^{\frac{1}{3}}\,|X_{11}|^{15}\,E\,|X_{11}|^3 + E^{\frac{2}{3}}\,|X_{11}|^{12} + E^{\frac{5}{12}}\,|X_{11}|^{12}\,E^{\frac{1}{6}}\,|X_{11}|^{18}]\}.$$

Using (14) and (13), we may write

$$D\left(F^{XTX^t}, F^{XT_\theta X^t}\right) \to 0 \ \ and \ \ D\left(F^{\hat{X}T_\theta \hat{X}^t}, F^{\hat{X}T\hat{X}^t}\right) \to 0 \ \ a.s.$$

Furthermore, by Lemma 1, we get

$$D\left(F^{XTX^t}, F^{\hat{X}T\hat{X}^t}\right) \to 0 \ \ a.s. \tag{19}$$

For $\hat{B}_N$ and $\tilde{B}_N$ defined by relations (3) and (4), we have from Lemma 2.5 of [8],

$$\left\| F^{\hat{B}_N} - F^{\tilde{B}_N} \right\| \to 0. \tag{20}$$

Let $\bar{\bar{X}}_{ij} = \tilde{X}_{ij} - \bar{X}_{ij}$. Hence,

$$\bar{\bar{X}}_{ij} = \tilde{X}_{ij} 1_{(|X_{ij}| \geq \ln N)} + E\tilde{X}_{ij} 1_{(|X_{ij}| < \ln N)}, \ \ \bar{\bar{X}} = \left(\frac{1}{\sqrt{N}} \bar{\bar{X}}_{ij}\right)$$

where $\tilde{X}_{ij}$ and $\bar{X}_{ij}$ are defined by the relations (4), (5), respectively.
Then, from Cauchy–Schwartz inequality, we can show that the squared distance

$$D^2\left(F^{\tilde{X}T_\theta\tilde{X}^t}, F^{\bar{X}T_\theta\bar{X}^t}\right)$$

is bounded by

$$\frac{1}{N}\left\{\theta^2 tr\left(\bar{\bar{X}}\bar{\bar{X}}^t\right)^2 + 4\left[\theta^4 tr\left(\bar{\bar{X}}\bar{\bar{X}}^t\right)^2 tr\left(\bar{X}\bar{X}^t\right)^2\right]^{\frac{1}{2}}\right.$$
$$\left. +4\left[\left[\theta^4 tr\left(\bar{\bar{X}}\bar{\bar{X}}^t\right)^2 tr\left(\bar{X}\bar{X}^t\right)^2\right]^{\frac{1}{2}} \theta^2 tr\left(\bar{\bar{X}}\bar{\bar{X}}^t\right)^2\right]^{\frac{1}{2}}\right\}.$$

Therefore, in order to show that almost surely

$$D\left(F^{\tilde{X}T_\theta\tilde{X}^t}, F^{\bar{X}T_\theta\bar{X}^t}\right) \to 0, \tag{21}$$

it suffices to verify that

$$\theta^4\frac{1}{N}tr\left(\bar{\bar{X}}\bar{\bar{X}}^t\right)^2 \to 0, \frac{1}{N}tr\left(\bar{X}\bar{X}^t\right)^2 = O(1) \;\; a.s. \tag{22}$$

Since $E\left(\bar{\bar{X}}_{11}\right) = 0$ and $\bar{\bar{X}}_{ij} = \tilde{X}_{ij}1_{(|X_{ij}|\geq \ln N)} + E\tilde{X}_{ij}1_{(|X_{ij}|<\ln N)}$, we have

$$E\left|\bar{\bar{X}}_{11}\right|^2 \leq KE\left|X_{11}\right|^2 1_{(|X_{11}|\geq \ln N)} \to 0. \tag{23}$$

For $p \geq 4$,

$$E\left|\bar{\bar{X}}_{11}\right|^p \leq K\left(N^{\frac{p-4}{2}}E\left|X_{11}\right|^4 1_{(\ln N\leq|X_{11}|<\sqrt{N})} + 1\right). \tag{24}$$

By dominated convergence theorem, we get

$$E\left|\bar{X}_{11}\right|^2 \to E\left|X_{11}\right|^2 = \gamma. \tag{25}$$

For $p \geq 4$ and definition of rv's $\bar{X}_{11}$, we have

$$E\left|\bar{X}_{11}\right|^p \leq K(\ln N)^{p-2}. \tag{26}$$

From (15), (23), (24), $E(|X_{11}|^4 1_{(\ln N\leq|X_{11}|<\sqrt{N})}) \leq NE\left|X_{11}\right|^2$ and relation (17), we may write

$$E\left[\frac{1}{N}\theta^4 tr\left(\bar{\bar{X}}\bar{\bar{X}}^t\right)^2\right] \le K\theta^4\left[E^{\frac{2}{3}}\left|X_{11}\right|^2 1_{(|X_{11}|\ge \ln N)} + \frac{1}{N}\right] \to 0.$$

Also (18) gives

$$var\left(\frac{1}{N}\theta^4 tr\left(\bar{\bar{X}}\bar{\bar{X}}^t\right)^2\right) \le K\theta^8\left[\frac{1}{N^{7/6}}E^{1/6}\left|X_{11}\right|^4 1_{(\ln N\le|X_{11}|<\sqrt{N})} + \frac{1}{N^2}\right],$$

where the latter bound is summable by (16).

Hence, we obtain $\frac{1}{N}\theta^4 tr\left(\bar{\bar{X}}\bar{\bar{X}}^t\right)^2 \to 0$ a.s.

Now it remains to show that $\frac{1}{N}tr\left(\bar{X}\bar{X}^t\right)^2 = O(1)$ a.s. Using (17), (25) and (26), we find

$$K\{-\frac{(\ln N)^{\frac{8}{3}}}{N^2}\} \le E\left[\frac{1}{N}tr\left(\bar{X}\bar{X}^t\right)^2\right] - \frac{n}{N}\left(\frac{n}{N}+1-\frac{2}{N}\right)E^2\left|\bar{X}_{11}\right|^2$$
$$\le K\{\frac{n}{N}\frac{(\ln N)^2}{N} + \frac{(\ln N)^{\frac{8}{3}}}{N^2} + \frac{(\ln N)^{\frac{4}{3}}}{N}\}.$$

Consequently, $E\left[\frac{1}{N}tr\left(\bar{X}\bar{X}^t\right)^2\right] - \frac{n}{N}\left(\frac{n}{N}+1-\frac{2}{N}\right)E^2\left|\bar{X}_{11}\right|^2 \to 0$, and, $E\left[\frac{1}{N}tr\left(\bar{X}\bar{X}^t\right)^2\right] \to \gamma^2\left[c\left(c+1\right)\right].$

Concerning the variance, by (18), (25) and (26), we may obtain

$$var\left(\frac{1}{N}tr\left(\bar{X}\bar{X}^t\right)^2\right) \le K\frac{(\ln N)^{17/3}}{N^3},$$

which is summable. Then, (22) is verified from which (21) follows. This result with (14) allow us to write, $D\left(F^{\bar{X}T\bar{X}^t}, F^{\tilde{X}T\tilde{X}^t}\right) \to 0\ a.s.$

From (19) and (20), in order to prove $D\left(F^{XTX^t}, F\right) \to 0$ a.s, it suffices to verify that, $D\left(F^{\bar{X}T\bar{X}^t}, F\right) \to 0\ a.s.$ For this aim, we shall show that for any $z \in \mathbb{C}^+$, $m_{F^{\bar{X}T\bar{X}^t}}(z) \to m_F(z)\ a.s.$

Let $z \in \mathbb{C}^+$ and $\bar{B}_N = \bar{X}T\bar{X}^t$, the sequence $\{F^{\bar{B}_N}\}$ satisfies the assumptions of Lemma 2.8 of [8]. So $\exists m > 0$ such that

$$\inf_N F^{\bar{B}_N}[-m, m] > 0,\ \delta = \inf_N \Im\mathrm{m}\left(m_{F^{\bar{B}_N}}(z)\right) > 0\ a.s.$$

Write $\bar{B}_N - zI = (x-z)I + \bar{X}T\bar{X}^t - xI$, and then

$$(x-z)^{-1} - m_{\bar{B}_N}(z) = \frac{1}{N}\sum_{j=1}^{n}\frac{\tau_j}{1+\tau_j m_{F^{\bar{B}_N}}(z)}d_j, \tag{27}$$

where

$$d_j = d_j^N = \frac{1+\tau_j m_{F^{\bar{B}_N}}(z)}{1+\tau_j \bar{q}_j^t \left(\bar{B}_{(j)} - zI\right)^{-1} \bar{q}_j} \bar{q}_j^t \left(\bar{B}_{(j)} - zI\right)^{-1} \left((x-z)^{-1} I\right) \bar{q}_j$$
$$-\frac{1}{N} tr \left(\bar{B}_N - zI\right)^{-1} \left((x-z)^{-1} I\right),$$

with $\bar{q}_j$ denote the jth column of $\bar{X}$, and $\bar{B}_{(j)}$, x, $x_{(j)}$ are defined by relations (8) and (9).

Lemma 3.1 of [8] and assumption (d) of the Theorem 1 permit us to obtain

$$\max_{j\le n} \max \left[\beta_1, \beta_2, \beta_3\right] \to 0 \ \ a.s \tag{28}$$

where

$$\beta_1 = \left| \left\| \bar{q}_j \right\|^2 - 1 \right|,$$
$$\beta_2 = \left| \bar{q}_j^t \left(\bar{B}_{(j)} - zI\right)^{-1} \bar{q}_j - \frac{1}{N} tr \left(\bar{B}_{(j)} - zI\right)^{-1} \right|,$$
$$\beta_3 = \left| \bar{q}_j^t \left(\bar{B}_{(j)} - zI\right)^{-1} \left(\left(x_{(j)} - z\right) I\right)^{-1} \bar{q}_j - \frac{1}{N} tr \left(\bar{B}_{(j)} - zI\right)^{-1} \left(\left(x_{(j)} - z\right) I\right)^{-1} \right|.$$

Lemma 2.6 of [8] gives us, $\max_{j\le n} \max[\mid \gamma_1 \mid, \mid \gamma_2 \mid] \to 0$, where $\gamma_1 = m_{F^{\bar{B}_{(j)}}}(z) - m_{F^{\bar{B}_N}}(z)$, $\gamma_2 = m_{F^{\bar{B}_N}}(z) - \bar{q}_j^t \left(\bar{B}_{(j)} - zI\right)^{-1} \bar{q}_j$.

So that for N large enough, we have, $\max_{j\le n} \max[|\Im m\gamma_1|, |\Im m\gamma_2|] < \frac{\delta}{2}$.

Then, for $j, l \le n$,

$$\left| \frac{1+\tau_j m_{F^{\bar{B}_N}}(z)}{1+\tau_j^t \bar{q}_j \left(\bar{B}_{(j)} - zI\right)^{-1} \bar{q}_j} - 1 \right| < \frac{2}{\delta} \mid \gamma_2 \mid,$$

and

$$\left| \frac{\tau_l}{1+\tau_l m_{F^{\bar{B}_N}}(z)} - \frac{\tau_l}{1+\tau_l m_{F^{\bar{B}_{(j)}}}(z)} \right| \le \frac{2}{\delta^2} \mid \gamma_1 \mid .$$

Therefore,

$$\max_{j\le n} \max\left[\left| \frac{1+\tau_j m_{F^{\bar{B}_N}}(z)}{1+\tau_j^t \bar{q}_j \left(\bar{B}_{(j)} - zI\right)^{-1} \bar{q}_j} - 1 \right|, \left| x - x_{(j)} \right| \right] \to 0. \tag{29}$$

Using Lemmas 2.6, 2.7 of [8] and (28), (29), we may have

$$\max_{j \leq n} d_j \to 0.$$

Since

$$\left| \frac{\tau_j}{1 + \tau_j m_{F^{\bar{B}_N}}(z)} \right| \leq \frac{1}{\delta},$$

we may conclude from (27) that

$$(x - z)^{-1} - m_{\bar{B}_N}(z) \to 0.$$

Hence, the relation (11) is satisfied.

Finally, using (6), the proof of Theorem 1 is now complete.

Acknowledgements The authors would like to thank the Editor and anonymous referees for insightful comments improving the presentation of this paper.

References

1. Bai, Z.D., Silverstein, J.W.: Spectral Analysis of Large Dimensional Random Matrices, 2nd edn. Springer (2010)
2. Bosq, D.: Linear processes in function spaces. Theory and Applications. Lecture Notes in Statistics, vol. 149. Springer (2000)
3. Boutet de Monvel, A., Khorunzhy, A.: Limit theorems for random matrices with correlated entries. Markov Process. Relat. Fields **4**(2), 175–197 (1998)
4. Grenander, U., Silverstein, J.W.: Spectral analysis of networks with random topologies. SIAM J. Appl. Math. **32**(2), 499–519 (1977)
5. Jonsson, D.: Some limit theorems for the eigenvalues of a sample covariance matrix. J. Multivar. Anal. **12**, 1–38 (1982)
6. Khettab, Z., Mourid, T.: Eigenvalues empirical distribution of covariance matrices of AR processes. A simulation study. Annales de l'ISUP, 60, fasc 1–2, 3–22 (2016)
7. Marcenko, V.A., Pastur, L.A.: Distribution of eigenvalues in certain sets of random matrices. Mat. SB (N. S.). **72**(114), 507–536 (1967)
8. Silverstein, J.W., Bai, Z.D.: On the empirical distribution of eigenvalues of a class of large dimensional random matrices. J. Multivar. Anal. **54**(2), 175–192 (1995)
9. Silverstein, J.W., Bai, Z.D., Couillet, R., Debbah, M.: Eigen-inference for energy estimation of multiple sources. IEEE Trans. Inf. Theor. **57**(4), 2420–2439 (2011)
10. Silverstein, J.W., Choi, S.I.: Analysis of the limiting spectral distribution of large dimensional random matrices. J. Multivar. Anal. **54**(2), 295–309 (1995)
11. Wachter, K.W.: The limiting empirical measure of multiple discriminant ratios. Ann. Stat. **8**, 937–957 (1980)
12. Yin, Y.Q.: Limiting spectral distribution for a class of random matrices. J. Multivar. Anal. **20**, 50–68 (1986)

References

Part II
Computational Intelligence Methods for Time Series

Deep Learning for Detection of BGP Anomalies

Marijana Cosovic, Slobodan Obradovic and Emina Junuz

Abstract The Internet uses Border Gateway Protocol (BGP) for exchange of routes and reachability information between Autonomous Systems (AS). Hence, BGP is subject to anomalous traffic that can cause problems with connectivity and traffic loss. Routing Table Leak (RTL), worm and power outage events are considered anomalous in the sense that they can disrupt the Internet routing and cause slowdowns of varying severity, which leads to packet delivery reliability issues. Deep learning, a subfield of machine learning, could be applied in detection of BGP anomalies. Studying RTL, worm, and power outage events are of interest to network operators and researchers alike. In this paper, we consider datasets of several events, all of which caused large-scale Internet outages. We use artificial neural network (ANN) models based on a backpropagation algorithm for anomalous event classification.

Keywords Machine learning · Deep learning · Anomaly detection · BGP Sampling

1 Introduction

The AS-level Internet topology is a structure in which autonomous systems (AS), collections of routers with same routing policies, are represented by nodes, while the connection between the nodes are data paths used for exchanging reachability infor-

M. Cosovic (✉) · S. Obradovic
Faculty of Electrical Engineering, University of East Sarajevo, Istocno Sarajevo, Bosnia and Herzegovina
e-mail: marijana.cosovic@etf.unssa.rs.ba

S. Obradovic
e-mail: slobo.obradovic@gmail.com

E. Junuz
Faculty of Information Technology, Dzemal Bijedic University, Mostar, Bosnia and Herzegovina
e-mail: emina@edu.fit.ba

I. Rojas et al. (eds.), *Time Series Analysis and Forecasting*,
Contributions to Statistics, https://doi.org/10.1007/978-3-319-96944-2_7

mation between the ASs. Each AS is uniquely represented by an autonomous system number (ASN). As the number of ASs increased over time in the Internet, BGP was defined to improve existing External Gateway Protocol (EGP) and its drawbacks in terms of hierarchical structure that limited efficient expansion of the Internet. The latest version of the BGP protocol is BGP-4 defined in [1]. BGP is a routing protocol that connects different domains and it is used for routing in networks composed of a large number of ASs. It is used for routing among autonomous systems, and its latest version allows for Classless Inter-Domain Routing (CIDR), path aggregation, incremental additions, better filtering capabilities, and determining the routing policy. The Routing Information Service (RIS) project was initiated in 2001 by the Réseaux IP Européens Network Coordination Centre (RIPE NCC). RIPE NCC belongs to one of the five Regional Internet Registries (RIRs) that manage allocation and registration of IP addresses and ASNs. The scope of the RIS project is collecting and storing routing data from Remote route collectors (RRC) positioned predominantly at Internet exchange points. RRCs are software routers, released on the Linux platform that collect routing information. RRCs are positioned in all five RIRs but a majority of them are located within the RIPE NCC domain. Presently, 18 RRCs are active and using Quagga routing software for collecting raw data that is stored in MRT routing information export format [2] to two different type of files: all BGP packets created every 5 mins and a complete BGP routing table that is created every 8 hr [3]. PyBGPDump, a library written in Python is used to convert MRT into ASCII format. The quality of the data is inspected after the conversion, as missing and corrupt data may occur.

By studying BGP packets files and, in particular, by extracting BGP update messages from them, as they contain important reachability information, we can study connectivity disruption in the Internet during anomalous events. We investigated several types of anomalous events, namely, routing table events, worms, and power outage events. RTL events are in general initiated by router misconfigurations and, although not malicious in nature, can cause connectivity problems and traffic loss. Worm and power outage events can also contribute to connectivity and traffic loss issues. All of the events considered in this study were globally visible events. We extracted 15 features from BGP update messages on a minute-level: features related to volume of BGP messages and AS-PATH features related to AS-PATH attribute. For the duration of the anomalous event, we label the class feature with one (anomaly present), while the time before and after the anomalous event we label the class feature with zero (anomaly not present). In this way, we obtained a labeled feature matrix for each of the events. Routing data, extracted from the BGP update messages, could be considered as time series data since data points are indexed in time order.

Machine learning techniques have been employed in anomaly classification tasks [4–7]. Deep learning, part of machine learning, has been used extensively in voice and image recognition, language modeling, and information retrieval, amongst others, and has impacted the wide range of information processing tasks [8]. Detection of anomalies in time series data has employed deep learning techniques in the past. ANNs are systems that can be trained to recognize patterns in data and classify anomalous from regular data instances [9, 10]. Routing data could be used to analyze past anomalous events and aid in classification of future anomalous events.

The paper is organized as follows. In Sect. 2, we describe ANNs. Introduction of anomalous events, such as particular RTL events, worm events and power outage event are discussed in Sect. 3. In addition, extraction of BGP features from the datasets concludes Sect. 3. Classification methodology and used performance measures are discussed in Sect. 4. We conclude with Sect. 5.

2 ANN—Deep Learning

Artificial Neural Networks (ANN) are originally developed to mimic basic biological systems and to learn based on examples in the way humans do. In essence, neural networks learn gradually from the interdependence of data input properties. This interdependence can be linear or nonlinear in nature. Application of ANN has been present in the anomaly detection field [7, 9, 10]. When used in supervised learning neural network needs labeled input data; hence it is known in advance which class the data belongs to. Based on a comparison between the output of the neural network and the target function, during the training process, ANN adjusts the weights as shown in Fig. 1.

Artificial neural networks can be classified as Feedforward or Feedbackward structures, depending on the direction of propagation of the information. The Feedbackward structure of neural networks refers to the spread of information backwards. When the input vector is applied to the input layer of the neural network, it propagates through the network throughout all its layers, and it generates output values by using the output layer of the network. The output values are compared with a desired target function, and for each of the neurons in the output layer the difference is calculated. Further information about these differences propagates backwards until all the neurons in the neural network are affected by the difference between the original and the target output value. The value of the weighting factors is determined by the optimization technique (typically a minimizing of the loss function with respect to

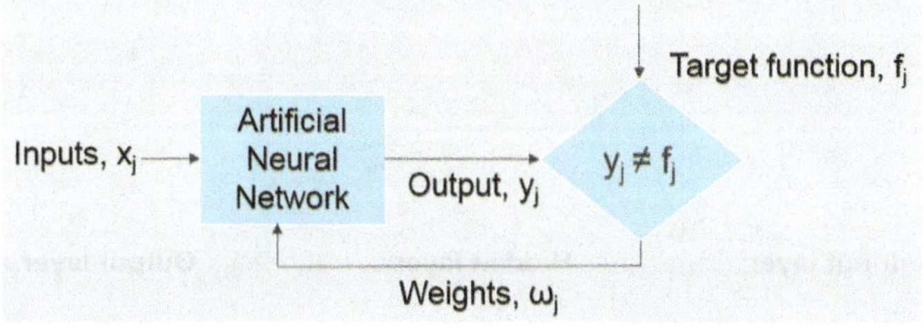

Fig. 1 Artificial neural network

the weights in the network), which determines the weighting factors such that the loss function is minimized.

ANN are simple mathematical methods made up of basic processing elements called neurons. The structures of neural networks differ in the number of layers used. Between the first and last layers of neural networks, there are hidden layers: usually one hidden layer in simpler networks and more hidden layers in complex neural networks.

The architecture of the neural network is engaged in specific neuronal connectivity as a whole. Usually, the number of neurons in the input layer is equal to the number of features (number of columns in the feature matrix). Each neuron has one input, and all the outputs are connected to all neurons of the next layer, as shown in Fig. 2. When using a neural network for classification, the output layer can have one or more neurons, depending on whether it is binary or multi-class classification. The most commonly used functions for the output neuron modeling are sigmoid or normalized exponential [11] functions.

Perceptron is a neuron model type developed in the original neural networks, in which each neuron has a number of inputs (x_j) associated with corresponding weight factors (ω_j), which show the effect of a particular input on the output. Thus, the output neuron classifies information by comparing the value of the sum (1) and the threshold value, which is a parameter of the neuron.

$$\sum_j \omega_j x_j \tag{1}$$

Modeling of neurons with a perceptron has the following disadvantage: a small change in the weight factor of any perceptron can lead to a sudden change in its output. This, in turn, can lead to a complicated change in the rest of the network, which may be difficult to control. The most commonly used artificial neuron model, which solves the aforementioned problem, is the sigmoid neuron, shown by the following expression:

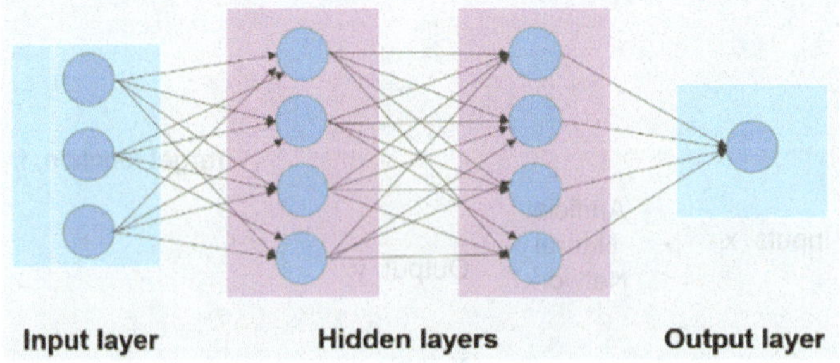

Fig. 2 Architecture of ANN with four layers: one input layer, two hidden layers, and one output layer

$$\frac{1}{1+exp\left(-\sum_j \omega_j x_j - b\right)} \tag{2}$$

where ω_j are weighting factors, x_j are input neurons, and b is bias. It turns out that a change in the output of sigmoid neurons linear function changes the weighting factors and bias. In this way, it is easier to determine how changes in weighting factors and bias may influence the change of the output neuron; hence, the neural network could be considered more resilient to changes of data and the ability to learn.

3 Anomalous Events

The BGP routing system is subject to frequent incidents that result in significant interruptions of Internet connectivity. This can be observed in BGP update messages. In this paper, we consider the following routing table leak events: Routing Leak AS9121 [12], AWS Route Leak [13], Telecom Malaysia AS4788 Route Leak [14], and Indosat Routing Table Leak [15], all of which showed an increased number of announced IP prefixes throughout the duration of the events. We also consider the Slammer [16] and Code Red I [17] worm events, as well as the Moscow power blackout event [18].

BGP Datasets

We obtain datasets from the RIPE NCC that collects Internet routing data by using Routing Information Service (RIS) Remote Route Collectors (RRC) positioned in various locations throughout the world. The effects of all the events considered in this paper and presented in Table 1 caused globally visible connectivity issues. We have used routing updates collected at two RRCs located in CIPX, Geneva and VIX, Vienna. We used BGP update messages during the occurrence of the routing leak, worm, and power outage events stored in MRT format described in [2]. In order to create a feature matrix, we observed BGP update messages during a five day period, including two days before and two days after the actual event.

Duration of the actual events lasted between 79 min, in the case of the AS9121 RTL event, and 869 min, in the case of the Slammer worm event. The rest of the anomalous events have durations that fall between those two values (Table 1). Python code was written in order to extract features from the dataset that is a collection of BGP update messages after MRT to ASCII conversion for each of the events. For example, Table 2 shows that on June 12, 2015 at 16:05:02 UTC, AS12350 (192.65.185.157) announced that the address prefixes 177.155.50.0/23, 177.155.52.0/23, 201.46.160.0/19, and 201.46.232.0/21 were available. The path by which the above prefixes were available was AS-PATH: 12350 174 6762 262589 262589 262589 262589 262589 28615. The

Table 1 BGP anomalous events

Dataset	Regular class	Anomaly class	Number of features
AS9121 RTL	7121	79	15
AWS RTL	7085	115	15
Malaysian Telecom RTL	7018	182	15
Indosat RTL	7050	150	15
Slammer	6331	869	15
Code Red I	6600	600	15
Moscow power outage	7031	169	15

Table 2 BGP update message during Telecom Malaysia AS4788 routing leak

FIELD:	VALUE:
TIME:	06/12/15 16:05:02
TYPE:	BGP4MP/MESSAGE/Update
FROM:	192.65.185.157 AS12350
TO:	192.65.185.40 AS12654
ORIGIN:	IGP
AS-PATH:	12350 174 6762 262589 262589 262589 262589 262589 28615
NEXT_HOP:	192.65.185.157
MULTI_EXIT_DISC:	0
ANNOUNCE	177.155.50.0/23 177.155.52.0/23 201.46.160.0/19 201.46.232.0/21

original autonomous system was AS28615, while AS262589, AS6762, and AS174 were transit autonomous systems. BGP update messages traveled from the original, via transit, to the ultimate AS. On the other hand, the data transmitted were traversed by the sequence of ASs defined by the AS-PATH attribute path (from left to right). Fifteen volume and AS-PATH features were extracted from BGP messages on a minute-level during the 5-day period, hence producing a feature matrix of 7200×15 in size.

The volume features that we observed are the number of BGP messages announcing new routes, the number of BGP messages withdrawing already existing routes, the number of announced IP prefixes (Figs. 3, 5, 6, 7, 8, 10 and 11), the number of withdrawn IP prefixes (Fig. 12), the number of duplicate announced messages, the number of duplicates withdrawn messages, the number of implicitly withdrawn messages, the number of BGP messages which NLRI originates from the Exterior Gateway Protocol (EGP), the number of BGP messages which NLRI originates from the Interior Gateway Protocol (IGP), and the number of BGP messages which NLRI originates from unknown sources.

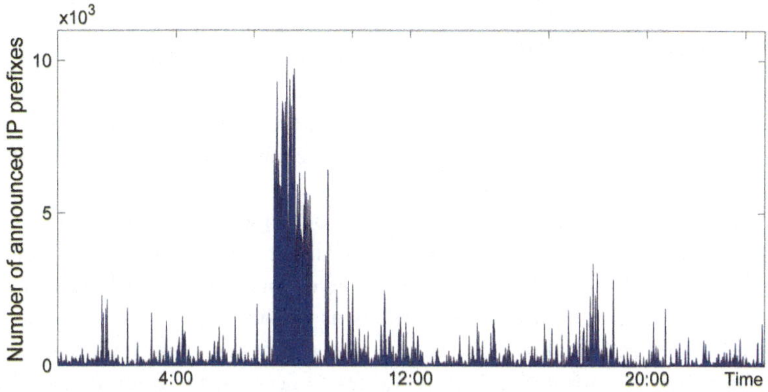

Fig. 3 Number of announced network layer reachability information (NLRI) prefixes during AS9121 Routing Leak Event as observed on RIPE Route Collector rrc04, CIPX

Duplicate announcements and withdrawal messages are defined as BGP update messages that announce the same combination of IP prefix and AS-PATH attribute that has previously been announced. Implicit withdrawal implies that the same IP prefix has been announced with a different AS-PATH attribute, hence it is an implicit withdrawal of a previous announcement (same IP prefix but different AS-PATH).

The features we computed based on the AS-PATH attribute are: the average length of the AS-PATH attribute, the maximum length of the AS-PATH attribute (Fig. 6), the average length of each unique AS-PATH attribute, the average edit distance, and the maximum edit distance (Figs. 4 and 9). While extracting information from the AS-PATH attribute, we considered regular and unique AS-PATHs. We also considered AS-PATHs as a string of ASNs (autonomous system number) and computed the similarity of two adjacent AS-PATHs by finding their edit distance [19].

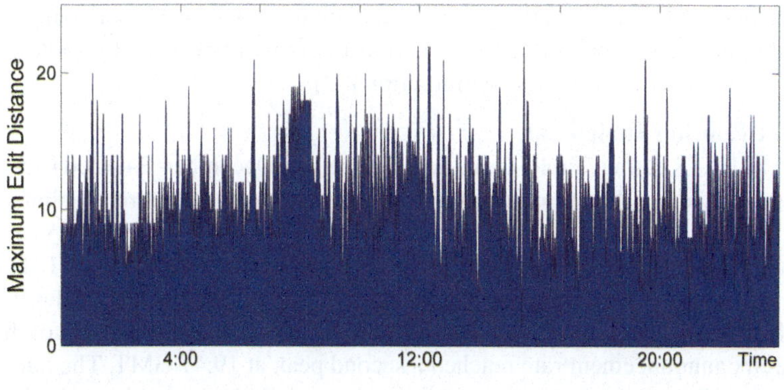

Fig. 4 Maximum edit distance during AS9121 Routing Leak Event as observed on RIPE Route Collector rrc04, CIPX

Features belong to three types: continuous, categorical, and binary. All of the volume features belong to the continuous type, since features may have an infinite number of values. On the other hand, features derived from the AS-PATH attribute may have a finite number of values and hence, are categorical. The class feature is of the binary type: given volume and AS-PATH features, we either have anomalous instances or not.

We have labeled all 7200 time instances (described by 15 features) as either belonging to anomalous or regular class in accordance with the information regarding the beginning, duration, and end of each of the events. We have referred to several sources in order to label our data as correctly as possible.

Considering that global routing tables increased in size from the time of the first event, we needed to normalize feature values to account for Internet size growth. Normalization is done such that each feature vector has a zero mean and a standard deviation of one [20]. We also performed feature discretization for the features of the continuous type prior to training the neural network. We did not encounter any missing data during the seven events observed, although we did have an increased number of outliers in the case of the Indosat RTL and Code Red I datasets, which can be observed in Fig. 7 and Fig. 10, respectively.

Routing Table Leak Events

Many of the events that cause connectivity issues are classified as routing leaks. It is often unclear what is meant by that term. Based on research of actual events on the Internet, which can be of use to network operators and Internet users, the authors in [21] define routing leaks as a propagation of announced paths beyond the intended scope. This means that the BGP path announcement from one AS to another in some way violates the routing agreements between a sending AS, a receiving AS or any transit AS. The consequence of routing leaks is traffic redirection through a path not originally planned, and thus, various malicious attacks from analyzing data to eavesdropping could be performed. The most common reasons why routing leaks occur are errors in the router's configuration [22].

AS9121 Routing Table Leak

The AS9121 Routing Table Leak took place on December 24, 2004. AS9121 announced to other ASs through BGP sessions that were used to reach almost 70% of all prefixes, which at that time amounted to more than 106 k prefixes. As a result, the data of tens of thousands of networks were either lost or diverted. AS9121 started to announce prefixes to its neighbors around 9:20 GMT, and the event lasted until just after 10:00 GMT. AS9121 continued announcing prefixes for the rest of the day. The prefix announcement rate reached a second peak at 19:47 GMT. The number of announced IP prefixes during the routing leak event is shown in Fig. 3. An increase was observed in the number of withdrawn IP prefixes, as well. Besides the increase in the number of announced/withdrawn prefixes, the maximum edit distance (the

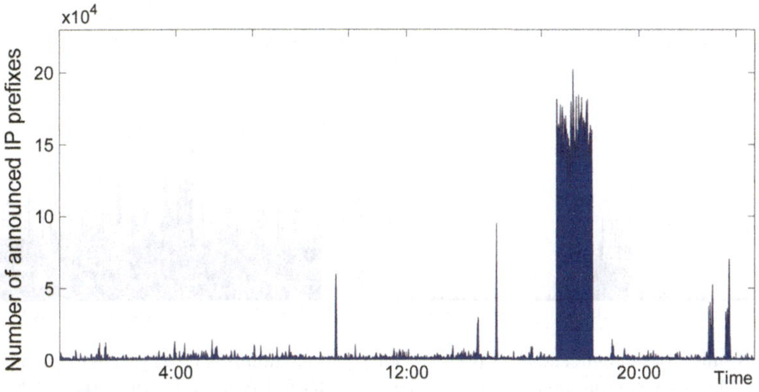

Fig. 5 Number of announced NLRI prefixes during AWS Routing Leak Event as observed on RIPE Route Collector rrc04, CIPX

measure of similarity between two AS-PATH attributes) increased during the duration of the event, as can be observed in Fig. 4. This could indicate that the choice of the paths differed from the common ones, and it was a sign of disruption between commonly connected ASs.

AWS Route Leak
The AWS Route Leak started at 17:10 UTC on April 22, 2016, and affected a large number of ASs and prefixes. Loss of traffic and connectivity were present since networks with high traffic prefixes, such as Google, Amazon, and Twitter, were affected, amongst others.

The event occurred due to maintenance issues on Innofield AG (AS 200759) that is connected to Swiss Internet eXchange (SwissIX). Innofield AG normally announces one IPv4 and IPv6 prefix to SwissIX. During maintenance reactivation of BGP sessions, AS 200759 distributed prefixes belonging to Amazon as belonging to private AS 65021. Prefix announcements were propagated through AS 6939 Hurricane Electric (HE) that peers at SwissIX. This resulted in a redirection of traffic passing through HE to a private AS, and hence, it compromised the reachability of Amazon AS. Since the event was widespread and likely caused by a misconfigured route optimizer, we observed an increase in announced IP prefixes at CIPX, as shown in Fig. 5.

Telecom Malaysia Route Leak
The Malaysian Telecom (AS 4788) leaked one-third of all IP prefixes in the global routing table to the backbone provider Level3 (AS 3549).

The event, triggered by routers misconfiguration at Telecom Malaysia, started on June 12, 2015 at 8:43 UTC and lasted until 11:45 UTC. Level3 (AS 3549) propagated traffic from its peers and customers via Telecom Malaysia, which was not capable of handling the traffic volume, resulting in major packet loss and performance degradation. The performance degradation was especially pronounced between the Asia Pacific region and the rest of the Level 3 network. Figure 6 shows an increased

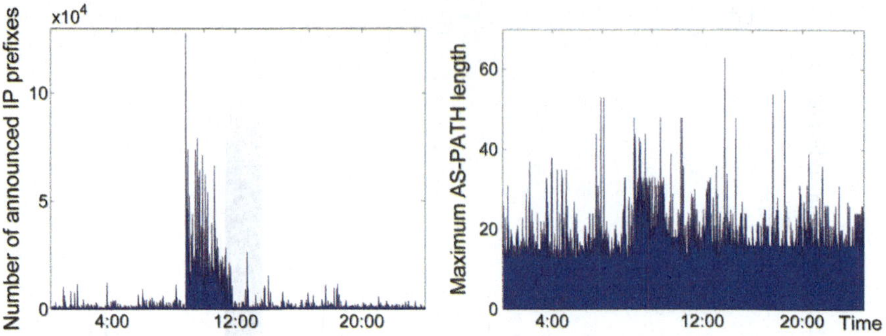

Fig. 6 Number of announced NLRI prefixes (left) and maximum AS-PATH length (right) during Telecom Malaysia Routing Leak Event as observed on RIPE Route Collector rrc04, CIPX

number of announced IP prefixes (left) and also an increase of maximum AS-PATH length (right) for the duration of the route leak event.

Indosat Routing Table Leak

The Indosat routing table leak occurred on April 2, 2014. At the time of the event, the global routing table consisted of nearly half a million routes. AS 4761 (Indosat) leaked around 320,000 routes, which happened during scheduled maintenance, starting at 18:25 UTC. The reason behind Indosat originating prefixes that were not assigned to it is assumed to be that BGP was redistributed with bad upstream filtering. This inadvertent error had an impact that was observed on various route collectors through an increase of announced IP prefixes, as shown in Fig. 7. Several hundreds of those prefixes were widely accepted, and services of some networks such as Akamai, a leading content delivery network (CND) and cloud service provider, were disrupted.

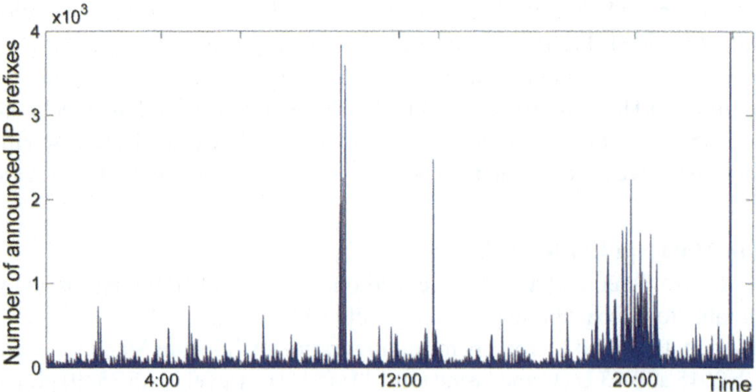

Fig. 7 Number of announced NLRI prefixes during Indosat Routing Leak Event as observed on RIPE Route Collector rrc04, CIPX

Worm Events

The Slammer worm is a single packet UDP scanning worm (404 bytes) that attacked MS SQL server and MS SQL server desktop addition on January 25, 2003. It spread worldwide in less than 10 min by sending copies of itself to random IP addresses. The main reason behind this rapid spread was the result of a bandwidth-limited scanner: each copy of the worm could scan at the maximum rate that the processor and network bandwidth could support. Depending on the upload bandwidth, every Slammer copy could be sending infectious packets at the maximum rate, hence the rapid spreading in which the number of infected machines doubled every 8.5 s. The number of announced IP prefixes during the Slammer worm is shown in Fig. 8, while the maximum edit distance increase can be observed in Fig. 9.

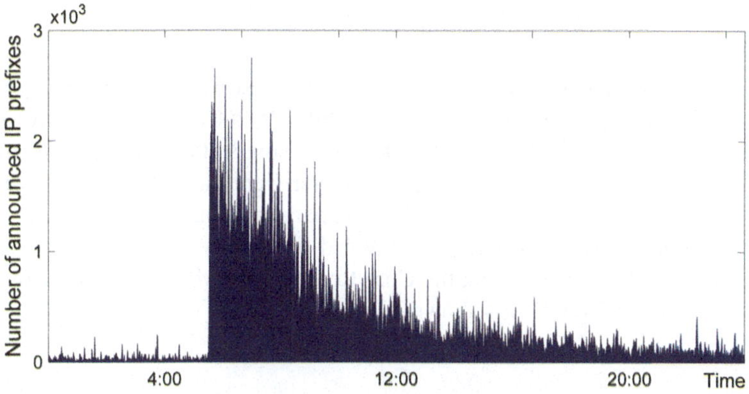

Fig. 8 Number of announced NLRI prefixes during Slammer worm event as observed on RIPE Route Collector rrc04, CIPX

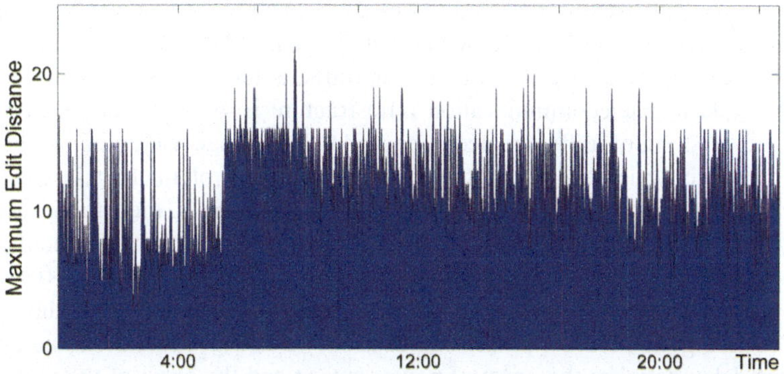

Fig. 9 Maximum edit distance during Slammer worm event as observed on RIPE Route Collector rrc04, CIPX

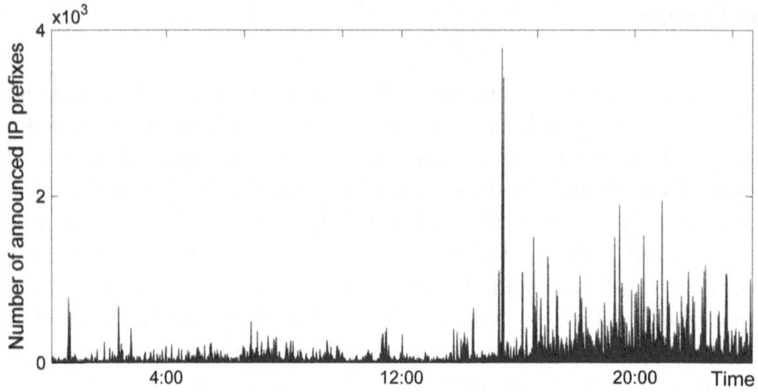

Fig. 10 Number of announced NLRI prefixes during Core Red I worm event as observed on RIPE Route Collector rrc04, CIPX

The Code Red I worm was released through IIS servers in June 2001, but the peak of infected computers was observed on July 19, 2001. The worm spread itself by creating a sequence of random IP addresses using static seed for generation of new IP addresses. Since every infected computer went through the same list of IP addresses, fewer systems were infected in comparison to the Slammer worm spread, and the number of infected machines doubled only every 40 min. The number of announced NLRI prefixes during the Core Red I worm event as observed on RIPE Route Collector rrc04, CIPX is shown in Fig. 10.

Power Outage Events

The Power outage event considered in this study is a Moscow power blackout that occurred on May 25, 2005. Moscow Internet eXchange (MSK-IX) was shut down at that time. Considering that 80% of Russian traffic at the time was passing through MSK-IX and that telecommunication infrastructure is greatly centralized around Moscow, rerouting of traffic created congestion. The outage had an impact on wide scale connectivity issues, hence, there was a disruption of Internet service: even though the data centers of main Russian websites had power, their traffic was still going through MSK-IX. Figure 11 shows an increased number of announced IP prefixes in BGP update messages during the power outage, as observed at RIS remote route collector in Vienna Internet eXchange (VIX). The number of withdrawn IP prefixes during the Moscow power outage is shown in Fig. 12, and we can observe the time delay between the onset of power outage and the onset of the increase in number of withdrawn IP prefixes in BGP messages.

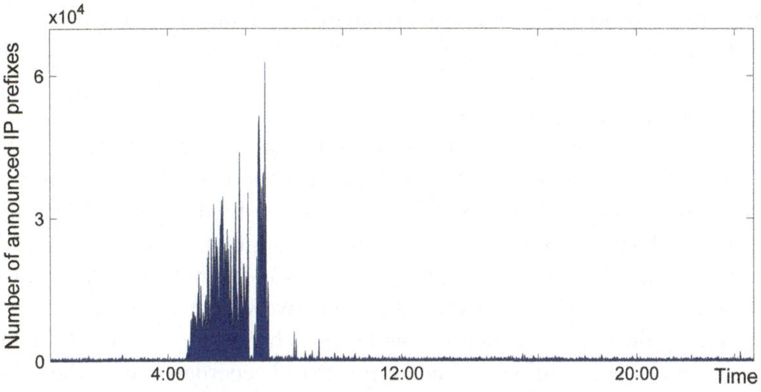

Fig. 11 Number of announced NLRI prefixes during Moscow power blackout as observed on RIPE Route Collector rrc05, VIX

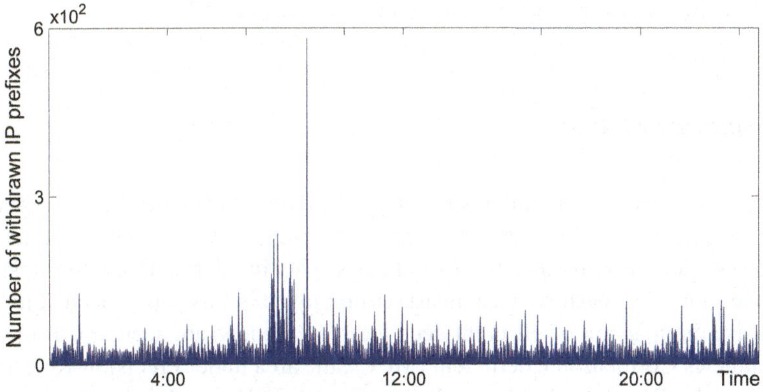

Fig. 12 Number of withdrawn NLRI prefixes during Moscow power blackout as observed on RIPE Route Collector rrc05, VIX

4 Classification of Anomalous Events

Methodology

We used the Keras Python library [23] with the Theano backend for development and evaluation of deep learning models. Models, based on a backpropagation algorithm for training of fully connected multilayer perceptron (MLP) neural networks, are defined as sequences of layers: an input layer, hidden layers, and an output layer. The shape of the input data needs to be specified only for the first layer in the sequence. In Keras, using Dense class is one of the ways to define fully connected layers. Network weights can be initialized to random numbers using either uniform or Gaussian distribution. Use of appropriate activation function allows for better training of the network [24]. Traditionally, sigmoid and tanh activation functions are used, but the authors in [24] have shown that better performance can be achieved using a rectifier activation function. In the output layer, we use a sigmoid function as we are dealing with binary classification.

We use 10-fold cross validation for determining accuracy on the test dataset, and as we increase the number of hidden layers beyond two, classification accuracy decreases. We found that a neural network with two hidden layers is the optimal model for the anomalous datasets considered: routing table leak, worm, and power outage. Using either too few or too many neurons in the hidden layers may result in problems of underfitting and overfitting, respectively. General guidelines are used for determining the number of neurons within each hidden layer. We selected neural network architecture based on trial and error, but in accordance with the following general guidelines: the number of neurons in hidden layers should be between the sizes of input and output layers, and they should be the sum of two-thirds of the input layer neurons and output layer neurons. Hence, we trained the neural network with two dense hidden layers with 15 and 10 neurons, respectively.

Performance Measures

The goal of binary classification is to categorize data into two different classes: regular or anomalous. In most cases, the number of anomalous instances is a fraction of regular instances, and as such, the cost of classifying regular or anomalous instances is not the same. The performance measures employed in this paper, needed for comprehensive comparison of different deep learning models, are accuracy, F-measure, the Matthews Correlation Coefficient (MCC), the area under Precision-Recall (PR), the area under Receiver Operating Characteristics (ROC), and time taken to build a model. Accuracy, considering our datasets are highly imbalanced (Table 1), might not be the most accurate performance measure. This is due to the fact that misclassification would have different costs associated with points belonging to either the regular or anomalous class. Accuracy is defined as the ratio of points belonging to the regular/anomalous class that are classified as regular/anomalous and the total number of points in the dataset. In order to define F-measure, we first define recall (R) as the ratio of detected anomalous points and all points labeled as anomalous. On the other hand, precision (P) is a ratio of detected anomalous points and all anomalous points. Specificity (S) is a ratio of detected regular points and all regular points; hence it is a measure of how many regular instances are identified as regular. F-measure is given as a double ratio of the product of P and R and the sum of P and R. MCC is given by (3), where N is the number of all points and TP is the number of data points classified as anomalous.

$$MCC = \frac{TP/N - PR}{\sqrt{PR(1-P)(1-R)}} \tag{3}$$

The PR curve is more often used when there is a class imbalance problem [25] because both precision and recall measures are defined by focusing on the number of detected anomalous points. ROC curve represents the relationship between recall and specificity measure, and as such, remains the same regardless of the baseline prior probability of the anomalous class. This is reflected in the reported results (Tables 3, 4 and 5) in which the area under the PR curve is often smaller than the area under the ROC curve.

Table 3 Performance measures of the original anomalous events

Dataset	Acc	F-measure	MCC	ROC	PR	Time (s)
AS9121 RTL	0.99375	0.945	0.942	0.998	0.946	10.55
AWS RTL	0.99431	0.808	0.807	0.961	0.848	10.95
Malaysian Telecom RTL	0.9925	0.852	0.848	0.979	0.883	10.58
Indosat RTL	0.93056	0.753	0.707	0.897	0.802	10.65
Slammer	0.95986	0.834	0.811	0.976	0.916	10.57
Code Red I	0.94542	0.586	0.582	0.887	0.628	10.65
Moscow power outage	0.99639	0.920	0.919	0.974	0.923	8.76

Table 4 Performance measures of anomalous events using oversampling techniques

Dataset	Acc	F-measure	MCC	ROC	PR	Time (s)
AS9121 RTL	0.99816	0.998	0.996	0.999	0.999	22.02
AWS RTL	0.99167	0.992	0.983	0.994	0.984	20.95
Malaysian Telecom RTL	0.98953	0.990	0.979	0.995	0.994	21.12
Indosat RTL	0.92087	0.923	0.844	0.958	0.940	21.19
Slammer	0.93854	0.940	0.878	0.977	0.967	19.63
Code Red I	0.89932	0.899	0.799	0.954	0.947	20.43
Moscow power outage	0.98962	0.990	0.979	0.997	0.997	18.1

Table 5 Performance measures of anomalous events using undersampling techniques

Dataset	Acc	F-measure	MCC	ROC	PR	Time (s)
92AS9121 RTL	0.98734	0.987	0.975	0.999	0.999	0.28
AWS RTL	0.96087	0.960	0.923	0.979	0.986	0.36
Malaysian Telecom RTL	0.95055	0.950	0.901	0.975	0.981	0.29
Indosat RTL	0.88333	0.878	0.770	0.927	0.948	0.31
Slammer	0.94131	0.941	0.883	0.983	0.979	2.73
Code Red I	0.9075	0.901	0.821	0.947	0.964	1.93
Moscow power outage	0.9645	0.964	0.929	0.986	0.989	0.48

Classification Results

We used a neural network with two hidden layers and obtained the performance measure values shown in Table 3. Accuracy is not the best approach to compare classification of different events, as the datasets are highly imbalanced. Hence, we used performance measures as introduced in Sect. 4.2. In addition, time taken to build a model is added in the results. Table 1 shows that amongst all RTL events, Malaysian Telecom RTL has the largest set of data labeled as anomalous—182 compared to the AS 9121 RTL event in which only 79 instances are labeled as anomalous. The Indosat RTL event shows the worst performance of all RTL datasets, and we can contribute that to noise in the dataset (Fig. 7). Slammer, followed by the Code Red I dataset, has the largest number of instances belonging to the anomaly class amongst all datasets. Noise in the Code Red I dataset, as shown in Fig. 10, might be the reason behind poor performance measures of Code Red I presented in Table 3. We used undersampling and oversampling techniques as in [26] to balance regular and anomalous instances in all datasets. In the case of oversampled and undersampled datasets, their imbalance ratio is around 1, meaning the classes are balanced; hence, accuracy and F-measure are approximately the same values.

Oversampling techniques are algorithms that create additional instances of the class that is represented by a smaller number of instances in the dataset. We used six oversampling techniques, namely, Synthetic Minority Oversampling Technique (SMOTE), Support Vector Machine (SVM)-SMOTE, Borderline1-SMOTE, Borderline2-SMOTE, Adaptive Synthetic Sampling (ADASYN), and Random Oversampling (ROS) algorithms. By using balancing techniques of the datasets, we achieved better performance measures, as shown in Table 4. The best results were achieved using the SVM-SMOTE oversampling technique for AS9121 RTL, AWS RTL, Indosat RTL, Slammer, Code Red I and Moscow dataset, while the Malaysian Telecom RTL dataset, when oversampled by ROS algorithm, had the best performance measure that was better by a small margin than when oversampled by the SVM-SMOTE algorithm.

Undersampling techniques are algorithms that remove instances from the dataset that belong to the more represented class. We used ten undersampling algorithms, namely, Near Miss-1, Near Miss-2, Near Miss-3, Tomek Links, Cluster Centroids, One-sided selection, Random undersampling (RUS), Edited Nearest Neighbors, Neighborhood Cleaning Rule, and Condensed Nearest Neighbors. By using undersampling balancing techniques of the datasets, we achieved better performance measures, as shown in Table 5. When comparing Tables 4 and 5, the values of performance measures (F-measure, MCC, and ROC) are greater in the case of oversampling techniques for most datasets, and this is due to possible overfitting. Also, when datasets are oversampled, additional points from the anomalous class are added into the original dataset, hence, the area under the PR curve increases, as can be observed in Table 4.

The best results were achieved using the RUS undersampling technique for AS9121 RTL, AWS RTL, Slammer, and Code Red I datasets, while Code Red I, Indosat RTL and Malaysian Telecom RTL datasets, when undersampled by the Near Miss-1 algorithm, had the best performance measure, which was only better by a small margin than when undersampled by the RUS algorithm.

5 Conclusion

We have developed a model for anomaly detection based on artificial neural networks with two hidden layers, which are optimal because performance indices deteriorated with additional hidden layers. We used a cross-validation technique to determine the number of neurons in each of the layers. Balancing techniques (dataset oversampling and undersampling) were employed, as the original datasets are highly imbalanced. Classification of the Indosat RTL and Code Red I datasets achieved the worst performance measures, possibly due to noise in the datasets. Similar performance measures on those datasets propagated when undersampling and oversampling techniques were used. We concluded that employing volume and AS-PATH features extracted from BGP update messages could lead to the reliable classification of anomalous events.

References

1. Rekhter, Y., Li, T., Hares, S.: A Border Gateway Protocol 4 (BGP-4). http://ietf.org/rfc/rfc4271 (2006). Accessed 20 June 2017
2. Manderson, T.: Multi-threaded routing toolkit (MRT) border gateway protocol (BGP) routing information export format with geo-location extensions. rfc6397.txt (2011). Accessed 20 November 2017
3. RIPE RIS raw data. http://www.ripe.net/data-tools/stats/ris/ris-raw-data

4. Ćosović, M., Obradović, S., Trajković, Lj.: Performance evaluation of BGP anomaly classifiers. In: Proceedings of the International Conference on Digital Information, Networking and Wireless Communication, pp. 115–120 (2015)
5. Cosovic, M., Obradovic, S., Trajkovic, L.J.: Classifying anomalous events in BGP datasets. In: Proceedings of the 29th Annual IEEE Canadian Conference on Electrical and Computer Engineering (CCECE 2016), pp. 697–700 (2016)
6. Cosovic, M., Obradovic, S.: Ensemble methods for classifying BGP anomalies. Ind. Technol. **4**(1), 12–20 (2017)
7. Ćosović, M., Obradović, S., Junuz, E.: Deep learning for detection of BGP anomalies. In: Proceedings of International Work-Conference on Time Series (ITISE 2017), pp. 487–498 (2017)
8. Deng, L., Yu, D.: Deep learning: methods and applications. Found. Trends Signal Process. **7**(3–4), 197–387 (2014)
9. Dau, H.A., Ciesielski, V., Song, A.: Anomaly detection using replicator neural networks trained on examples of one class. In: Proceedings of the 10th International Conference on Simulated Evolution and Learning, pp. 311–322 (2014)
10. Jadidi, Z., Muthukkumarasamy, V., Sithirasenan, E., Sheikhan, M.: Flow-based anomaly detection using neural network optimized with GSA algorithm. In: Proceedings of the 33rd IEEE International Conference on Distributed Computing Systems Workshops (ICDCSW'13), pp. 76–81 (2013)
11. Bishop, C.M.: Pattern Recognition and Machine Learning. Information Science and Statistics. Springer-Verlag New York Inc., Secaucus, NJ, USA (2006)
12. Popescu, A.C., Premore, B.J., Underwood, T.: Anatomy of a Leak: AS9121. https://www.nanog.org/meeting-archives/nanog34/presentations/underwood.pdf (2005). Accessed 20 November 2017
13. AWS Route Leak-North American Network Operators Group Mailing List. https://mailman.nanog.org/pipermail/nanog/2016-April/085410.html (2016). Accessed 20 June 2016
14. Telecom Malaysia AS4788 Route Leak-North American Network Operators Group Mailing List. https://mailman.nanog.org/pipermail/nanog/2015-June/076187.html (2015). Accessed 20 June 2016
15. Indosat Routing Table Leak-North American Network Operators Group Mailing List. https://mailman.nanog.org/pipermail/nanog/2014-April/065920.html (2014). Accessed 20 June 2016
16. Moore, D., Paxson, V., Savage, S., Shannon, C., Staniford, S., Weaver, N.: Inside the Slammer Worm. IEEE Secur. Priv. **1**(4), 33–39 (2003)
17. Schauer, R.C.: The mechanisms and effects of the Code Red worm. https://www.sans.org/reading-room/whitepapers/dlp/mechanisms-effects-code-red-worm-87 (2001). Accessed 20 November 2017
18. Moscow Power Blackout-North American Network Operators Group Mailing List. https://www.nanog.org/mailinglist/mailarchives/old_archive/2005-05/msg00650.html (2005). Accessed 20 June 2016
19. Levenshtein, V.I.: Binary codes capable of correcting deletions, insertions and reversals. Doklady Akademii Nauk SSSR **163**(4), 845–848 (1965)
20. LeCun, Y., Bottou, L., Orr, G.B., Müller, K.-R.: Effiicient BackProp. In: Montavon, G., Orr, G.B., Müller, K.-R. (eds.) Neural Networks: Tricks of the Trade. LNCS, vol. 7700, pp. 9–48. Springer-Verlag, London, UK (1998)
21. Sriram, K., Montgomery, D., McPherson, D., Osterweil, E., Dickson, B.: Problem Definition and Classification of BGP Route Leaks. https://www.rfc-editor.org/rfc/rfc7908.txt (2016)
22. Mahajan, R., Wetherall, D., Anderson, T.: Understanding BGP misconfiguration. In: Proceedings of the Conference on Applications, Technologies, Architectures, and Protocols for Computer Communications (SIGCOMM'02), pp. 3–16 (2002)
23. Chollet, F.: Keras. https://github.com/fchollet/keras (2016)
24. Nair, V., Hinton, G.E.: Rectified linear units improve restricted Boltzmann machines. In: Proceedings of 27th International Conference on Machine Learning, pp. 807–814 (2010). Accessed 20 November 2017

25. Davis, J., Goadrich, M.: The relationship between precision-recall and ROC curves. In: Proceedings of 23rd International Conference on Machine Learning, pp. 233–240 (2006). Accessed 20 November 2017
26. Ćosović, M., Obradović, S.: BGP anomaly detection with balanced datasets. Tehnički vjesnik/Technical Gazette **25**(3) (2018)

Using Scaling Methods to Improve Support Vector Regression's Performance for Travel Time and Traffic Volume Predictions

Amanda Yan Lin, Mengcheng Zhang and Selpi

Abstract Long queues often happen on toll roads, especially at the tollgates. These create many problems including having an impact on the regular roads nearby. If travel time and traffic volume at the tollgates can be predicted accurately in advance, this would allow traffic authorities to take appropriate measures to improve traffic flow and the safety of road users. This paper describes a novel combination of scaling methods with Support Vector Machines for Regression (SVR) for travel time and tollgate volume prediction tasks, as part of the Knowledge Discovery and Data Mining (KDD) Cup 2017. A new method is introduced to handle missing data by utilising the structure of the road network. Moreover, experiments with reduced data were conducted to evaluate whether the conclusions from combining scaling methods with SVR could be generalised.

Keywords Travel time prediction · Traffic volume prediction · Tollgate · SVR Time series analysis · SVR with scaling · Support vector regression

1 Introduction

Traffic jams are common scenes in most roads including toll roads or controlled access roads. The tollgates, in particular, are well known as bottleneck, especially during rush hours and holidays. Reliable methods to predict future traffic flow and

A. Y. Lin · M. C. Zhang · Selpi (✉)
Department of Mechanics and Maritime Sciences, Chalmers University of Technology, SE-412 96 Göteborg, Sweden
e-mail: selpi@chalmers.se

M. C. Zhang
e-mail: mengchengz@outlook.com

A. Y. Lin
e-mail: l.amanda0821@gmail.com

A. Y. Lin · M. C. Zhang
Department of Computer Science and Engineering, Chalmers University of Technology, SE-412 96 Göteborg, Sweden

I. Rojas et al. (eds.), *Time Series Analysis and Forecasting*,
Contributions to Statistics, https://doi.org/10.1007/978-3-319-96944-2_8

demands are important for traffic management authorities and road users. With precise predictions, the traffic regulators can decide how to deal with the problems (e.g. to open more tollgates or divert traffic at upstream intersections) and road users can plan their routes better. This paper is an extended version of [1], where we combined Support Vector Machine for Regression (SVR) with scaling methods for predicting travel time and traffic volume (as part of a competition in Knowledge Discovery and Data Mining (KDD) Cup 2017 [2]) for a given road and tollgate during rush hours, knowing the previous two-hour data and some days before. While the base of the work here is the same as in [1], a new piece of work (i.e. testing the generalisation of our methods) is added here. Furthermore, we elaborate on our own method to fill in the missing data by utilising the road network topology. For completeness purpose, we include the relevant parts of the base work here.

Travel time is the time taken from a designated start point to a designated end point. Traffic volume is a record of the number of vehicles at a designated point. Travel time and volume calculations depend on many stochastic factors, such as weather condition, holidays, time of day and season, thus making the tasks of predicting travel time and traffic volume challenging to date.

SVR is a version of Support Vector Machine (SVM) for regression that was proposed in 1996 by Drucker et al. [3]. SVR is chosen here due to past researches that have shown good performances using SVR in different areas, including financial time series forecasting [4], stock market price forecasting [5], real-time flood stage forecasting [6] and also travel time prediction [7].

The rest of this paper is arranged as follows. Section 2 describes the data and the objectives of this work. Section 3 describes the related work. In Sect. 4, the methods used are introduced. We describe and discuss the results of our experiments for travel time prediction and traffic volume prediction in Sects. 5 and 6, respectively. We test if the conclusions from Sects. 5 and 6 could be generalised in Sect. 7. The final conclusions are presented in Sect. 8.

2 Data and Work Objectives

The data used here are from the KDD Cup 2017. It consists of four types, i.e. road network topology, time-stamped records of actual vehicles driving from intersections to tollgates (called vehicle trajectories data), traffic volume at tollgates and weather data. The road network is represented as a sequence of road links and implemented as a directed graph Fig. 1. The network includes three intersections (A, B, C) and three tollgates (1, 2, 3). These make up ten routes. Only data from vehicles using Amap navigation software was included in the vehicle trajectories data [2]. Therefore, there was quite a lot of missing data in the provided data set.

The objectives of this work are to address the following tasks:

- Task 1—Travel time prediction: Given training data described above for the period from 19 July to 24 October, predict the average travel time for each route during

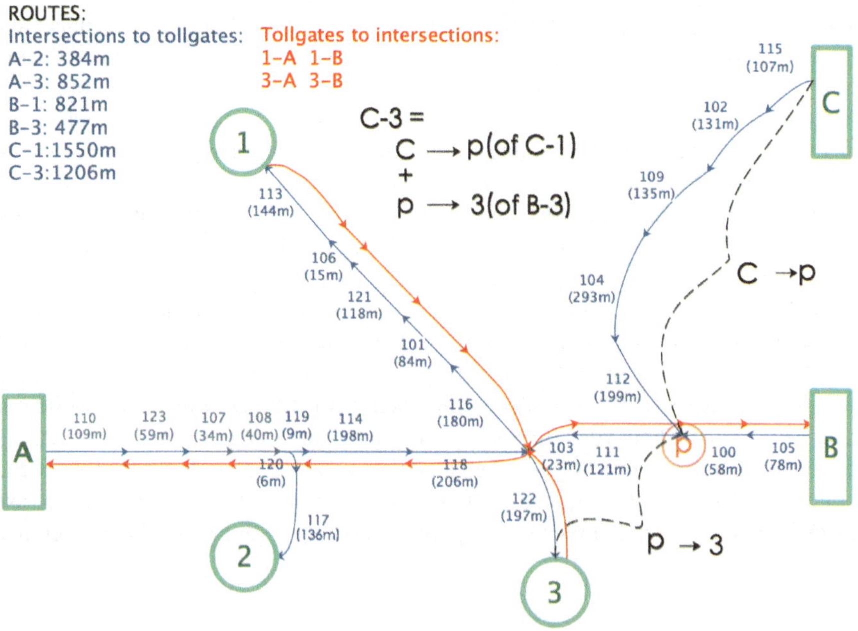

Fig. 1 The link representation of road network. Each route is composed by a sequence of links and each link is represented by an arrow. The value without parentheses over a link represents the unique ID of the link and the value in parentheses represents the length of the link

rush hours (08:00–10:00 and 17:00–19:00), per 20-min interval, for the period from 25 October to 31 October.

- Task 2—Traffic volume prediction: Given training data described above for the period from 19 September to 24 October, estimate the volume for each of the five tollgate-direction pairs (Tollgate 1-entry, Tollgate 1-exit, Tollgate 2-entry, Tollgate 3-entry and Tollgate 3-exit) during rush hours, per 20-min interval, for the period from 25 October to 31 October.
- Task 3—Test the generalisation of the methods used, i.e. testing if the conclusions from Tasks 1 and 2 still hold if we use different data as input.

3 Related Work

Traffic flow prediction, a well-known problem in traffic network, has been studied by many researchers. Both statistical (data-driven) and analytical approaches (model-based) had been tried for such predictions (see a recent review in [8]). The statistical approach uses time series data consisting variables such as travel times, speeds and volumes as input and predict the near future travel time based on historical traffic patterns. This approach assumes that the current or near future travel time will have similar pattern as historical travel time, while the analytical approach deduces the travel time from traffic conditions. The traffic conditions in turn are predicted from traffic propagation on the network by using traffic simulators. The statistical approach is suitable to be used when there are good amount of historical data, while the analytical approach can be applied to the situation with changes in input factors, for example, adding additional networks [8]. Compared with analytical approach, an obvious advantage of statistical approach is that there are lots of ready-to-use software packages, and the approach does not need much expertise about traffic flow modelling [9].

SVR belongs to the statistical approach and is a data-driven method. An application of SVR for highway travel time prediction has been studied by Wu et al. in [7]. There exist two main differences between data sets in Wu et al.'s paper and in our project, one is that they collected the data from different highways, while our data were collected between different intersections and tollgates, the other is that in our data, we have special holidays and lots of missing data, but they avoided special holidays and set the data loss rate within some threshold values. In addition, we use feature scaling as a data preprocessing step, which was not included in Wu et al.'s work.

4 Methods

Experiments using SVR with and without scaling methods were conducted. The scaling methods investigated include standard scaling, min-max scaling and robust scaling. A combination of our own method (called complementary) and linear interpolation was used to fill in the missing data. The use of different combinations of features was tested. Cross-validation was used to measure the predictive performance of each model built using different scaling methods and feature sets. Generalisation of the methods was tested.

Support Vector Regression

The Support Vector Regression (SVR) uses the same principles as the Support Vector Machine for Classification (SVC). The goal of SVR is to find a function, with at most ϵ deviation from the actual target y. The problem can be written as a convex optimisation problem

$$\text{minimize } \frac{1}{2}\|w\|^2 \quad \text{subject to } y_i - \langle w, x_i \rangle - b \leq \epsilon$$
$$\langle w, x \rangle + b - y_i \leq \epsilon$$

If the problem is not feasible, slack variables ξ_i, ξ_i^* are introduced. The formulation becomes

$$\text{minimise } \frac{1}{2}\|w\|^2 + C\sum_{i=1}(\xi_i + \xi_i^*) \quad \text{subject to } y_i - \langle w, x_i\rangle - b \le \epsilon + \xi_i$$
$$\langle w, x\rangle + b - y_i \le \epsilon + \xi_i^*$$
$$\xi_i, \xi_i^* \ge 0$$

where the constant $C > 0$ is penalty parameter. More about SVR can be found in [10, 11]. In this project, we used the SVR implementation from Scikit-learn library in Python [12].

Scaling Methods

Scaling is a way to systematically alter all the values in a data set. The simplest method, min-max scaling, is rescaling the data to a fixed range, usually [0, 1] or [−1, 1]. For a given data set X, a min-max scaling is typically done via the following equation:

$$lb + \frac{X - min(X)}{max(X) - min(X)}(ub - lb),$$

where lb is a lower bound of the range and ub is an upper bound [13].

One common and widely used scaling method is standard scaling. The idea of standard scaling is to make the values of each feature in the data have zero mean and unit variance, according to

$$\frac{X - mean(X)}{standard\ deviation(X)}.$$

Another scaling method is robust scaling, which is based on the median and the interquartile range. If the data set X contains many outliers, robust scaling often gives better results [14]. Robust scaling is defined as

$$\frac{X - median(X)}{IQR},$$

where IQR is interquartile range [14].

Error Measurements and Validation Method

Mean Absolute Percentage Error (MAPE) has been chosen by KDD cup team to evaluate the predictions. The MAPE is defined as

$$MAPE = \frac{1}{R}\sum_{r=1}^{R}(\frac{1}{T}\sum_{t=1}^{T}|\frac{d_{rt} - p_{rt}}{d_{rt}}|). \quad (1)$$

For Task 1 (travel time prediction), d_{rt} and p_{rt} are the actual and predicted average travel time for route r during time window t. For Task 2 (volume prediction), R is the number of tollgate-direction pairs (1-entry, 1-exit, 2-entry, 3-entry and 3-exit), T is the number of time windows in the testing period and d_{rt} and p_{rt} are the actual and predicted traffic volume for a specific tollgate-direction pair r during time window t.

Cross-validation was used to assess the predictive performance of our models.

5 Travel Time Prediction

To build a good model for Task 1, we addressed the sub-task to estimate the average travel time, per 20 mins interval, from designated intersections to tollgates, for the hours 08:00–10:00 and 17:00–19:00 during 18 to 24 October, with training data from 19 July to 17 October. In order to test our models, the previous two-hour data of the period to be predicted were used as test data.

Quite a lot of data points were missing in the data set. Before running experiments, the missing data were filled in by applying our own 'Complementary' method as shown in Fig. 1 and then linear interpolation. 'Complementary' is a method to fill in missing data in a route with the relevant part of other route(s) data. For instance, if there is missing data for a specific time window in route C-3, we gather part of that specific time window data from C-1 to get the data from Intersection C to point p (C → p) and part of data in route B-3 to get the data from point p to Tollgate 3 (p → 3) to fill in the missing part in C-3, see Fig. 1. 'Complementary' was applied to the data in routes B-1, B-3 and C-1 in the same way. Since there are not many missing data in routes A-2 and A-3, the missing parts were only filled in by linear interpolation.

We assumed that, in the morning and afternoon, the travel time of every given route is independent of each other, and as a result, we applied the same prediction procedure on each route in the morning and afternoon separately. After iterative trial and error experiments with different parameter values chosen randomly, Radial Basis Function (RBF) was selected as our kernel function, with parameters $\gamma = 0.005$ and $\varepsilon = 0.5$. Furthermore, parameter C was chosen based on

$$max(|\bar{y} + 3\sigma_y|, |\bar{y} - 3\sigma_y|) \quad (2)$$

where $\bar{y}$ and σ_y are the mean and standard deviation of the y values from training data [15]. It has been found that SVR with RBF is less sensitive to data preprocessing methods such as scaling [13].

Several cross-validation experiments, for example, testing different scaling methods, different number of training data and different feature sets were conducted. Time window position and the previous two-hour travel time were treated as two basic features during the process. *Time window position*: since the prediction is for the rush hours (the definitions of rush hours are 08:00–10:00 and 17:00–19:00), every 20 mins interval, the rush hours are divided into six 20-min time window. For instance, for the rush hour in the afternoon, 17:00–17:20 is the first time window position, 17:20–17:40 is the second and so on. *Previous two-hour travel time*: it means the two-hour travel time data before the rush hours. For example, the previous two-hour travel time for the rush hours in the afternoon is the data from 15:00 to 17:00. They are divided into six 20-min time window as well.

Obviously, the travel time is a product of dynamic interplay of traffic demand and traffic supply [16]. High traffic flow denotes high traffic demand. The factors, including temporal effects, such as daily pattern, weekly pattern and holiday have influences on the traffic demand [7]. The factors, for example, crashes, road works, weather and so on have influences on the traffic supply. For this reason, we added some extra features into the prediction one by one. The predictive performance of every resulting model was evaluated by comparing the results from validation and prediction phases. We show extra features that can capture the traffic demand as follows. *Special days*: holidays, weekends or working days. *Tollgate volume*: it means the traffic volume at the tollgate in the target route. For instance, when predicting the travel time in route A-2, the tollgate volume means the volume at Tollgate 2 (shown in Fig. 1). *Adjacent tollgate volume*: it means the traffic volume at the tollgate adjacent to route to be predicted. When two routes are from the same intersection and then go to different tollgates, while one of them is the route to be predicted, another one will become the adjacent route. For instance, route A-2's adjacent tollgate volume is the volume at Tollgate 3.

The predictive performances of using SVR combined with different scaling methods are presented in Table 1. This table also shows the results of the experiments using

Table 1 Average MAPE from cross-validation experiments with basic features, using two sets of training data (all means 19/7 to 17/10 and part means 19/9 to 17/10). Test set is from 18/10 to 24/10

Scaling method	Validation using all data	Prediction using all data	Validation using part of data	Prediction using part of data
Robust scaling	0.2302	0.1886	0.1901	0.2073
Standard scaling	0.2296	0.1902	0.1888	0.2083
Min-Max scaling [0, 1]	0.2276	0.1935	0.1811	0.1928
No scaling	0.2464	0.2081	0.1977	0.2001

two different amounts of training data ('all' means training data from 19/7 to 17/10 and 'part' means training data from 19/9 to 17/10).

In Table 1, one can see that using fewer weeks data for training gives better validation results, but worse prediction results. This also means that our experiments did not show anything conclusive about the influence of season on the travel time prediction (note that the period 19/7 to 18/9 is a summer season). Similarly, our experiments (not shown here due to space) suggest that most of the weather-related features did not increase predictive performance of our models. If any, only temperature was worth adding.

The best experimental result from the travel time prediction task is achieved by applying robust scaling with the two basic features (the previous two-hour travel time and time window position). Table 1 also shows that using scaling method gives better predictive performance compared to no scaling. Robust scaling seems to be particularly good for time series with more varying patterns (that include summer season), while min-max scaling seems to be particularly good for time series with more similar patterns.

6 Traffic Volume Prediction

In order to build a good model for Task 2, we addressed this sub-task: estimate the average volume for every tollgate-direction pair, per 20 mins interval, during rush hours (08:00–10:00 and 17:00–19:00) from 18 October to 24 October using training data from 19 September to 17 October.

We assumed that, in the morning and afternoon, the volume at a given tollgate-direction pair is independent of each other, and as a result, we applied the same prediction procedure on each tollgate-direction pair in the morning and afternoon separately. MAPE defined in Eq. 1 was used to calculate the average error for every tollgate-direction pair. We applied SVR for the volume prediction as well. After iterative trial and error experiments with different parameter values chosen randomly, Radial Basis Function (RBF) was selected as our kernel function, with parameters $\gamma = 0.01$ and $\epsilon = 0.01$. Furthermore, parameter C was selected by Eq. 2.

Similar feature selection strategy, as in Sect. 5, is used for this task. Time window position and the previous two-hour volume were treated as two basic features during the process. Time window position is the same as in Sect. 5 and the previous two-hour volume means the two-hour volume data before the rush hours to be predicted.

The performances of combining different scaling methods with SVR are presented in Table 2. For the volume prediction, applying SVR combined with a scaling method gives a huge improvement to the result compared with only using SVR. And again, it appears that robust scaling is particularly good for time series with more varying patterns. Note that the period from 1 October to 7 October is a big holiday period in China and it is widely known that the traffic volume is unusual during that period.

Traffic volume may depend on many factors, including time of day, day of week, holiday, weather, etc. For this reason, an additional feature called special days to

Table 2 Average MAPE from cross-validation experiments with features: time window position and previous two-hour volume. Training data are from 19/9 to 17/10. Test data are from 18/10 to 24/10.

Scaling method	Validation result	Prediction of test data
Robust scaling	0.2710	0.1472
Standard scaling	0.2717	0.1502
Min-Max scaling [0, 1]	0.3467	0.1526
No scaling	1.0374	0.3128

capture the holiday and weekend effect was added. Other features, extracted from the provided volume data, including the number of vehicles with ETC and the number of vehicles having vehicle model, n ($n \in [0, 7]$), were also tested in our experiments (not shown here due to space). The predictive performance increases when we add special days to the feature set. The best performance (with average MAPE 0.2691 in validation phase and 0.1436 in prediction phase) is from an experiment, where special days are included in the feature set, suggesting that the feature special days are very important for traffic volume prediction.

7 Generalisation

Based on the experimental results from the previous sections, we conclude that: (i) SVR with a scaling method performs better compared to without scaling, (ii) robust scaling is especially good for time series with varying patterns and (iii) min-max scaling is especially good for time series with similar patterns. These conclusions could depend on the provided input data. Here, we want to test if the conclusions could be generalised, if different traffic data sets are used. Due to the lack of other traffic data set, we analyse the following question instead:

Do these conclusions still hold if some of the data had been missing?

If some of the data had been missing, we start from slightly different data. The basic idea to address this question is: randomly delete some values from the original data (pretend those values were missing) and repeat the same experiment with reduced input data. The procedure can be summarised as follows:

1. Delete $p\%$ of the original data randomly.
2. Fill in the originally missing data and the deleted data using complementary and linear interpolation.
3. Take the data after Step 2, for each of the three scaling methods, run the experiment with a fixed feature set and a fixed SVR setting. For simplicity reason, we use the basic feature set (time window position and the previous two-hour travel

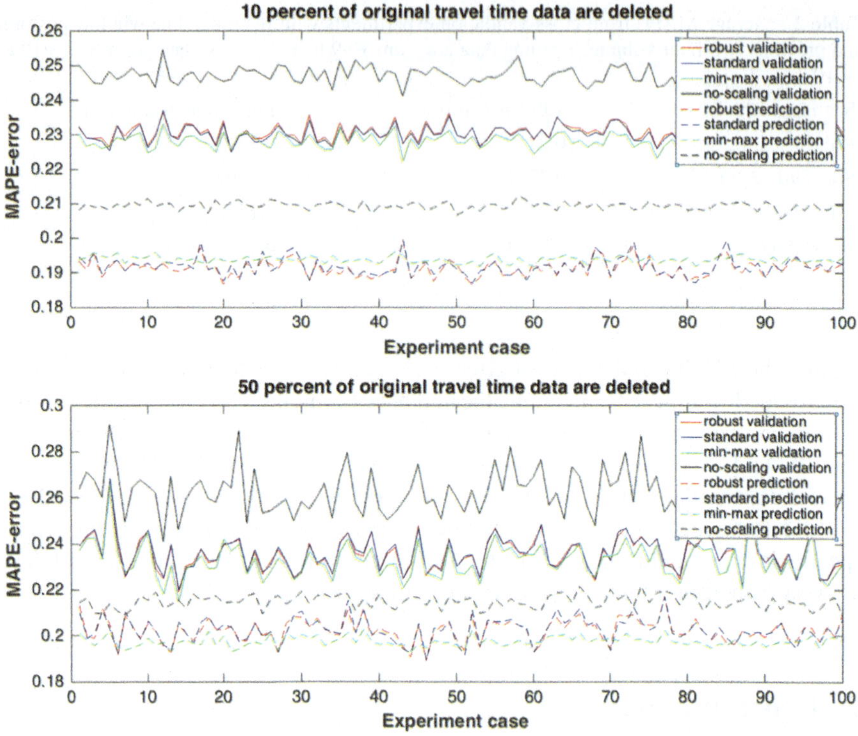

Fig. 2 Validation and prediction error for 100 experiments with 10% (top) and 50% (bottom) deleted data for the generalisation of Task 1 (travel time prediction)

time), RBF-kernel and SVR parameters, $\epsilon = 0.5$, $\gamma = 0.005$. The output from this step is a table similar to Table 1, but with changed values.

4. Repeat Step 1 to Step 3 100 times.

Task 1 was investigated with five levels of deletion (10, 20, 30, 40 and 50%). For brevity, only the results from 10 and 50% are reported, here, in Fig. 2.

The results from Task 1, Fig. 2, show that the performance of no scaling (black) is noticeably worst compared to that of the other methods for both validation (solid lines) and prediction (dashed lines). For validation, the performances of robust scaling (solid red) and standard scaling (solid blue) are very similar, while min-max scaling (solid green) is slightly better than the other scaling methods. For prediction, the performances of robust scaling (dashed red) and standard scaling (dashed blue) deteriorate more than min-max scaling (dashed green) as more data are deleted. A possible explanation of this is that as more data are deleted, the more outliers disappear and are replaced with smoother values (since we use complementary and linear interpolation to fill in the deleted data). This suggests that min-max scaling is especially good for time series with similar patterns.

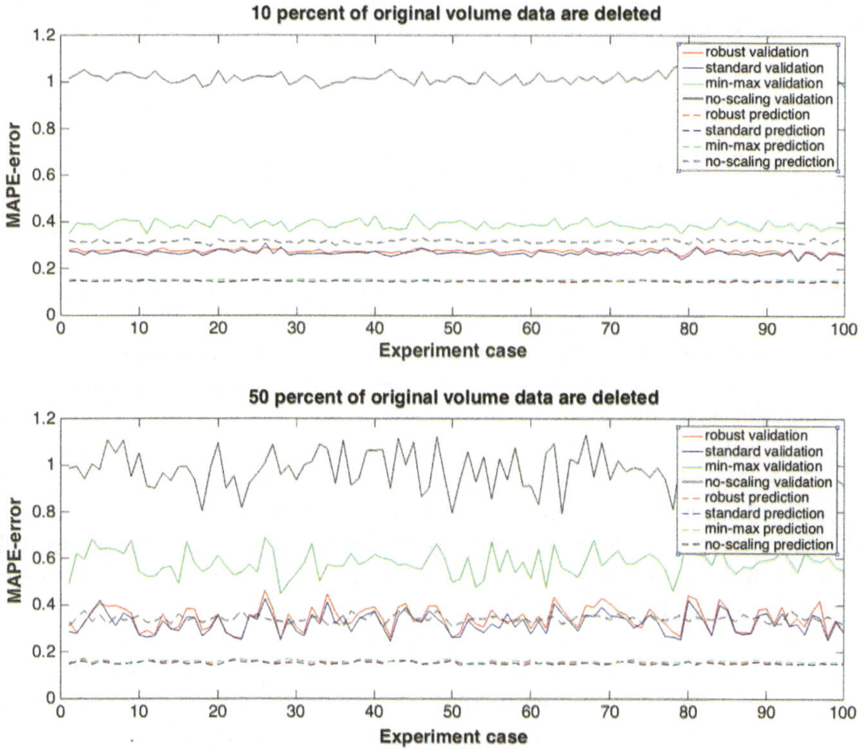

Fig. 3 Validation and prediction error for 100 experiments with 10% (top) and 50% (bottom) deleted data for the generalisation of Task 2 (volume prediction)

A modified generalisation procedure was applied for Task 2, with the same five levels of deletion. The modifications include: In Step 2, only linear interpolation is used to fill in the originally missing data and the deleted data; In Step 3, use time window position and the previous two-hour volume as feature set and set SVR parameters, $\epsilon = 0.01$, $\gamma = 0.01$.

The results from Task 2, Fig. 3, show that, in all cases, the performance of no scaling (black) is worst compared to that of the other methods for both validation (solid lines) and prediction (dashed lines). The performances of robust (red) and standard scaling (blue) are very similar in both validation and prediction. Unlike the results of generalisation from Task 1, the validation performance of min-max scaling (solid green) deteriorates more than the other scaling methods as more data are deleted. For prediction, the performances of all three scaling methods are very similar. A possible explanation for this is that the original pattern (varying pattern) of volume data are mostly preserved after deletion and filling in process.

Based on the results of this section, the conclusions (i) and (iii) still hold. Regarding conclusion (ii), it seems that, for the data with varying pattern, the robust and standard scaling performs very similar and slightly better than min-max scaling.

8 Conclusion

This paper demonstrated the application of SVR with scaling methods for travel time and tollgate volume predictions in rush hours. The impact of using three different scaling methods (robust scaling, standard scaling and min-max scaling) with SVR predictor was investigated. Furthermore, experiments to test if the conclusions from Sects. 5 and 6 still hold if reduced data are used as input were conducted. Our results suggested that SVR combined with a scaling method provides a more accurate prediction than without scaling, especially for volume prediction task. Min-max scaling was found to be particularly good for time series with more similar patterns. The performances of robust scaling and standard scaling were found to be pretty similar, and they seemed to perform slightly better than min-max scaling for time series with varying patterns.

Features that capture different travel time/volume influencing factors were analysed in the experiments. Although adding the features tollgate volume and adjacent tollgate volume have been found to increase the performance in some of our experiments for travel time prediction, but it is not always the case. The feature special days was found to be useful for volume prediction. Weather-related features were not found to be that useful in our experiments.

When our model was applied to Task 1, the mean absolute percentage error of the travel time prediction is around 0.19, which differs by only 0.02 from the best result obtained by other contestants (this is a competition task, the best prediction result was announced). Similarly, when our model was applied to Task 2, the mean absolute percentage error of the volume prediction is around 0.144, which differs by only 0.03 from the best result. We conclude that SVR combined with a scaling method can still provide a reasonable performance for travel time and traffic volume predictions, even when the training data contain many outliers (like holiday data) and no deep analysis of the data was applied.

Acknowledgements S. acknowledges strategic funding support from Chalmers Area of Advance Transport while writing this paper.

References

1. Lin, A.Y., Zhang, M.C., Selpi.: Combining support vector regression with scaling methods for highway tollgates travel time and volume predictions. In: Proceedings of International Work-Conference on Time Series Analysis. ITISE (2017)
2. KDD2017, https://tianchi.aliyun.com/competition/information.htm?spm=5176.100067.5678.2.ru0ea4&raceId=231597. Last accessed 15 Mar 2017
3. Drucker, H., Burges, C.J., Kaufman, L., Smola, A., Vapnik, V.: Support vector regression machines. Advanc Neural Inf. Process. Syst. **9**, 155–161 (1997)
4. Lu, C.J., Lee, T.S., Chiu, C.C.: Financial time series forecasting using independent component analysis and support vector regression. Decision Support Syst. **47**(2), 115–125 (2009)
5. Yeh, C.Y., Huang, C.W., Lee, S.J.: A multiple-kernel support vector regression approach for stock market price forecasting. Exp. Syst. Appl. **38**(3), 2177–2186 (2011)

6. Yu, P.S., Chen, S.T., Chang, I.F.: Support vector regression for real-time flood stage forecasting. J. Hydrol. **328**(3), 704–716 (2006)
7. Wu, C.H., Ho, J.M., Lee, D.T.: Travel-time prediction with support vector regression. IEEE Trans. Intell. Trans. Syst. **5**(4), 276–281 (2004)
8. Oh, S., Byon, Y.J., Jang, K., Yeo, H.: Short-term travel-time prediction on highway: a review of the data-driven approach. Transp. Rev. **35**(1), 4–32 (2015)
9. Van Lint, J.W.C., Hoogendoorn, S.P., van Zuylen, H.J.: Accurate freeway travel time prediction with state-space neural networks under missing data. Transp. Res. Part C Emerg. Technol. **13**(5), 347–369 (2005)
10. Smola, A.J., Schlkopf, B.: A tutorial on support vector regression. Stat. Comput. **14**(3), 199–222 (2004)
11. Basak, D., Pal, S., Patranabis, D.C.: Support vector regression. Neural Inf. Process. Lett. Rev. **11**(10), 203–224 (2007)
12. SVR. http://scikit-learn.org/stable/modules/generated/sklearn.svm.SVR.html. Last accessed 15 Mar 2017
13. Crone, S.F., Guajardo, J., Weber, R.: The impact of preprocessing on support vector regression and neural networks in time series prediction. In: DMIN, pp. 37–44 (2006)
14. RobustScaler. http://scikit-learn.org/stable/modules/generated/sklearn.preprocessing.RobustScaler.html. Last accessed 15 Mar 2017
15. Cherkassky, V., Ma, Y.: Practical selection of SVM parameters and noise estimation for SVM regression. Neural Netw. **17**(1), 113–126 (2004)
16. van Lint., J.W.C.: Reliable travel time prediction for freeways: bridging artificial neural networks and traffic flow theory. In: TRAIL Research School (2004)

Part III
Dimensionality Reduction and Similarity Measures in Time Series

Linear Trend Filtering via Adaptive LASSO

Matúš Maciak

Abstract Linear trend filtering methods are popular due to their overall simplicity—the model is linear in each segment and there are typically only few segments considered. These segments are defined by unique points where the trend changes its direction—so-called changepoints. In this paper, we consider an innovative estimation approach for such models. Our proposal is based on recent developments in the atomic pursuit techniques: we present an estimation algorithm based on the adaptive LASSO penalty and we introduce a fully data-driven method which can be effectively used to fit the continuous linear trend models. Some statistical properties are discussed and the empirical performance is compared with respect to other competitive LASSO-based techniques.

Keywords Linear trend filtering · Joinpoint regression · Regularization · Lasso Adaptive lasso · Changepoints · Oracle properties

1 Introduction

In general, the trend filtering models can be viewed in terms of some convex minimization problem where one minimizes some objective function (usually an L_2-norm loss) together with some specific penalty—typically the sum of the absolute kth order differences of the consecutive parameter estimates (for some reasonable choice of $k \in \mathbb{N}$). The resulting model is then formed as a linear combination of the k degree spline functions such that the overall fit is continuous (for $k > 0$). A specific scenario is obtained for $k = 1$. This choice leads to the linear trend filtering case: the resulting

M. Maciak (✉)
Department of Probability and Mathematical Statistics, Charles University,
Sokolovská 82, 186 75 Prague, Czech Republic
email: maciak@karlin.mff.cuni.cz
URL: http://karlin.mff.cuni.cz/~maciak

I. Rojas et al. (eds.), *Time Series Analysis and Forecasting*,
Contributions to Statistics, https://doi.org/10.1007/978-3-319-96944-2_9

estimate is piecewise linear and, in addition, it is continuous over the whole domain. Such models are also commonly known as joinpoint regression models (segmented regression, sequential linear models or linear trend models respectively).

The key advantage of the linear trend filtering models is their overall simplicity while still offering a quite reasonable amount of flexibility when fitting some data (see, for instance, [1] for an overview). This is also the reason why these models gained so much popularity in last decades: statistical models which can be interpreted in a simple and straightforward way are needed especially in situations when the final conclusions drawn from the model are about to be explained to non-statisticians or, more generally, to people with a lack of mathematical skills. Indeed, the linear trend model can be, in each of its phase, interpreted as an ordinary linear regression model. The model phases are put together at the points where the slope of the regression line changes, but it the overall continuity property is still preserved. The points where the slope breaks are usually referred to as *changepoints* (sometimes also transition points, breakpoints or joinpoints, respectively). The linear trend changes in order to adapt for the existing structural changes in the underlying dependence.

There are various methodological approaches for estimating the piecewise linear model and for performing statistical inference in such model. Standard techniques are based on the L_2-norm minimization with some prior knowledge on the right number of changepoints, or slope breaks, respectively. Theoretical results suggest that it might be advantageous to know the number of changepoints in advance; this knowledge can improve the overall performance and it also provides a more convenient theoretical background for proving certain statistical properties. Moreover, the prior knowledge on the number of changepoints in the model also yields a better rate of convergence.

In practical situations, however, the number of changepoints is rarely known in advance and some model selection techniques need to be used to choose one final model from a set of plausible ones. Usually, a statistician deals with a whole sequence of (nested) models with different number of changepoints and the selection can be performed, for instance, by the means of some permutation tests [2–5], or the likelihood ratio tests [6, 7] by evaluating some goodness-of-fit criterion (usually BIC). An alternative approach can be seen in adopting the Bayesian framework instead (see [8, 9]).

All these techniques, however, require a multistage approach where, in the first stage, one needs to estimate the set of plausible models while the model selection is performed later, in the next stages. A convenient approach would be estimating and selecting the final model at once—in just one single stage. This can be obtained by using some ideas of the sparse signal processing methods involving atomic pursuit techniques and adopting, for instance, the k order differences common for the trend filtering [10, 11], the total variation penalty [12, 13], or the LASSO regularization [14, 15] for the changepoint detection and estimation. These methods based on the sparsity principle share the same idea: the estimated model is heavily over-parametrized and the data themselves are used to choose only a small subset of relevant parameters while shrinking all remaining parameters exactly to zero. This is performed exclusively by utilizing the nature of the L_1-norm (see Fig. 1 for an illustration). The objective function used in the minimization problem is combined

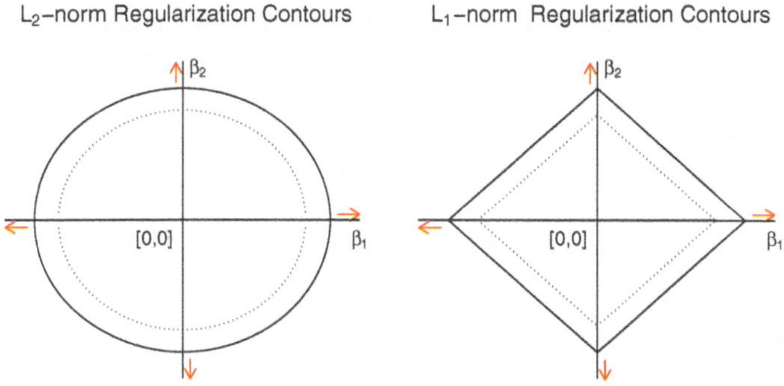

Fig. 1 Standard contours of the L_2-norm penalty and the L_1-norm penalty. The smooth favor of the L_2-norm is responsible for selecting all parameters to be nonzero while the sharp edges of the L_1-norm creates the background for selecting just some nonzero parameters and shrinking remaining ones exactly to zero

with the L_1-norm penalty which penalizes the magnitude of each estimated parameter (see [16, 17]).

On the other hand, the standard LASSO problem is well known for having a tendency to choose slightly more parameters in the final model than necessarily needed [18, 19]. This also implies that the LASSO estimation is not oracle consistent in general scenarios (see, for instance, [20]), but there were some oracle consistent modifications proposed (for instance, an elastic net by [21] and adaptive LASSO by [20]) to avoid this inconsistency issue and to improve the overall selection performance of the LASSO estimated models.

In this paper, we use the adaptive LASSO modification and we improve the original idea presented in [14] for the LASSO based changepoint detection and estimation in piecewise linear models. We introduce the adaptive LASSO trend filtering approach and we show that the final model can be obtained within the same time costs as the standard LASSO based model. The adaptive based trend filtering can be also used to consistently estimate the true slope breaks (i.e., consistent changepoint detection and estimation) in the overall trend and it can be shown to satisfy the oracle properties.

In addition, by utilizing the complete solution paths as elucidated by [22] we can, in an easy and straightforward manner, obtain the whole set of all possible models (beginning with a simple linear trend with no breaks up to a perfect linear interpolation overall unique data points), and by using the recent advances in the post-selection inference (see, for instance, [23, 24]) we can test for the significance of the final model. The model selection step is thus performed via the L_1 regularization and both, the changepoint location detection and the changepoint magnitude estimation are performed in a data-driven manner in just one single step. Finally, the proposed modeling approach can be easily extended to handle some qualitative restrictions imposed on the final estimate (in a similar manner as in [15]), which can be easily accounted for by adding some set of straightforward linear constraints. This turns out to be useful especially in economics and econometric modeling.

The paper is organized as follows: in the next section, we describe the underlying model and we propose the adaptive LASSO trend filtering approach. Some statistical properties are presented in Sect. 3 and finite sample properties are investigated in Sect. 4 where the performance of the proposed adaptive LASSO estimation approach is put in contrast with the standard LASSO methods and the obtained results are discussed.

2 Adaptive LASSO in Trend Filtering

Let $\{(X_i, Y_i);\ i = 1, \ldots, n\}$ be a random sample drawn from some unknown population, with the join distribution function $F_{(X,Y)}$, such that X_i's have some compact domain, say, an interval $(0, 1)$, and they are all unique. Without the loss of generality, we may assume that $X_i < X_{i+1}$, for all $i = 1, \ldots, n-1$. Alternatively, one can also consider the X_i values to be some specific time points in some given period of time, denoting the times when actual observations Y_i's are taking place. In the following, however, we will refer to the general scenario only. Let, moreover, $\{\xi_i\}_{i=1}^{n-1}$ be a sequence of unobserved points from the same domain as the X_i values, such that $X_i < \xi_i \leq X_{i+1}$, for every $i = 1, \ldots, n-1$. We assume that the overall dependence structure between Y and X can be expressed as a piecewise linear model with the linear pieces joining together at the points ξ_i's only, however, not necessarily all of them. Under these assumptions, we can express the underlying piecewise linear model as

$$Y_i = a_i + b_i X_i + \varepsilon_i, \qquad \text{for } i = 1, \ldots, n, \tag{1}$$

where we assume independent random error terms $\varepsilon_i \sim N(0, \sigma^2)$, for an unknown constant $\sigma^2 > 0$, while the overall continuity condition can be formulated as

$$a_i + b_i \xi_i = a_{i+1} + b_{i+1} \xi_i, \qquad \text{for } i = 1, \ldots, n-1. \tag{2}$$

Considering the model in (1)–(2), we would like to estimate the unknown parameters $a_i, b_i \in \mathbb{R}$, for $i = 1, \ldots, n$, and we want to detect the locations, if there are some, where the overall slope breaks. In particular, we need to specify those ξ_i's, for which $b_i \neq b_{i+1}$.

The sparsity principle in Model (1)–(2) is employed throughout the parameters b_i, for $i = 1, \ldots, n$, as we assume that $b_i = b_{i+1}$ holds for all, but some small subset of indexes from $\mathcal{I} = \{1, \ldots, n-1\}$. If there is some location ξ_i, for which $b_i \neq b_{i+1}$, then we introduce a changepoint in the model: it holds that $\xi_i = \frac{a_i - a_{i-1}}{b_{i+1} - b_i}$ and the linear trend changes at this location from b_i to b_{i+1} in order to adjust for the underlying structural change in the data. The intercept parameters, values a_i, for $i = 1, \ldots, n$, are then determined by the requirement on the overall continuity over the whole domain of interest (i.e., interval $(0, 1)$).

Using the standard properties of the L_1-based regularization, and LASSO penalty in particular, we can define the corresponding parameter estimates as the solutions of the minimization problem

$$\begin{array}{ll}\text{Minimize} & \frac{1}{n}\sum_{i=1}^{n}(Y_i - a_i - b_i X_i)^2 + \lambda_n \sum_{i=1}^{n-1}|b_{i+1} - b_i|,\\ a_i, b_i \in \mathbb{R} & \\ \text{w.r.t.} & X_\ell < \frac{a_\ell - a_{\ell+1}}{b_{\ell+1} - b_\ell} \leq X_{\ell+1}, \text{ for } \ell \in \{i;\ b_{i+1} - b_i \neq 0\}.\end{array} \tag{3}$$

Note, that for $b_{i+1} = b_i$ we are not estimating any location ξ_i, as there is no change in the overall trend at this point. Thus, we only minimize (3) with respect to optimal changepoint locations $\xi_i = \frac{a_i - a_{i+1}}{b_{i+1} - b_i}$, for which $b_{i+1} \neq b_i$.

The penalty term in (3) can be interpreted as a penalty of the total variation type (see [12, 25, 26]); $\lambda_n > 0$ represents a classical regularization parameter, here controlling the number of changepoints in the final estimate. For $\lambda_n \to 0$, we expect changepoints to occur in every ξ_i, for $i = 1, \ldots, n-1$, resulting in an interpolating piecewise linear curve with possible slope changes between every two neighboring observations; for $\lambda_n \to \infty$, on the other hand, an overall linear trend over the whole domain of interest is produced and no changepoints are present, thus $b_i = b_{i+1}$, for all $i = 1, \ldots, n-1$.

The minimization problem in (3) is well defined, but, unfortunately, it is not convex. It can be still solved using some optimization toolboxes though, but it can get quite complex and time consuming if the number of observations or changepoints grows. Moreover, the standard optimization methods may end up in a local minimum instead of the global minimum and some caution is always necessary when interpreting the results in such non-convex problem. The reason of this non-convexity is the assumption we made about the changepoint locations: we assume that $\xi_i \in (X_i, X_{i+1}]$, for $i = 1, \ldots, n-1$, and thus, it can be easily seen that the design matrix of the model depends on these location parameters, which are also subjects to the minimization in (3).

An intuitive way around is to assume a slightly simplified version of the model: the overall trend can only change in the actual observational points X_i's. For instance, we may define $\xi_i = X_{i+1}$, for $i = 1, \ldots, n-1$, which now simplifies the initial minimization problem and brings it back to a standard convex optimization scenario. The potential regression segments are now only determined by two neighboring observations, possibly having $n-1$ segments at most joining together at the design points $X_2, \ldots, X_{n-1}$. This scenario was already proposed in [14] and further investigated in [15]. As the authors pointed out, this restriction might be slightly limiting in some finite sample cases, but this limitation becomes negligible and vanishes as the sample size increases. Once we assume that the possible changepoint locations can only occur at the observational points X_i's, then the minimization problem in (3) can be expressed in terms of a standard LASSO problem

$$\underset{\boldsymbol{\beta} \in \mathbb{R}^n}{\text{Minimize}} \ \frac{1}{n} \|\boldsymbol{Y} - \mathbb{X}\boldsymbol{\beta}\|_2^2 + \lambda_n \sum_{j=2}^{n-1} |\beta_j|, \tag{4}$$

where $\boldsymbol{Y} = (Y_1, \ldots, Y_n)^\top$ is the response vector, for the parameter vector we have $\boldsymbol{\beta} = (\beta_0, \ldots, \beta_{n-1})^\top = (a_1, b_1, (b_2 - b_1), \ldots, (b_{n-1} - b_{n-2}))^\top \in \mathbb{R}^n$, and the design matrix takes the form

$$\mathbb{X} = \begin{pmatrix} 1 & X_1 & 0 & 0 & \ldots & 0 \\ 1 & X_2 & 0 & 0 & \ldots & 0 \\ 1 & X_3 & (X_3 - X_2) & 0 & \ldots & 0 \\ \vdots & \vdots & \vdots & \ddots & \ddots & \vdots \\ 1 & X_n & (X_3 - X_2) & (X_4 - X_3) & \ldots & (X_n - X_{n-1}) \end{pmatrix}. \tag{5}$$

The regularization parameter $\lambda_n > 0$ is, as we already mentioned, used to control the number of changepoints appearing in the final model. The design matrix $\mathbb{X}$ can be decomposed into two parts: the first two columns corresponds with a standard design matrix used for an ordinary linear regression with the corresponding intercept and slope parameters $(a_1, b_1) \in \mathbb{R}^2$ and the remaining columns of $\mathbb{X}$ are used to model hypothetical trend breaks in the final model. If there is some changepoint detected at some location, the corresponding column of the design matrix is used to model the break and the involved parameter estimate become nonzero.

The minimization problem in (4) is convex, and thus, it can be effectively solved using some standard optimization tools. Another approach, for example, is to use the LARS algorithm proposed in [22], and to obtain the whole solution paths for all $\lambda_n > 0$. On the other hand, this problem also has some limitations, which are evident especially when it comes to the theoretical properties of the estimates obtained by (4). The standard LASSO penalty in (4) is well known for over-fitting the final model and, in general, it does not provide a consistent selection in terms of the sparsity—more nonzero parameters are present in the model than the true number of nonzero values (see [18]).

In order to improve the selection performance of the model we adopt the idea presented in [20], and we introduce an analogous minimization problem where the estimate for $\boldsymbol{\beta} = (\beta_0, \ldots, \beta_{n-1})^\top$ is now defined as

$$\widehat{\boldsymbol{\beta}} = \underset{\boldsymbol{\beta} \in \mathbb{R}^n}{\text{Argmin}} \ \frac{1}{n} \|\boldsymbol{Y} - \mathbb{X}\boldsymbol{\beta}\|_2^2 + \lambda_n \sum_{j=2}^{n-1} \frac{|\beta_j|}{|\widehat{\beta}_j^{(LS)}|}, \tag{6}$$

where, in addition to (4), we are using a specific scaling factor for each parameter β_j, $j = 2, \ldots, n-1$, in order to make the shrinkage effect (in terms of the relative magnitude) same for all elements. The scaling factor is given by the corresponding ordinary least squares estimate $\widehat{\beta}_j^{(LS)}$ of β_j, for each $j = 2, \ldots, n-1$. Recall, that we again do not penalize for the overall intercept and slope parameters, which are

$\beta_0, \beta_1 \in \mathbb{R}$. The ordinary least squares estimates are well defined and they are unique: the total number of parameters equals to the total number of observations, and thus, the ordinary least squares yield a perfect data interpolating fit.

From the theoretical point of view, it is not strictly required to use the least squares estimates in (6) for rescaling the β_j parameters in the LASSO penalty. The adaptive LASSO was shown (see [20] for details) to preserve its selection consistency and oracle properties with any $\sqrt{n}$-consistent estimates of β_j's when these are used instead of the least squares estimates in (6). Especially in situations, where the standard least squares estimates cannot be calculated (for instance, if the number of observations is less than the number of parameters, or the data matrix is not regular), it might be suitable to use some alternative techniques instead (for instance, the estimates obtained by the elastic net minimization approach).

Computational Aspects

As already pointed out, the minimization formulated in (6) is a convex optimization problem and therefore, a global minimum can be found by using the same efficient algorithms, which are used to solve the standard LASSO problems. In the following, we propose a straightforward algorithm to perform the adaptive LASSO trend filtering as defined in (6).

1. Calculate the ordinary least squares estimate for $\boldsymbol{\beta} = (\beta_0, \ldots, \beta_{n-1})^\top \in \mathbb{R}^n$ as $\widehat{\boldsymbol{\beta}}^{(LS)} = \left(\mathbb{X}^\top\mathbb{X}\right)^{-1}\mathbb{X}\boldsymbol{Y}$, where $\mathbb{X}$ is defined in (5).
2. Define a new design matrix $\widetilde{\mathbb{X}} = (\boldsymbol{X}_0, \boldsymbol{X}_1, w_2\boldsymbol{X}_2, \ldots, w_{n-1}\boldsymbol{X}_{n-1})$, where $w_j = |\widehat{\beta}_j^{(LS)}|$, for $j = 2, \ldots, n-1$, and $\boldsymbol{X}_0, \ldots, \boldsymbol{X}_{n-1}$ are the columns of $\mathbb{X}$.
3 Solve the standard LASSO minimization formulated as

$$\widetilde{\boldsymbol{\beta}} = \underset{\boldsymbol{\beta} \in \mathbb{R}^n}{\text{Argmin}} \ \frac{1}{n}\left\|\boldsymbol{Y} - \widetilde{\mathbb{X}}\boldsymbol{\beta}\right\|_2^2 + \lambda_n \sum_{j=2}^{n-1} |\beta_j|. \tag{7}$$

4 Calculate the final estimates for $\beta = (\beta_0, \ldots, \beta_{n-1})^\top$ as $\widehat{\beta}_j = \widetilde{\beta}_j$, for $j = 0, 1$, and $\widehat{\beta}_j = \widetilde{\beta}_j w_j$, for $j = 2, \ldots, n-1$.

The minimization in (7) does not yet fully correspond with the standard LASSO problem as defined in [17], where the author considered the same vector of parameters in the objective function and also in the penalty term. In our situation, we do not want to penalize for the first two parameters (the overall intercept and slope) and thus, only a sub-vector of $\boldsymbol{\beta} \in \mathbb{R}^n$ plays the role in the penalty term in (7). The problem can be, however, easily transformed to exactly fit the LASSO problem defined in [17]. By splitting the design matrix into two parts as $\widetilde{\mathbb{X}} = (\widetilde{\mathbb{X}}_1, \widetilde{\mathbb{X}}_2)$, where $\widetilde{\mathbb{X}}_1$ consists of the first two columns of $\widetilde{\mathbb{X}}$, and the corresponding vector of parameters is split accordingly, $\boldsymbol{\beta} = (\boldsymbol{\beta}_{(1)}^\top, \boldsymbol{\beta}_{(2)}^\top)^\top$, such that $\boldsymbol{\beta}_{(1)} = (a_1, b_1)^\top = (\beta_0, \beta_1)^\top$, then it only needs some basic algebra calculations to show that $\widetilde{\mathbb{X}}\widetilde{\beta} = \widetilde{\mathbb{X}}_1\widetilde{\beta}_{(1)} + \widetilde{\mathbb{X}}_2\widetilde{\beta}_{(2)}$, for $\widetilde{\beta} = (\widetilde{\beta}_{(1)}^\top, \widetilde{\beta}_{(2)}^\top)$ being the solution of (7), can be equivalently expressed as $\mathbb{H}Y + (\mathbb{I} - \mathbb{H})\widetilde{\mathbb{X}}_2\widetilde{\beta}_{(2)}$, where $\mathbb{H} = \widetilde{\mathbb{X}}_1\big(\widetilde{\mathbb{X}}_1^\top\widetilde{\mathbb{X}}_1\big)^{-1}\widetilde{\mathbb{X}}_1$, and $\widetilde{\beta}_{(2)}$ is now the solution of the minimization problem

$$\widetilde{\beta}_{(2)} = \underset{\beta \in \mathbb{R}^{n-2}}{\text{Argmin}} \; \frac{1}{n}\left\|(\mathbb{I} - \mathbb{H})Y - (\mathbb{I} - \mathbb{H})\mathbb{X}_2\beta_2\right\|_2^2 + \lambda_n\|\beta_2\|_1,$$

which now fully corresponds with the standard LASSO problem as discussed in [17]. Technical details can be found in [14].

3 Theoretical Properties

The theoretical properties of the adaptive LASSO are mostly well known and details can be found in [20]. Some theoretical results related to the changepoint detection and estimation can be also found in [27]. In this session, we summarize some important facts and we provide additional technical details related to the trend filtering models and are, otherwise, not found elsewhere.

Let us recall, that X_j, for $j = 0, \ldots, n-1$, denote the corresponding columns of the designs matrix $\mathbb{X}$, defined in (5). For the vector of parameter estimates $\widehat{\beta}$ given by (6), and the vector of the corresponding signs $\hat{s} \in [-1, 1]^{n-2}$, with $\hat{s}_j = sign(\widehat{\beta}_{j+1})$, for $j = 1, \ldots, n-2$, the KKT optimality conditions can be expressed in a straightforward way as

$$\begin{aligned}
&X_j^\top(\mathbb{X}\widehat{\beta} - Y) = 0, && \text{for } j = 0, 1; \qquad (8)\\
&X_j^\top(\mathbb{X}\widehat{\beta} - Y) + \frac{\lambda_n}{|\widehat{\beta}_j^{(LS)}|}\hat{s}_{j-1} = 0, && \text{for } j = 2, \ldots, \text{n} - 1;\\
&\hat{s}_{j-1} = sign(\widehat{\beta}_j), && \text{if } \widehat{\beta}_j \neq 0, \text{ for j} = 2, \ldots, \text{n} - 1;\\
&\hat{s}_{j-1} \in [-1, 1], && \text{if } \widehat{\beta}_j = 0.
\end{aligned}$$

Using the KKT conditions above and applying the same idea as used in Theorem 1 in [14], we can prove the consistency property of the adaptive LASSO estimate of the underlying linear trend in a classical L_2-norm sense. In addition, inheriting the properties of the adaptive LASSO approach, we can also show that the obtained estimates preserve the oracle properties, which, beside the asymptotic normality property of the nonzero estimates, means that the consistent selection is achieved, meaning that

$$P\big[\forall j \in \{0, \ldots, n-1\};\; sign(\widehat{\beta}_j) = sign(\beta_j^o)\big] \to 1, \quad \text{for } n \to \infty,$$

for some vector of the true parameters $\boldsymbol{\beta}^o = (\beta_0^o, \ldots, \beta_{n-1}^o)^\top \in \mathbb{R}^n$, and the optimal choice of the regularization parameter $\lambda_n > 0$, which needs to fulfill $\lambda_n \to \infty$ and $\lambda_n/\sqrt{n} \to 0$, for $n \to \infty$.

Reflecting this selection consistency property and the mean consistency in the L_2-norm sense back onto the initial model scenario defined in (1) and (2), we can conclude that the changepoints in the model (i.e., breaks in the overall linear trend) are estimated consistently as the sample size tends to infinity. The theoretical properties guarantee both—the consistent estimation of the true number of changepoints with their locations and the consistent recovery of the overall underlying trend.

The asymptotic normality of the vector of estimated parameters can be expressed as

$$\sqrt{n}[\widehat{\boldsymbol{\beta}}_{\mathcal{A}} - \boldsymbol{\beta}^o_{\mathcal{A}}] \xrightarrow[n\to\infty]{D} N\left(\mathbf{0}, \sigma^2 \mathbb{C}_{\mathcal{A}}^{-1}\right), \tag{9}$$

where $\mathcal{A} = \{j;\ \beta_j^o \neq 0\}$ stands for the set of indexes for all nonzero parameters in the true parameter vector $\boldsymbol{\beta}^o = (\beta_0^o, \ldots, \beta_{n-1}^o)^\top \in \mathbb{R}^n$, $\widehat{\boldsymbol{\beta}}_{\mathcal{A}}$ and $\boldsymbol{\beta}^o_{\mathcal{A}}$ are the corresponding sub-vectors of $\widehat{\boldsymbol{\beta}}$ and $\boldsymbol{\beta}^o$ with elements which correspond to indexes in $\mathcal{A}$, and the matrix $\mathbb{C}_{\mathcal{A}}$ is composed of the rows and columns of $\mathbb{C} = \lim_{n\to\infty} \frac{1}{n}\mathbb{X}^\top\mathbb{X}$, where again, the indexes of the rows and columns are in the set $\mathcal{A}$.

Due to the selection consistency property, it can be easily seen that for all other parameter estimates $\widehat{\beta}_j$, where $j \in \{0, \ldots, n-1\} \setminus \mathcal{A}$, we easily obtain that

$$\widehat{\beta}_j \xrightarrow{P} 0 \quad \text{for } n \to \infty.$$

The asymptotic normality in (9) can be used to perform statistical tests about the vector of parameter estimates or it can be utilized for constructing the corresponding confidence intervals.

Alternatively, one can also adopt the residual bootstrap approach (see [28] for further details) to mimic the asymptotic normal distribution in (9) and to avoid standard plug-in techniques which are known for their poor performance in some situations with a rather slow convergence and insufficient approximation. From the practical point of view, the proposed model can be easily extended to account for additional shape restrictions which might be convenient in econometric modeling and finance models. Such restrictions (for instance, monotonicity) are easily enforced by using a set of well defined linear constraints. These linear constraints can be easily implemented into the estimation algorithm discussed in Sect. 2 with no additional computational costs.

4 Simulations

For the simulation purposes, we consider a simple situation for a linear trend model with two existing breaks: the first changepoint occurs at $x_1 = 0.2$ and the second at $x_2 = 0.7$. The underlying function takes the form

$$f(x) = 2x\mathbb{I}_{\{x\in(0,0.2)\}} + (0.6 - x)\mathbb{I}_{\{x\in[0.2,0.7)\}} + (x/2 - 0.45)\mathbb{I}_{\{x\in[0.7,1)\}}, \quad (10)$$

where the unit interval (0, 1) is considered to be the domain for x. The data were simulated according to the model in (1), with independent error terms being normally distributed with the zero mean parameter and variance $\sigma^2 > 0$.

In order to investigate various signal-to-noise scenarios, we considered three different values for the variance parameter $\sigma^2 \in \{\frac{1}{4}, \frac{1}{2}, 1\}$ and four sample sizes for $n \in \{50, 100, 500, 1000\}$. For each combination of the variance and sample size, we obtained 100 Monte Carlo simulations and four different models were fitted: the first model with the standard LASSO approach and the prior knowledge on two existing changepoints with the corresponding regularization parameter denoted as $\lambda_{(2)}$; the second model based on the standard LASSO approach with the asymptotically optimal value of $\lambda_n = n^{-1}(\log n)^{5/2}$, denoted as λ_{AS}; the third model, again based on the standard LASSO approach, however, with the regularization parameter defined by the minimum cross-criterion, denoted as λ_{CV}; finally, the last model is based on the proposed adaptive LASSO approach with the regularization parameter $\lambda_n = o(\sqrt{n})$ (see [20]), denoted as λ_{AD}. All four models are compared with respect to three different quantities: the mean estimation bias $\frac{1}{n}\sum_{i=1}^{n}(f(X_i) - \boldsymbol{x}_i^\top\widehat{\boldsymbol{\beta}})$ and the mean squared error (MSE) $\frac{1}{n}\sum_{i=1}^{n}(f(X_i) - \boldsymbol{x}_i^\top\widehat{\boldsymbol{\beta}})^2$ to assess the performance with respect to the conditional mean estimation and the changepoint detection rate $\frac{1}{2}\sum_{k=1}^{2}|x_k - \hat{x}_k|$ to judge the performance with respect to the consistency of the changepoint recovery. Let us just recall that $\widehat{\boldsymbol{\beta}}$ is the sparse vector estimate defined by (6) and $\boldsymbol{x}_i$ is the ith row of the model matrix $\mathbb{X}$ in (5).

The simulation results for the estimation bias and the mean squared error quantity are given in Table 1 and the results for the number of estimated changepoints (the set $\widehat{\mathcal{A}}_n$) and the changepoint detection rate are given in Table 2.

Considering Table 1, it is clear that the proposed adaptive LASSO approach for the linear trend filtering performs at the same quality as the other three LASSO based techniques. The mean squared error quantity converges towards zero in all four cases and, in addition, the observed convergence rate is roughly at the same level. The mean estimated bias is, as expected close to zero and even identical for all four models—this is however, not the case. The tabulated values differ at a smaller magnitude which is not reflected in the table.

On the other hand, if we focus on the results in Table 2, the adaptive LASSO evidently outperforms the other three standard LASSO approaches. Moreover, the standard LASSO approaches, beside the model with λ_{AS}, are worse even from the practical point of view. Indeed, the model with the regularization parameter $\lambda_{(2)}$ uses a prior knowledge about two existing changepoints and the model with λ_{CV} requires a multistage approach where, in the first few stages, we need to fit various models for different values of $\lambda > 0$, and later, once the minimum value of the cross-validation criterion is obtained, the final model can be fitted. The standard LASSO model with λ_{AS} is, same as the adaptive LASSO approach, a single stage method—the asymptotically optimal value of the regularization parameter can be easily obtained given the sample size. On the other hand, all standard LASSO approaches perform poorly with

Table 1 The simulation results for four different LASSO models: the standard LASSO approach (SLASSO) given by (4) with three different choices of the regularization parameters (value $\lambda_{(2)}$ taken under the prior knowledge of two changepoints in the model; the asymptotically optimal value $\lambda_{AS} = n^{-1}(\log n)^{5/2}$ and the regularization parameter λ_{CV} given by the minimum cross-validation criterion) and the proposed adaptive LASSO method (ALASSO) defined in terms of the minimization problem in (6) with the asymptotically optimal value $\lambda_{AD} = o(\sqrt{n})$. The empirical performance with respect to the conditional mean estimation is compared with respect to the estimation bias quantity and the means squared error quantity where both are averaged over 100 independent Monte Carlo simulations. The corresponding standard error terms are given in brackets

σ^2	n	SLASSO with $\lambda_{(2)}$		SLASSO with λ_{AS}		SLASSO with λ_{CV}		ALASSO with λ_{AD}	
		Est.Bias	MSE	Est.Bias	MSE	Est.Bias	MSE	Est.Bias	MSE
$\sigma^2 = 0.25$	50	−0.007 *(0.069)*	0.020 *(0.012)*	−0.007 *(0.069)*	0.021 *(0.009)*	−0.007 *(0.069)*	0.016 *(0.009)*	−0.007 *(0.069)*	0.021 *(0.009)*
	100	0.000 *(0.048)*	0.013 *(0.006)*	0.000 *(0.048)*	0.013 *(0.005)*	0.000 *(0.048)*	0.010 *(0.005)*	0.000 *(0.048)*	0.014 *(0.005)*
	500	0.001 *(0.019)*	0.007 *(0.003)*	0.001 *(0.019)*	0.003 *(0.002)*	0.001 *(0.019)*	0.003 *(0.002)*	0.001 *(0.019)*	0.003 *(0.002)*
	1000	0.002 *(0.016)*	0.006 *(0.003)*	0.002 *(0.016)*	0.002 *(0.001)*	0.002 *(0.016)*	0.002 *(0.001)*	0.002 *(0.016)*	0.002 *(0.001)*
$\sigma^2 = 0.50$	50	−0.010 *(0.097)*	0.039 *(0.025)*	−0.010 *(0.097)*	0.034 *(0.020)*	−0.010 *(0.097)*	0.028 *(0.017)*	−0.010 *(0.097)*	0.034 *(0.020)*
	100	0.000 *(0.068)*	0.021 *(0.010)*	0.000 *(0.068)*	0.021 *(0.010)*	0.000 *(0.068)*	0.016 *(0.009)*	0.000 *(0.068)*	0.021 *(0.010)*
	500	0.001 *(0.026)*	0.009 *(0.003)*	0.001 *(0.026)*	0.007 *(0.004)*	0.001 *(0.026)*	0.005 *(0.003)*	0.001 *(0.026)*	0.005 *(0.003)*
	1000	0.003 *(0.023)*	0.007 *(0.003)*	0.003 *(0.023)*	0.005 *(0.002)*	0.003 *(0.023)*	0.003 *(0.001)*	0.003 *(0.023)*	0.003 *(0.001)*
$\sigma^2 = 1.00$	50	−0.014 *(0.138)*	0.074 *(0.050)*	−0.014 *(0.138)*	0.063 *(0.045)*	−0.014 *(0.138)*	0.048 *(0.035)*	−0.014 *(0.138)*	0.064 *(0.046)*
	100	0.000 *(0.097)*	0.039 *(0.022)*	0.000 *(0.097)*	0.041 *(0.023)*	0.000 *(0.097)*	0.026 *(0.017)*	0.000 *(0.097)*	0.039 *(0.021)*
	500	0.002 *(0.037)*	0.011 *(0.005)*	0.002 *(0.037)*	0.016 *(0.008)*	0.002 *(0.037)*	0.009 *(0.004)*	0.002 *(0.037)*	0.011 *(0.006)*
	1000	0.004 *(0.032)*	0.009 *(0.003)*	0.004 *(0.032)*	0.012 *(0.005)*	0.004 *(0.032)*	0.005 *(0.002)*	0.004 *(0.032)*	0.007 *(0.003)*

Table 2 The simulation results for four LASSO: standard LASSO with three different regularization parameters (value $\lambda_{(2)}$ taken under the prior knowledge of two changepoints in the model; the asymptotically optimal value $\lambda_{AS} = n^{-1}(\log n)^{5/2}$ and the regularization parameter λ_{CV} given by the minimum cross-validation criterion) and the adaptive LASSO with the asymptotically optimal value $\lambda_{AD} = o(\sqrt{n})$. In the first three columns, the estimated number of changepoints over 100 Monte Carlo simulations is given in terms of the minimum, median, and maximum ($[m|m|m]$). The standard LASSO model with $\lambda_{(2)}$ always contains exactly two changepoints, therefore, it is omitted. In the remaining columns, the changepoint detection error detection rate calculated as $\frac{1}{2}\sum_{k=1}^{2}|x_k - \hat{x}_k|$ is given together with the corresponding standard errors in brackets

σ^2	n	Changepoints $\lvert\hat{\mathcal{A}}_n\rvert$			Changepoint error detection rate			
		λ_{AS}	λ_{CV}	λ_{AD}	Model $\lambda_{(2)}$	Model λ_{AS}	Model λ_{CV}	Model λ_{AD}
$\sigma^2 = 0.25$	50	[0|1|2]	[0|2|6]	[0|0|3]	0.078 *(0.045)*	0.128 *(0.069)*	0.073 *(0.041)*	0.120 *(0.047)*
	100	[0|2|5]	[0|3|7]	[0|1|4]	0.076 *(0.041)*	0.063 *(0.037)*	0.065 *(0.041)*	0.101 *(0.060)*
	500	[4|7|13]	[2|6|12]	[2|3|7]	0.044 *(0.032)*	0.038 *(0.023)*	0.035 *(0.024)*	0.036 *(0.023)*
	1000	[5|12|24]	[3|7|19]	[2|4|8]	0.036 *(0.019)*	0.024 *(0.015)*	0.027 *(0.021)*	0.027 *(0.019)*
$\sigma^2 = 0.50$	50	[0|1|4]	[0|2|6]	[0|1|3]	0.092 *(0.045)*	0.100 *(0.060)*	0.079 *(0.035)*	0.114 *(0.059)*
	100	[0|2|5]	[0|3|5]	[0|1|5]	0.089 *(0.046)*	0.076 *(0.042)*	0.072 *(0.040)*	0.078 *(0.040)*
	500	[4|8|15]	[1|5|12]	[2|3|7]	0.060 *(0.044)*	0.041 *(0.023)*	0.047 *(0.035)*	0.046 *(0.032)*
	1000	[7|14|28]	[3|8|16]	[2|4|8]	0.043 *(0.025)*	0.025 *(0.016)*	0.036 *(0.026)*	0.035 *(0.027)*
$\sigma^2 = 1.00$	50	[0|2|4]	[0|1|5]	[0|1|4]	0.097 *(0.043)*	0.105 *(0.049)*	0.095 *(0.045)*	0.103 *(0.040)*
	100	[0|3|5]	[0|2|5]	[0|2|5]	0.100 *(0.048)*	0.089 *(0.049)*	0.077 *(0.042)*	0.087 *(0.045)*
	500	[5|9|22]	[0|5|13]	[2|3|7]	0.077 *(0.053)*	0.034 *(0.019)*	0.061 *(0.048)*	0.052 *(0.030)*
	1000	[11|19|38]	[2|8|22]	[2|3|9]	0.058 *(0.034)*	0.021 *(0.012)*	0.047 *(0.033)*	0.036 *(0.021)*

respect to the number of estimated changepoints. The LASSO methods are common for their tendency to overfit the true model which can be also observed from Table 2—the number of estimated changepoints increases as the sample size increases. However, the adaptive LASSO seems to be able to overcome this inconvenient property and the number of estimated changepoints much closely correspond with the true number of changepoints (there are two true changepoints in the model in (10)).

The performance of the error detection rate seems to be at the same scale for all four methods but it would be natural to expect smaller error detection rates if there are more changepoints being estimated in the model (the error detection rate only considers two location estimators with the smallest distance from the true locations

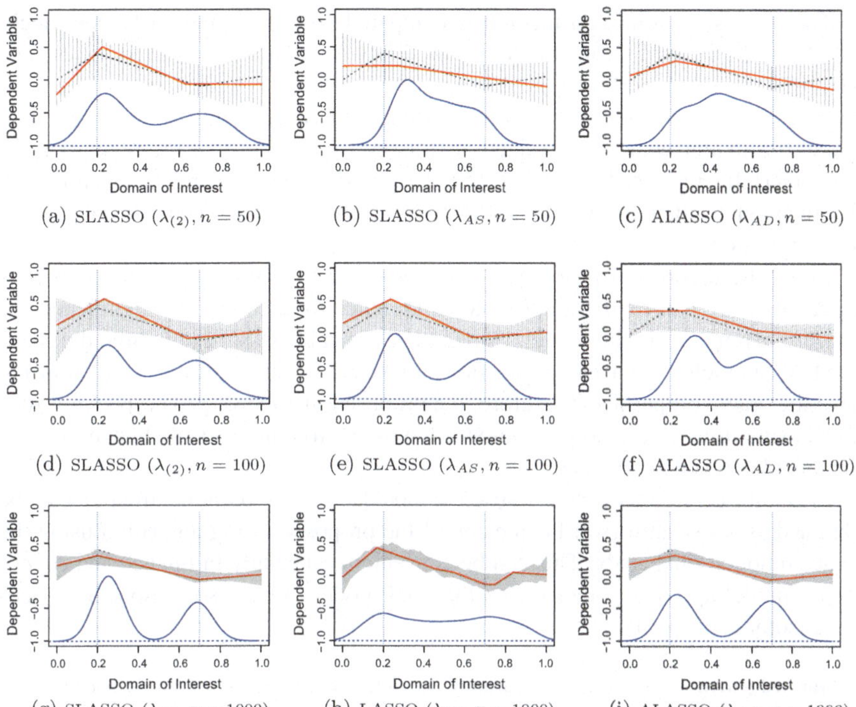

(a) SLASSO ($\lambda_{(2)}, n = 50$) (b) SLASSO ($\lambda_{AS}, n = 50$) (c) ALASSO ($\lambda_{AD}, n = 50$)

(d) SLASSO ($\lambda_{(2)}, n = 100$) (e) SLASSO ($\lambda_{AS}, n = 100$) (f) ALASSO ($\lambda_{AD}, n = 100$)

(g) SLASSO ($\lambda_{(2)}, n = 1000$) (h) LASSO ($\lambda_{AS}, n = 1000$) (i) ALASSO ($\lambda_{AD}, n = 1000$)

Fig. 2 Finite sample performance for three different models: the standard LASSO model (SLASSO) fitted with a prior knowledge that there are only two breaks in the slope is always provided in the first figure column (the corresponding regularization parameter is denoted as $\lambda_{(2)}$). In the middle column, there is again the standard LASSO model, however, with the asymptotically optimal value of the regularization parameter and finally, in the last column, there is the adaptive LASSO model (ALASSO) with the asymptotically optimal regularization parameter $\lambda_{AS} = o(\sqrt{n})$. The pointwise range for the estimated models over 100 Monte Carlo repetitions is given by the gray regions and the blue solid lines represent a scaled density of the estimated changepoint locations

$x_1 = 0.2$ and $x_2 = 0.7$). This is however not what we observe in Table 2—the error detection rates for the model with λ_{AD} are, in general, smaller than those for the model with λ_{CV} and, in some cases, even smaller than those for the model with λ_{AS}. Therefore, we can also conclude that even the changepoint detection performance is better for the adaptive LASSO approach (see also an illustration in Fig. 2).

5 Conclusion

In this paper, we proposed a new approach to fit linear trend models. The idea is based on the sparse fitting modeling approach where we adopted the adaptive LASSO approach in order to achieve the model selection consistency in terms of the true number of changepoints being estimated in the final model.

The proposed estimation method is compared with the standard LASSO techniques on the base of various models and different qualitative characteristics. The adaptive LASSO outperforms the standard LASSO methods with respect to the number of estimated changepoints: while the standard LASSO tends to overfit the final model (no matter what is the selection strategy for the value of the regularization parameter), the proposed adaptive LASSO estimation can perform consistently with the true model and, moreover, the estimation is fully automatic and performed in just one single step.

Given the theoretical properties of the adaptive LASSO—its oracle properties especially—we can easily adopt various inference tools to test the significance of nonzero parameters in the model. Unlike the classical LASSO estimation, the adaptive LASSO yields a consistent selection and the constructed estimates are known to have the oracle properties. Therefore, the proposed methodology is especially suitable in situations where no prior knowledge on the true number of changepoints in the model is given in advance.

In addition, various shape-restricted models (e.g., monotone trends) can be obtained as a straightforward extension of the proposed fitting algorithm using the same computational costs. This can be convenient especially in finance and econometric modeling where one expects the final model to have some specific monotonic/isotonic properties.

Acknowledgements This work was partially supported by the research grant provided by the Czech Science Foundation, grant number P402/12/G097 and the Mobility Grant provided by the Ministry of Education, Youth and Sports in the Czech Republic, grant number 7AMB17FR030.

References

1. Kim, H.J., Fay, M.P., Yu, B.: Comparability of segmented line regression models. Biometrics **60**(4), 1005–1014 (2004)
2. Bosetti, C., Bertuccio, P., Levi, F., Lucchini, F., Negri, E., La Vecchia, C.: Cancer mortality in the European Union, 1970–2003, with a joinpoint analysis. Ann. Oncol. **16**, 631–640 (2008)
3. Kim, H.J., Yu, B., Feuer, E.J.: Selecting the number of change-points in segmented line regression. Stat. Sinica **19**, 597–609 (2009)
4. Kim, H.J., Fay, M.P., Feuer, E.J., Midthune, D.N.: Permutation tests for joinpoint regression with application to cancer rates. Stat. Med. **19**, 335–351 (2000)
5. Qui, D., Katanoda, K., Tomomi, M., Tomotaka, S.: A joinpoint regression analysis of long-term trends in cancer mortality in Japan (1958–2004). Int. J. Cancer **24**, 443–448 (2009)
6. Feder, P.: The log likelihood ratio in segmented regression. Ann. Stat. **3**, 84–97 (1975)
7. Hinkley, D.V.: Inference in two-phase regression. J. Am. Stat. Assoc. **66**(336), 736–743 (1971)
8. Carlin, B.P., Gelfand, A.E., Smith, A.F.M.: Hierarchical Bayesian analysis of changepoint problems. Appl. Stat. **41**, 389–405 (1992)
9. Martinez-Beneito, M., García-Donato, G., Salmerón, D.: A Bayesian joinpoint regression model with an unknown number of break-points. Ann. Appl. Stat. **5**(3), 2150–2168 (2011)
10. Kim, S.J., Koh, K., Boyd, S., Gorinevsky, D.: l_1 trend filtering. SIAM Rev. **51**, 339–360 (2009)
11. Tibshirani, R.J.: Adaptive piecewise polynomial estimation via trend filtering. Ann. Stat. **42**(1), 285–323 (2014)

12. Mammen, E., Van De Geer, S.: Locally adaptive regression splines. Ann. Stat. **25**(1), 387–413 (1997)
13. Rudin, L.I., Osher, S., Faterni, E.: Nonlinear total variation based noise removal algorithm. Phys. D: Appl. Phys. **60**, 259–268 (1992)
14. Maciak, M., Mizera, I.: Regularization techniques in joinpoint regression. Stat. Papers **57**(4), 939–955 (2016)
15. Maciak, M.: Testing shape constrains in lasso regularized joinpoint regression. In: Antoch, J., Jurečková, J., Maciak, M., Pešta, M. (eds.) Analytical Methods in Statistics, AMISTAT, Prague, pp. 105–122, November 2015 (2016)
16. Chen, S.S., Donoho, D.L., Saunders, M.A.: Atomic decomposition by basis pursuit. SIAM Rev. **43**(1), 129–159 (2001)
17. Tibshirani, R.: Regression shrinkage and selection via the LASSO. J. R. Stat. Soc. Ser. B **58**(1), 267–288 (1996)
18. Meinshausen, N., Bhlmann, P.: Variable selection and high-dimensional graphs with the lasso. Technical report, ETH Zrich (2004)
19. Meinshausen, N., Yu, B.: Lasso-type recovery of sparse representations for high-dimensional data. Ann. Stat. **37**(1), 246–270 (2009)
20. Zou, H.: The adaptive lasso and its oracle properties. J. Am. Stat. Assoc. **476**(101), 1418–1429 (2006)
21. Zou, H., Hastie, T.: Regularization and variable selection via the elastic net. J. R. Stat. Soc. B **67**(2), 301–320 (2005)
22. Efron, B., Hastie, T., Tibshirani, R.: Least angle regression. Ann. Stat. **32**, 407–499 (2004)
23. Lockhart, R., Taylor, J., Tibshirani, R.J., Tibshirani, R.: A significance test for LASSO. Ann. Stat. **2**(42), 413–468 (2014)
24. Tibshirani, R., Taylor, J., Lockhart, R., Tibshirani, R.: Exact post-selection inference for sequential regression procedures. J. Am. Stat. Assoc. **514**(111), 600–620 (2016)
25. Koenker, R., Ng, P., Portnoy, S.: Quantile smoothing splines. Biometrika **81**, 673–680 (1994)
26. Koenker, R., Mizera, I.: Penalized triograms: total variation regularization for bivariate smoothing. J. R. Stat. Soc. Ser. B **66**, 1681–1736 (2004)
27. Shen, J., Gallagher, C.M., Lu, Q.: Detection of multiple undocumented change-points using adaptive lasso. J. Appl. Stat. **46**(6), 1161–1173 (2014)
28. Chatterjee, A., Lahiri, S.N.: Bootstrapping lasso estimators. J. Am. Stat. Assoc. **106**(494), 608–625 (2011)

An Efficient Anomaly Detection in Quasi-Periodic Time Series Data—A Case Study with ECG

Goutam Chakraborty, Takuya Kamiyama, Hideyuki Takahashi and Tetsuo Kinoshita

Abstract Anomaly detection from a time series is an important problem with applications to find or predict the development of a fault in a system. Depending on the source of the data, it could be nonperiodic, quasi-periodic, and periodic. Modeling an aperiodic data to detect anomaly is difficult. A pure periodic data seldom happens in nature. Finding anomaly in quasi-periodic time series signals, for example, bio-signals like ECG, heart rate (pulse) data, are important. But, the analysis is computationally complex because of the need for proper window size selection and comparison of every pair of subsequences of window-size duration. In this paper, we proposed an efficient algorithm for anomaly detection of quasi-periodic time series data. We introduced a new concept "*mother signal*", which is the average of normal subsequences. Creation of the *mother signal* is the first step in the process. Finding deviations of subsequences of varied duration (due to quasi-periodicity) from *mother signal*, is the second step. When this distance crosses a threshold, it is declared as a discord. The algorithm is light enough to work in real-time on computationally weak platforms like a mobile phone. Experiments were done with ECG signals to evaluate the performance. It is shown to be computationally more efficient compared to existing works, and could identify discords with higher rate.

Keywords Quasi-periodic time series · Anomaly detection · Fundamental period
Clustering

G. Chakraborty (✉) · T. Kamiyama
Graduate School of Software & Information Science, Iwate Prefectural University, Takizawa, Japan
e-mail: goutam@iwate-pu.ac.jp

T. Kamiyama
e-mail: takuya@gmail.com

H. Takahashi · T. Kinoshita
Research Institute of Electrical Communication, Tohoku University, Sendai, Japan
e-mail: hideyuki@riec.tohoku.ac.jp

T. Kinoshita
e-mail: kino@riec.tohoku.ac.jp

I. Rojas et al. (eds.), *Time Series Analysis and Forecasting*,
Contributions to Statistics, https://doi.org/10.1007/978-3-319-96944-2_10

1 Introduction

The health of any running system is monitored by a set of sensors, the collected data from which is analyzed to ensure safety and/or predict malfunctioning. Depending on the system, the data could be aperiodic or periodic. There are studies on anomaly detection, both on periodic and nonperiodic signals. Various techniques are used, such as HMM based [1], prediction-based [2], similarity-based [3], window-based [4] and segmentation-based [5]. For aperiodic signals, piecewise regression [6, 7] and model-based clustering methods [8] are proposed. These models need a lot of training data to optimize the model parameters. They require long training using computationally heavy algorithms like expectation maximization.

For many systems, the data collected are quasi-periodic, where the period varies slightly over an average. Important bio-signals to monitor health, like ECG, pulse rate are quasi-periodic. Recently, systematic collection, storing and analysis of bio-signals is widely adapted for personalized healthcare, medical informatics, drug testing and a plethora of applications. Anomalies in bio-signal can detect/predict heart disease, pulse failure or other kinds of life-threatening situations. Present healthcare systems installed on mobile devices collect data continuously, to get analyzed at the end of the day on a different platform. A real-time analysis to create alarms for people vulnerable to heart-related problems, could save lives, avoid driving accidents, etc. This is not realized yet.

The main motivation of this work is to detect an anomaly in quasi-periodic time series signals in real-time, on computationally weak platforms like smartphones. We did experiment on ECG data because there are many works done, and labeled data available on the web. The approach is from the signal point of view. No physiological or medical knowledge is used.

Definition of Time Series Discords

The anomaly subsequences in a quasi-periodic signal are called discords. In quasi-periodic signals, as the period randomly varies over the average, to find anomaly, one needs to compare every possible pair of subsequences. Usually, from the domain knowledge, one fixes the subsequence window length. Time series discords are subsequences, which are maximally different to all the rest of the subsequences of the whole sequence. Discords could be detected by comparing every pair of subsequences, and identifying and ordering them, with the one having largest distance from its nearest (least distant) neighbor heading the list. Usually, we are interested to find the set of discords for which the distance from the normal subsequence is above some predefined threshold. We can find discords using brute force method which is computationally heavy with time complexity of $O(n^2)$, where n is the total number of subsequences possible out of the whole time series.

Let us consider a discrete time series consisting of T time-slots. Let us also consider that the subsequence length is m time-slots, where $m \ll T$. Thus, the original signal consists of $n = (T - m + 1)$ such subsequences. The ith subsequence starts at ith slot, where $0 \leq i \leq (T - m)$. If all possible pairs are to be compared, we need $n^2 \approx T^2$ comparisons because $m \ll T$ and, therefore, $n \approx T$. In previous works, the length of the subsequence was user defined [4]. In our work, the length m is computed from the signal, and is equal to the average of the fundamental period of the quasi-periodic time series [9].

Related Works

As the time series length gets longer, comparing all pairs to find discords is computationally complex. Related works are proposals of how to find discords efficiently. As ECG data is one of the most popular quasi-periodic signal, many of the works are with ECG signals. One of the early work was by Keogh [4]. They discretize the data value into three levels, assigning alphabets "A", "B", "C", and proposed a heuristic algorithm to find maximal distance subsequences. In Keogh's work, one need to fix parameters, and the result depends on them. Subsequently, parameter-free discord search algorithm for quasi-periodic signals were reported in [10, 11]. Recently, anomaly detection in ECG artifacts using Motif was proposed in [12]. In [13], algorithms for assessing normality of multivariate signals are implemented in R.

Of quasi-periodic signals, ECG is one most investigated and recently finds applications in many mobile devices. It is now common to attach wearable sensors to collect bio-signal and transfer to mobile device by Blue-tooth communication. Simple analysis, like instantaneous pulse rate is available. Real analysis of the recorded data is done off-line after uploading the data on a PC. For example, the device would collect ECG data over a day, and then it is analyzed offline for detecting discords.

Existing algorithms to find anomaly in quasi-periodic signals like ECG are computationally heavy. Yet, it is important to identify anomaly in bio-signals in real-time, especially for those who have medical history. A real-time identification and warning could avoid fatality. The aim of this work is to propose an algorithm which can run on a weak computational platform, with low memory requirement—for example, a smartphone.

Depending on the algorithm used, anomaly location and frequency would vary. The ground truth could be understood and identified only by the domain expert, in case of bio-signals by a health professional. Anomaly detected on the basis of algorithmic analysis may not tell the ground truth. In fact, some signal anomaly could be of no health concern, whereas the algorithm may miss some subtle problems. Yet, a warning is important even when there could be an occasional false-alarm.

The rest of the paper is as follows. In Sect. 2, the core idea behind the algorithm, creation of *mother signal* is explained. How it is actually done for ECG data is also explained. In Sect. 3, the algorithm to find discords is explained. Experimental data set and their results are elaborated. The paper is concluded in Sect. 4.

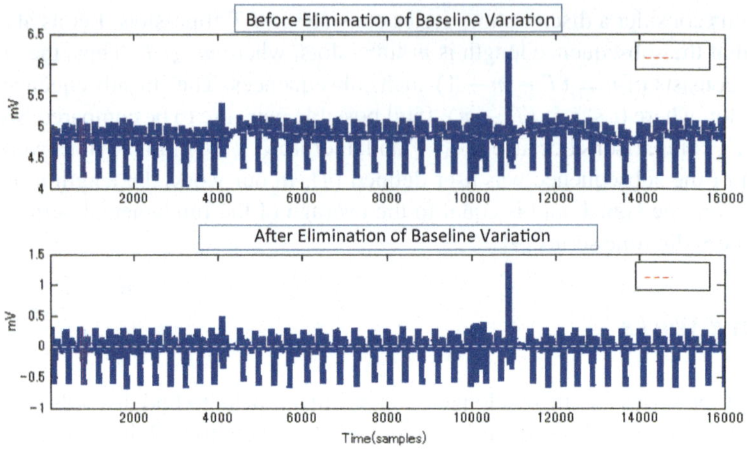

Fig. 1 The slowly moving baseline is shifted to constant value of zero

2 Proposed Idea—Mother Signal

What Is Mother Signal*?*

We proposed a new concept [14], we named "*mother signal*". As a first step, we set m equal to the fundamental period, average of most frequently occurring periodic signals. The fundamental assumption here is that normal (not discords) subsequences overwhelmingly outnumber subsequences with discords, which is true for any running system. *Mother signal* is the average of those normal subsequences. Once mother signal is known, discords are detected more efficiently. Even if we use exhaustive comparison with mother signal, the complexity is $O(mn)$ much less than $O(n^2)$. In other words, comparisons with mother signal will be much more efficient compared to brute force comparison of every possible pairs. The largest discord is the one whose distance is highest from the mother signal. Multiple discords can be detected as subsequences whose distances with mother signal exceed a predefined threshold. Otherwise, we can identify and list the first predefined number of discords in order of their distances from the mother signal.

From a given quasi-periodic time series, first, its fundamental period is determined. For that the slowly moving baseline is shifted to a constant level of zero, as shown in Fig. 1. Next, the peaks of the signal are identified, to find the span of different subsequences. This procedure is explained in Fig. 2. The time series signal, shown in Figs. 1 and 2, is an ECG signal. One peak to the next is one subsequence. The time duration of these subsequences vary over an average, except at places where there is a discord. We took a large number of samples to verify their distribution, and found that to be normal. The average of these periods is set as the value of m, the time duration of the so-called *mother signal*. For any two normal subsequences, though their durations may differ a little from m, if we normalize the durations to m and

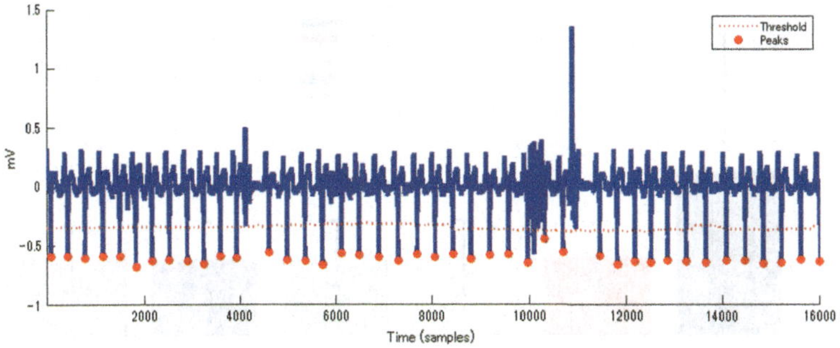

Fig. 2 Identifying the peaks to determine the periods

shift-and-rotate for maximum match (minimum Euclidean distance), their distance will be small. Only in case of discords the distance from *mother signal* would be large even when all possible shifts and rotations are tried.

Creation of the Mother Signal

In this section, we describe how the mother signal is created. As shown in Fig. 2, local minimum points are marked, and subsequences from one to its next are identified. The basic idea is to cluster those subsequences. The cluster with the maximum cardinality is considered to be consisting of normal subsequences. Their mean will form the *mother signal.*

For clustering, we need to measure the distance between any pair of subsequences. As they are of different length in time, ideally the distance measured by dynamic time warping (DTW), should be considered. We too did clustering using distances measured by DTW. But, DTW is computationally heavy. We proposed a faster solution. All subsequences are scaled to the same length, the average length of all subsequences. For length normalization, algorithms used for image compression and enlargement are tried. Euclidean distance between two subsequences were calculated after scaling them to same time duration using (a) Lanczos, (b) Bilinear, (c) Nearest Neighbor, (d) Piecewise Aggregate Approximation, and (e) Bicubic algorithms. The total computation time for clustering, using DTW to measure distance and Euclidean distance after scaling using different scaling algorithms, are shown in Fig. 3. Euclidean distance after scaling takes almost the same time, using different scaling algorithms. DTW requires nearly 100 times more computation time.

Clustering of Subsequences

The number of clusters is not known a-priori. As the motivation is to find the cluster with highest cardinality, the result is not affected much on the setting of the number of clusters. We tried different clustering algorithms as listed here: agglom-

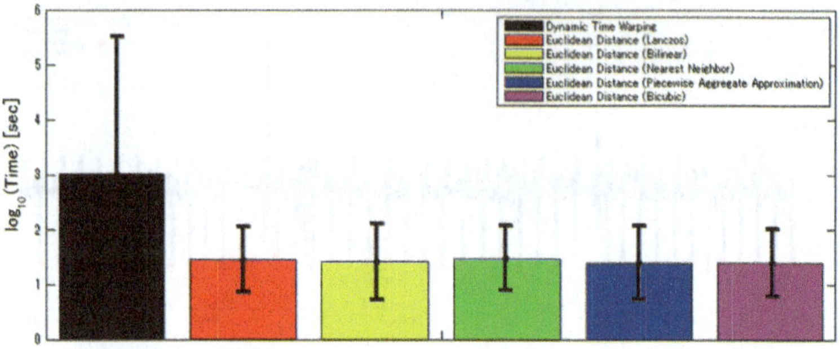

Fig. 3 Time required for clustering when different algorithms were used for measurement of distance between a pair of subsequences

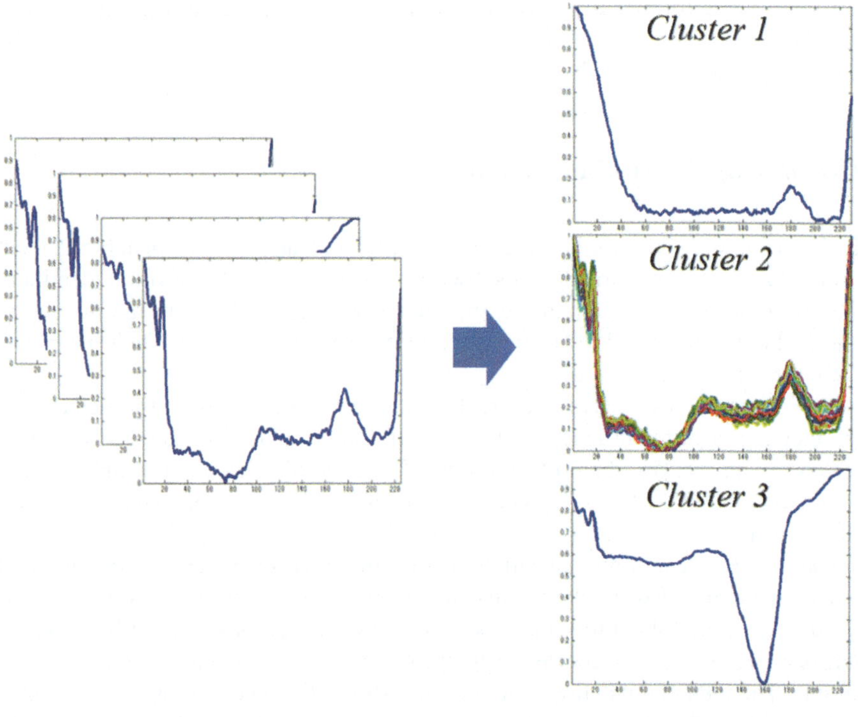

Fig. 4 Three clusters from 40 subsequences. Cardinality of cluster 2, which consists of normal subsequences, is the highest

erative clustering; k-means with K = 2, 3, 4, 5; X-means, SOM, DBSCAN. Both DBSCAN, X-means, where optimum cluster number is automatically set, resulted in four clusters. Three clusters are shown in Fig. 4, the result when only 40 subsequences were used for clustering. We included some discord portion. As shown in the figure, majority of signals fall in cluster 2, which we conclude is the ensemble of normal subsequences. The average of all members of this highest cardinality cluster is the *mother signal*. In case of ECG data, *mother signal* may change with time. In fact, with the same person, it will change with the person's activity. While we experimented with ECG data, we updated *mother signal* using latest ν subsequences, where $\nu = 40$. Once *mother signal* is ready, it is used for next incoming signal.

3 Proposed Algorithm, Experimental Data, and Results

The computational complexity for anomaly detection using brute force method is $O(n^2)$. Even all possible comparisons with mother signal will reduce the complexity to $O(mn)$. As $n \gg m$, comparison using mother signal is much more efficient. In this section, we propose a heuristic algorithm which will improve the efficiency even further.

The basic idea of proposed heuristic for reducing computation is as follows. Suppose, for ith subsequence, to find the minimum distance (i.e., maximum match) with mother signal, we need to shift-and-rotate the ith signal by p slots. We already found that using step-by-step shift-and-rotate comparison. In that case, for comparing $(i+1)$th subsequence with the mother signal, we need to shift-and-rotate $(i+1)$th subsequence by $(p+1)$ slots because $(i+1)$th subsequence is only one-slot shifted from the ith subsequence. In that way, for $(i+1)$th subsequence, the number of shift-rotate-compare will be reduced from m to 1. We can reduce the computation for distance comparison by m times. Through experiment, we found that 97% of times, this one-slot shift gave best match. But, it does not always happen, especially for discord subsequence. First, we shift $(i+1)$th subsequence by p-slots, $(p+1)$-slots and $(p+2)$-slots, and calculate the distance with mother signal. Ideally, $(p+1)$-slots shift will give the best match. If that does not happen, we need more elaborate searching for proper shift that would achieve maximum match with the mother signal. An efficient heuristic algorithm for this elaborate searching is proposed and explained below.

In Fig. 5, we show the distance when a normal signal is compared with the mother signal. The x-axis represents the shift and the y-axis is the distance. The change of distance with shift is smooth, with a prominent minimum at shift around 100 slots. The distance is symmetric on two sides of the minimum. Because of this shape, it is easy to find the minimum without comparing distances for all possible m shifts. We calculate distances for a few equally spaced shifts, and can easily converge to the minimum point (maximum match) in a few trials. The heuristic search algorithm to find minimum distance is explained in [14].

4 Experimental Data and Results

We used data from MIT-BIH Database [15], the detail of which is shown in Table 1. The algorithm complexity to find anomaly is basically the number of times we need to calculate distance between two subsequences. Distance function calculates Euclidean

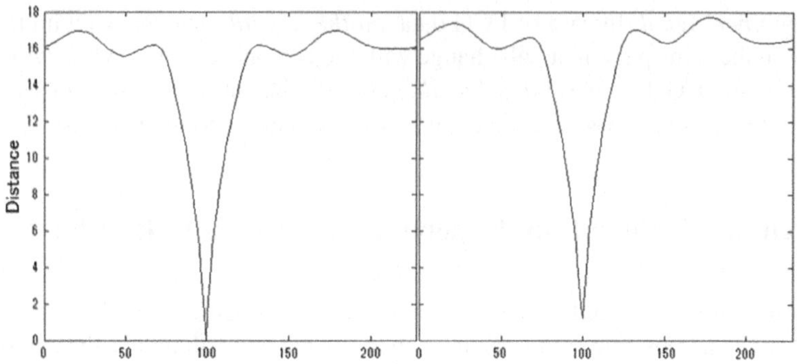

Fig. 5 Distance of two subsequences from mother, shown at different shifts. The graph is symmetrical on both sides of minimum. The one on the left is a closer match to mother signal, as distance minimum is almost zero

Table 1 Data used for experiments and Computation time

Data set	*T*	*m*	Proposed Algo	[16] (s = 20)	[16] (s = 100)	[4]
1	3751	228	**1.14**	2.48	6.9	8.8
2	3750	163	**0.85**	2.01	5.73	8.6
3	3750	251	**1.09**	2.06	5.77	9.03
4	3750	143	**0.95**	3.06	8.94	8.47
5	5400	351	**2.43**	4.77	12.67	15.46
6	5401	292	**1.56**	3.16	8.76	14.74
7	5400	399	**2.23**	2.84	8.55	16.85
8	16500	370	**8.28**	21.33	60.92	67.01

distance between two subsequences. Computation time for the proposed algorithm is at fourth column, and for other competitive algorithms, are shown in the last three columns of Table 1. For the proposed algorithm, time to generate mother signal is also included. We can see that the proposed algorithm is almost eight times faster compared to [16] (with parameter s = 100) and [4]. Though, [16] (s = 20) is faster compared to [16] (s = 100), in Table 2 we can see that its performance (F-score) is not good.

In Table 2, we show the average of F-score for all 8 data sets using our proposed algorithm and related works, reported in [4, 16]. This table verifies that, in spite of our algorithm being faster, its identification of discords (both precision and recall) are either the best, or very near to the best. In the last column, we show how F-score of our algorithm differs from the best result. Out of 8, for 4 data sets, it gave the best result. For the rest, it was near the best.

It is usual to evaluate any pattern recognition algorithm by precision and recall. In our case, the ground truth is not known. The decision of anomaly is solely based on the distance from the normal trend—both for the proposed method as well as

Table 2 Comparison of F-score for all eight Data sets

Data set	Proposed Algo	[16] (s = 20)	[16] (s = 100)	[4]	Difference with best
1	0.92	0.89	0.91	**0.94**	0.02
2	0.54	0.35	**0.58**	0.44	0.04
3	**0.88**	0.51	0.59	0.51	0.00
4	**0.95**	0.72	0.85	0.71	0.00
5	0.91	0.87	0.93	**0.95**	0.04
6	**0.76**	0.52	0.61	0.34	0.00
7	**0.72**	0.34	0.65	0.52	0.00
8	0.83	0.73	**0.84**	0.81	0.01

Table 3 Detection of PCV and SPVC discords

Data set	T	m	PVC count	SPVC count	PVC detection (%)	SPVC detection (%)	Error rate (%)
1	1,000,500	210	18	4	78	0	14
2	1,000,500	220	9	4	89	0	10
3	1,000,500	270	12	2	75	50	9
4	1,000,500	250	23	7	87	14	5
5	1,000,500	270	10	13	80	25	23
6	1,800,000	253	86	0	100	–	0
7	1,800,000	279	9	16	90	8	9
8	1,800,000	268	8	5	80	20	1

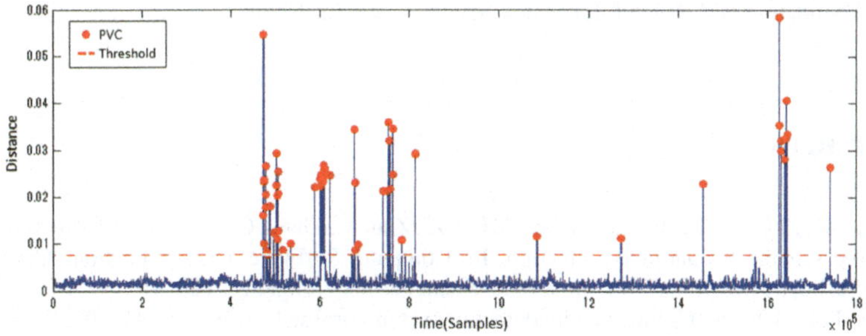

Fig. 6 Distance from mother signal and Discords for certain threshold value

brute force method. All the results in Table 2 are from signal shape perspective. The medical interpretation of discords could differ.

To evaluate the significance of our result in medical perspective, we did experiment with a different (much longer) data set. The data parameters and the result are shown in Table 3. These are labeled data, with two types of discords, PVC (Premature Ventricular Contractions and SVPC (Supraventricular Premature Contraction). The distance of subsequences, from mother signal is shown in Fig. 6. PVC discords are more prominent, and we could detect with very high accuracy. But, most of the SVPC discords were missed. It is also evident from Fig. 6 that though prominent discords form sharp spikes, setting proper threshold value to identify all discords, is difficult.

5 Conclusion

We could improve the efficiency of discord detection in periodic signals. We compared our algorithm to other kind of algorithms [16] proposed recently. Comparison with [12] is underway. The memory requirement is not analyzed. The parameters used in our algorithm, determine both the efficiency as well as the quality of the result. To find their optimum values for maximum efficiency without sacrificing quality is a challenge. We will investigate how the result changes with parameter values, for certain type of bio-signals. Finally, the most important aspect is to compare the results with ground truth. We hope to obtain more labeled data with discords, identified by health experts. Our algorithm works for quasi-periodic signals. The period changes over time, like pulse rate varies with the level of exhaustion or emotional state. We took care by using a window of $40 \times m$ (the last m) to find new m and the new *mother signal*. Selecting this window size is critical and application dependent. Discords with different waveforms have different meanings. For practical application, we need to analyze the discords to provide the user with their level of emergency and suggestions for action to be taken.

Acknowledgements Part of this work was carried out under the cooperative research project program of the research institute of electrical communication, Tohoku University, and grant from Sendai Foundation of Applied Information Sciences, Sendai, Japan.

References

1. Sha, W., Zhu, Y., Huang, T., Qiu, M., Ming, Z., Zhu, Y., Zhang Q.: A multi-order markov chain based scheme for anomaly detection. In: Computer Software and Applications Conference Workshops (COMPSACW), pp. 83–88. Shanghai Jiao TongPages (2013)
2. Ma, J., Perkins, S.: Online novelty detection on temporal sequences. In: KDD'03, Proceedings of the Ninth ACM SIGKDD International Conference on Knowledge Discovery And Data Mining, pp. 613–618 (2003)
3. Rebbapragada, U., Protopapas, P., Brodley, C.E., Alcock, C.: Finding anomalous periodic time series. **74**(3) 281–313 (2009)
4. Keogh, E., Lin, J., Fu, A.W., Van Herle, H.: Finding the unusual medical time series, algorithms and applications. EEE Trans. Inf. Technol. Biomed. **10**(3), 429–439 (2006)
5. Salvador, S., Chan, P.: Learning states and rules for detecting anomalies in time series. Appl. Intell. **23**(3), 241–255 (2005)
6. Brailovsky, V.L., Kempner, Y.: Application of Piece-wise regression to detecting internal structure of signal. Pattern Recognit. **25**(11), 1361–1370 (1992)
7. Ferrari-Trecate, G., Muselli, M.: A new learning method for piecewise linear regrassion. In: ICANN, Spain 28–30 Aug 2002

8. Chamroukhi, F.: Piecewise regression mixture for simultaneous functional data clustering and optimal segmentation. J. Classif. **33**(3), 374–411 (2016)
9. Basha, R., Ameen, J.R.M.: Unusual sub-sequence identifications in time series with periodicity. Int. J. Innov. Comput. Inf. Control (IJICIC) **3** (2007)
10. Luo W., Gallangher, M.: Faster and parameter free discord search in quasi-periodic time series. In: Huang, J.Z., Cao, L., Srivastave, J. (eds.) PAKDD 2011, Part II, LNAI 6635, pp. 135–145. Springer (2011)
11. Luo, W., Gallangher, M., Janet, W.: Parameter free search of time series discord. J. Comput. Sci. Technol. **28**(2), 300–310 (2013)
12. Sivaraks, H., Ratanamahantana, C.A.: Robust and accurate anomaly detection in ECG artifacts using time series motif discovery. Comput. Math. Methods Med. **2015**, 20. Article ID 453214. https://doi.org/10.1155/2015/453214
13. Korkmaz, S., Goksuluk, D., Zararsiz, G.: MVN: an R package for assessing multivariate normality. R J. **6**(2), 151–162 (2014)
14. Kamiyama, T., Chakraborty, G.: Real-time anomaly detection of continuously monitored periodic bio-signals like ECG. In: Otak, M., et. al. (ed.) LNAI , Springer 10091, pp. 418–427 (2017)
15. Goldberger, A.L., Amaral, L.A.N., Glass, L., Hausdorff, J.M., Ivanov, P.C.h., Mark, R.G., Mietus, J.E., Moody G.B., Peng, C.-K., Stanley, H.E.: PhysioBank, PhysioToolkit, and PhysioNet: components of a new research resource for complex physiologic signals. Circulation **101**(23), 215–220
16. Sugiyama, M., Borgwardt, K.M.: Rapid Distance-Based Outlier Detection via Sampling, NIPS (2013)

8. Chamroukhi, F.: Piecewise regression mixture for simultaneous functional data clustering and optimal segmentation. J. Class. 33(3), 374–411 (2016)
9. [illegible] (1967)
10. [illegible]
11. [illegible]
12. [illegible]
13. [illegible]
14. [illegible]
15. [illegible]

Similarity Analysis of Time Interval Data Sets—A Graph Theory Approach

Marc Haßler, Christian Kohlschein and Tobias Meisen

Abstract Comparison of entities, i.e., the measurement of their similarity, is a frequent, but challenging task in computer science. It requires a precise and quantifiable definition of similarity itself. Are two texts equal, if they overlap in a majority of their composing words? Does a pair of pictures resemble the same content? What defines the sameness of two songs? While certain distance-based approaches, e.g., Minkowski, make for a good starting point in defining similarity, there is no one-size-fits-all approach. In this work, we tackle a particularly interesting problem, namely, the definition of a similarity measure for comparing time interval data sets. Our approach regards the data sets as disjoint parts of a bigraph, thereby allowing for an application of methods from graph theory. We present both a formal definition of the similarity of two time intervals and our methods as well as concrete use-case from the medical domain, thus demonstrating the applicability for real-world scenarios.

Keywords Graph theory · Time interval data set · Similarity analysis · Medical data analysis · Distance measures

1 Introduction and Motivation

Time interval data occur in a huge variety of fields for example as part of process optimization in the production technology [1], within scheduling tasks in logistic or staff planning [2] or as labeled sleep data in the medical field [3]. All mentioned areas observe deviance in their recorded data due to different reasons, such as machine

M. Haßler (✉) · C. Kohlschein · T. Meisen (✉)
Institute of Information Management in Mechanical Engineering, RWTH Aachen University, Aachen, Germany
e-mail: marc.hassler@ima.rwth-aachen.de; marc.hassler@ima-zlw-ifu.rwth-aachen.de

C. Kohlschein
e-mail: christian.kohlschein@ima.rwth-aachen.de

T. Meisen
e-mail: tobias.meisen@ima.rwth-aachen.de; tobias.meisen@ima-zlw-ifu.rwth-aachen.de
URL: https://cybernetics-lab.de/

I. Rojas et al. (eds.), *Time Series Analysis and Forecasting*,
Contributions to Statistics, https://doi.org/10.1007/978-3-319-96944-2_11

failure, sick leave or other abnormalities. Regardless of the reason for the deviation, short reaction times must be guaranteed.

In many scenarios, such as manufacturing, medicine or production technology, analysts recognize deviations mostly based on their experience and start appropriate countermeasures. This procedure resembles a similarity analysis regarding past processes or situations. Recognitions like these can not often be part of the computer-aided similarity analysis because at the moment there are only few limited possibilities to quantify interval similarities. Examples for specific similarity analysis already exist within certain scientific research areas. Those methods include the area of text or image processing, where similarity analysis is used to optimize search algorithms [4], within biology to compare genes or gene groups [5, 6] or as a tool of audio recognition methods [7]. Research regarding time interval data sets gained importance in recent years [8–13]. However, basic considerations of similarities regarding time intervals are still missing in our opinion. While the similarity in the mentioned publications was derived from a sequence analysis and studies the existing data sets as a whole, in our approach, we deduce the similarity of the data set from the individual similarities between the underlying intervals.

Therefore, we concentrate solely on time intervals and at first construct a similarity measure to compare intervals with each other. This method allows for a detailed view on specific characteristics of the records saved in the time interval data set and a comparison even under big time offsets is possible. Further, our approach is able to measure the similarity of two data sets with well-known methods from graph theory [14]. To achieve our goal, we interpret the comparative data sets as the disjoint parts of an bigraph where each time interval is represented by a node within these parts and the weight of each edge represents the similarity measure of the corresponding intervals.

This chapter is an extended version of [15] and is structured as follows. In the chapter, we will give a brief overview of other approaches in the field of time interval data similarity. In Sect. 3 we construct the interval distance between two time intervals, which is then used in Sect. 4 to build the distance between two time interval data sets. At the end we discuss the results and give an outlook (Sect. 5), followed by a conclusion in Sect. 6. Throughout the whole paper we illustrate our results within the use case of similarity analysis between classified sleep data, where the main goal is the comparison of different nights of sleep. The used golden standard of sleep classification was provided by Rechtschaffen and Kales in 1968 [16] and is further developed by the American Academy of Sleep Medicine (AASM) [3] as the current standard in practice.

2 Related Work

The work of Kostakis et al. [8] regarding time interval data sets compares two data sets in relation to the correlation of the intervals within the respective data set. Within this method, a difference regarding the interval length is not considered as long as

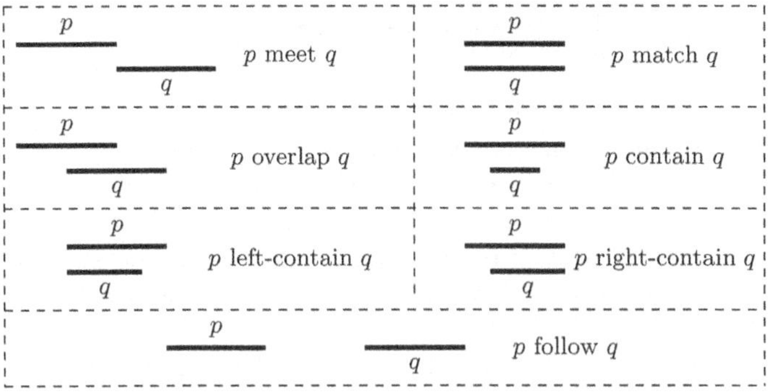

Fig. 1 Interval relation within a data set defined by Kostakis et al. [8]. Reprinted by permission of Springer-Verlag Berlin Heidelberg 2011

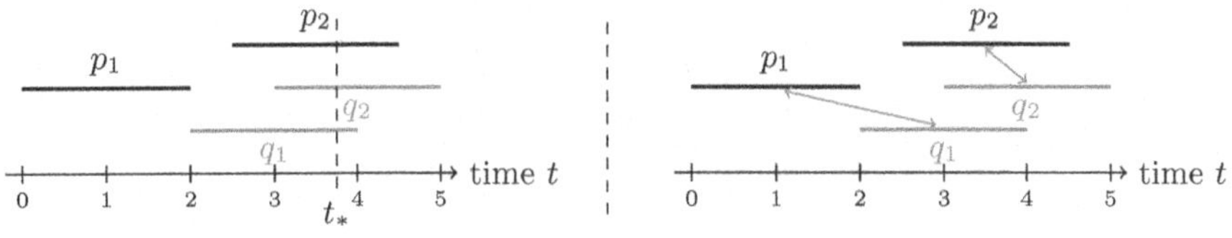

Fig. 2 *Left*: time-based view on data sets by Meisen [9]; *right*: our interval approach

it does not effect the correlation between the two intervals. The authors introduce seven interval correlations, which they used for their comparison (cf. Fig. 1).

Work from Meisen et al. [9] differentiate the similarity analysis into three distances, which are later combined to form the similarity measure. These distances are determined at a specific time t and are the following:

(1) "temporal order distance" compares the number of active intervals at time t.
(2) "temporal measure distance" matches the "value" of all intervals at time t.
(3) "temporal relation distance" analyzes the relation of all intervals at time t.

This approach takes into account the lengths of the individual intervals, but only considers the data set for each evaluation at a certain point in time. Therefore, even small time shifts in one of the datasets are completely changing the outcome of the analysis.

The previously mentioned methods can be described as static comparisons, as depicted in Fig. 2, yet global changes (like temporal displacements) are not regarded. Here, our method has a decisive advantage as we are able to allow global changes to be incorporated into the model by matching individual intervals.

3 Similarities Between Time Intervals

In order to make sure that two time intervals are comparable, we take a closer look at the construction of these intervals. They consist of a start point and an end point and any amount of metadata, such as device class or hourly cost to run, e.g., a specific process like introduced by Meisen et al. [9]. In this paper, we assume that the metadata is available in mathematical form and is thus comparable (cf. section "Metadata and Dealing with Deadlines"). We consider the following form for an interval p:

$$p := \left(s_p, e_p, M_{p_i} \mid i \in \mathbb{N}\right)$$
$$\text{or in } \textit{short form } p := \left(s_p, e_p\right)$$

where

$$s_p := \text{start point of the interval}$$
$$e_p := \text{end point of the interval}$$
$$M_{p_i} := \text{i-th metadata of the interval}$$

In our example of sleep data sets, each interval represent an occurring sleep phase. s_p and e_p describe the start and end point of the corresponding phase and M_p contains the information in which sleep phase the patient was during the interval period. That means we identify labeled sleep data as a time interval data set and visualize it like shown in Fig. 3, where the vertical lines represent the transition from one interval to the next.

The interval similarity analysis is divided into three parts. At first, the geometrical data of each interval, such as length or position on the time axis, is compared to generate geometrical distances between two intervals. In the second part, the metadata as well as the possibility to address deadlines or earliest starting time is added into

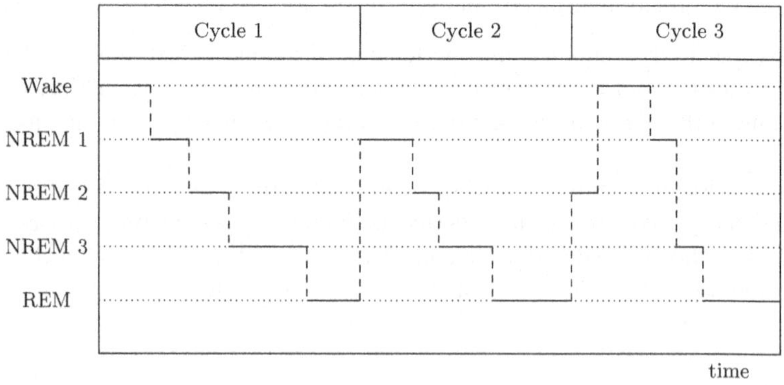

Fig. 3 Hypnogram that shows the characteristic sleep stages, referring to AASM standards [3]

the interval similarity. In the end, all of these information define a similarity measure for two time intervals.

Geometrical Analysis

In the first step, we use the information for each interval to generate several distances with the possibility to evaluate each characteristic differently. For two intervals $p = (s_p, e_p)$ and $q = (s_q, e_q)$ as well as a norm $|| \cdot ||$, we conclude the following geometrical attributes.

(1) Start point distance:

$$D_S(p, q) := \frac{||s_p - s_q||}{|| \max\{e_p, e_q\} - \min\{s_p, s_q\} ||} \tag{1}$$

(2) End point distance:

$$D_E(p, q) := \frac{||e_p - e_q||}{|| \max\{e_p, e_q\} - \min\{s_p, s_q\} ||} \tag{2}$$

(3) Lengths distance:

$$D_L(p, q) := 1 - \frac{\min\{||e_p - s_p||, ||e_q - s_q||\}}{\max\{||e_p - s_p||, ||e_q - s_q||\}} \tag{3}$$

(4) Overlap:

$$D_O(p, q) := 1 - \frac{||p \cap q||}{\min\{||e_p - s_p||, ||e_q - s_q||\}} \tag{4}$$

with the interval

$$p \cap q = \begin{cases} (\max\{s_p, s_q\}, \min\{e_p, e_q\}) & \text{for } \max\{s_p, s_q\} < \min\{e_p, e_q\} \\ 0 & \text{else} \end{cases}$$

(5) Gap:

$$D_G(p, q) := \begin{cases} \frac{\min\{||s_q - e_p||, ||s_p - e_q||\}}{|| \max\{e_p, e_q\} - \min\{s_p, s_q\} ||} & \text{for } ||p \cap q|| = 0 \\ 0 & \text{else} \end{cases} \tag{5}$$

Example 1 To visualize (Fig. 4) the geometrical attributes, we take a closer look at the intervals $p := (0, 10)$ and $q := (3, 7)$ and calculate their attributes:

(1) $||p|| = 10$, $||q|| = 4$ and $|| \max\{e_p, e_q\} - \min\{s_p, s_q\} || = 10$
(2) $p \cap q = (3, 7)$ and therefore $||p \cap q|| = 4$
(3) $||s_p - s_q|| = 3$, $||e_p - e_q|| = 3$ and $D_G(p, q) = 0$.

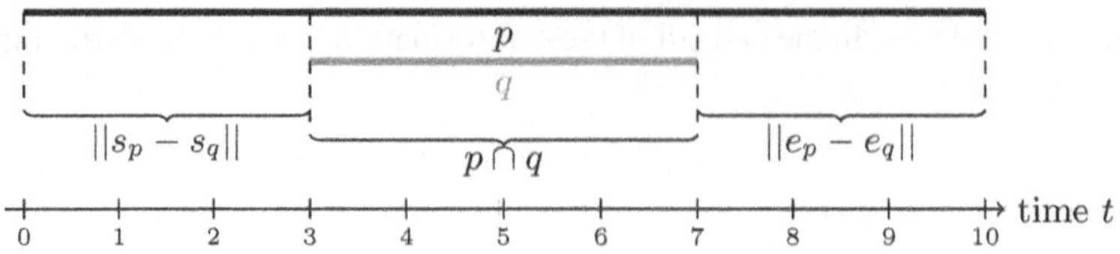

Fig. 4 Visualization of two intervals with their geometrical attributes

Metadata and Dealing with Deadlines

As previously stated, it is assumed that the metadata related to the considered time intervals p and q are in mathematically comparable form. That means that for every metadata i there is a continuous distance D_{M_i} with $0 < D_{M_i}(p, q) < 1$ available. The metadata can, for example, be used to identify, if two intervals are comparable or not. In our use case we use the appearing sleep phases to determine, if the corresponding time intervals should be compared.

A comparison criterion regarding interval deadlines is also added, which means that, if interval p is compared with q, we want to make sure that interval q does not end after p has ended. The same holds true for a start condition. The difference to the distances already mentioned is the lack of symmetry because within these distances the order of the intervals is crucial. For the mentioned criteria, we defined the following two distances:

$$D_{END}(p, q) := \begin{cases} \min\{1, ||e_q - e_p||\} & \text{for } e_q > e_p \\ 0 & \text{else} \end{cases} \tag{6}$$

$$D_{START}(p, q) := \begin{cases} \min\{1, ||s_p - s_q||\} & \text{for } s_p > s_q \\ 0 & \text{else} \end{cases} \tag{7}$$

Similarity of Two Time Intervals

With all the introduced distances, a distance measure for two time intervals is defined, where every characteristic is individually weighted. Therefore, it is adaptable to specific use cases because in case of machine failure the length of the failure could be more relevant than the time of appearance, while within staff planing the point of time where a shortage is located may be a higher priority. This measure is later used in Sect. 4 to calculate the similarity between two data sets. It can be defined as follows:

Definition 1 (*distance between time intervals*) For two intervals p and q as well as weight factors $\lambda_* \in \mathbb{R}_0^+$, the distance between the intervals is measured by calculating the weighted sum of distances:

$$S(p, q) := \sum_{i \in I} \lambda_i \cdot D_i(p, q) \tag{8}$$

Thus, the more similar the two intervals p and q are to each other, the smaller the value of $S(p, q)$ is. In the next step, two time interval data sets are compared and the similarity using this approach is evaluated.

4 Similarity Analysis Regarding Time Interval Data Sets

In this chapter, two interval data sets P and Q are compared. At first, the same cardinality for both P and Q is assumed, that means the number of intervals in each data set is the same. In section "How to Deal with Different Cardinality", a procedure for dealing with different cardinalities is introduced. Furthermore, the intervals in P are specified with p_i and q_i for Q. For the remainder of the paper, data sets are considered as disjoint partial sets of a complete, weighted bipartite graph, in which the edge weight between two nodes corresponds to the interval similarity measure S (cf. Definition 1). In Fig. 5 such visual transition from two hypnograms to the graph form is displayed.

Hence, the similarity of time interval data sets (STIDes) is equivalent to a perfect matching with minimal weight within our constructed bipartite graph.

Definition 2 (*STIDes approach*) Let P and Q be two time interval data sets, $p_i \in P$, $q_i \in Q$ and $|P| = |Q| = n$. Furthermore, Π is the set of permutations of a set with n elements and $\pi \in \Pi$. The similarity between P and Q is determined by the following distance measure:

$$S(P, Q) := \min_{\pi} \left\{ \sum_{i=1}^{n} S(p_i, q_{\pi(i)}) \right\}_{\pi \in \Pi} \tag{9}$$

Such minimization problems in bipartite graphs can be solved in polynomial time by using, for example, the Hungarian algorithm [14]. Our approach is, therefore, capable of calculating a similarity measure within polynomial time while being able

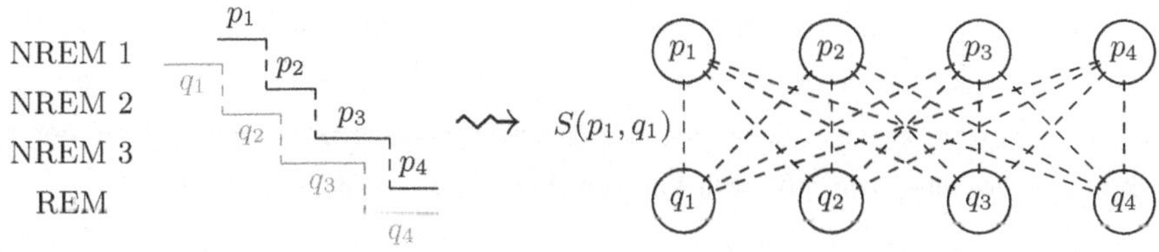

Fig. 5 Left: two datasets as hypnograms; right: representation as bipartite graph

to prioritize certain characteristics and measure similarities even with existing time shift. In the next part, we expand this static approach for a dynamic similarity search, which also includes rescaling and shifting possibilities.

Dynamic Changes Within One Data Set

Until now, the previous static approach has difficulties in determining realistic similarities as soon as one of the time interval sets has big temporal shifts. In Fig. 6, we recognize that the length of each sleep stage except the first one, as well as their sequence is identical for both data sets D_1 and D_2. That means only the time to fall asleep is much longer in data Set D_1.

If we can globally shift the second data set to the right, we get much more realistic distance values with our approach. A comparison with true-to-scale model data sets is not provided in the basic configuration either, however, the construction of our interval distances allows an extension of both desired properties. Therefore, we define two kinds of operators, where the Eq. 10 deals with temporal displacement and Eq. 11 is able to rescale the intervals of one data set.

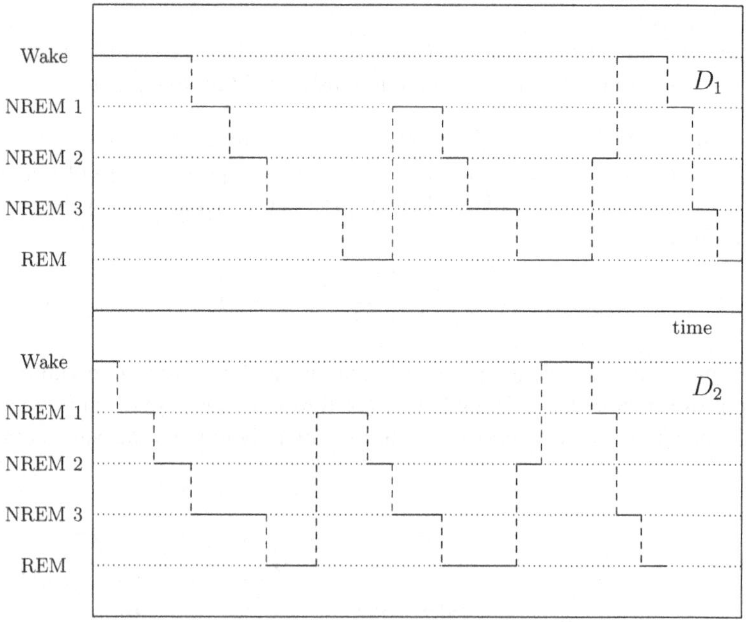

Fig. 6 Two data sets where only the first interval differ in their length

Definition 3 Let $p = (s_p, e_p)$ be an time interval in short form. Furthermore, let $v \in \mathbb{R}$ be a shift parameter and $s \in \mathbb{R}_+$ a scaling factor. The functions

$$p + v := (s_p + v, e_p + v) \tag{10}$$
$$s \cdot p := (s \cdot s_p, s \cdot e_p) \tag{11}$$

map an interval onto a new interval, hence we can integrate these functions into our similarity measure.

For the similarity analysis of our data sets, this means that we have to solve the following minimization problems:

Definition 4 Let P and Q be two time interval data sets, $p_i \in P$, $q_i \in Q$ and $|P| = |Q| = n$. Furthermore, let Π be the set of permutations of a set with n elements, $\pi \in \Pi$, $v \in \mathbb{R}$ a shift parameter and $s \in \mathbb{R}_+$ a scaling factor. The degree of similarity taking into account global displacement (12) or global scaling (13) can then be calculated with

$$S(P, Q + v) := \min_{\pi, v} \left\{ \sum_{i=1}^{n} S(p_i, q_{\pi(i)} + v) \right\}_{\pi \in \Pi, v \in \mathbb{R}} \tag{12}$$
$$S(P, s \cdot Q) := \min_{\pi, s} \left\{ \sum_{i=1}^{n} S(p_i, s \cdot q_{\pi(i)}) \right\}_{\pi \in \Pi, s \in \mathbb{R}_+} \tag{13}$$

Although an efficient solver of the above minimization problem is still part of our research, the solubility of the problem can be shown.

Lemma 1 (Existence of the Minimum) *Let the conditions of Definition 4 be satisfied. The following functions are then continuous with a global minimum.*

$$F_1(v) := \min_{\pi} \left\{ \sum_{i=1}^{n} S(p_i, q_{\pi(i)} + v) \right\}_{\pi \in \Pi} \tag{14}$$
$$F_2(s) := \min_{\pi} \left\{ \sum_{i=1}^{n} S(p_i, s \cdot q_{\pi(i)}) \right\}_{\pi \in \Pi} \tag{15}$$

For a detailed proof of the above lemma see [15], but the idea behind the proof is the concept of the extreme value theorem by Weierstrass in 1860 and therefore it was shown, that both functions are continuous and lower bounded.

How to Deal with Different Cardinality

Until now, we assumed that both data sets have equal cardinality, but realistically this will not often be the case in real life situations. If we look at our use-case of sleep data and we want to compare two different nights of sleep, the chances are minimal, that the same amount of sleep phases occur in different nights of sleep. Therefore, a option to deal with these cardinality differences is needed. Our approach, therefore, is filling the smaller data set with additional nodes, dubbed "dummy nodes", and set the edge weight of all edges connecting these dummy nodes to the maximum occurring edge weight. In Fig. 7 an example with one dummy node is shown. We then adjust the calculation within the STIDes approach and define the following.

Definition 5 Let P and Q be two interval data sets with $|P| > |Q|$. We then generate a set of dummy nodes D with $|D| = |P| - |Q|$, set the interval similarity $S(p_k, d_l) = \max_{i,j}\left\{S(p_i, q_j)\right\}$ for every $p_k \in P$ and $d_l \in D$. The adjusted similarity is then calculated as

$$S(P, \{Q \cup D\}) := \min_{\pi}\left\{\sum_{i=1}^{n} S(p_i, q_{\pi(i)})\right\}_{\pi\in\Pi} - |D| \cdot \max_{i,j}\left\{S(p_i, q_j)\right\} \quad (16)$$

The extent to which unmatched intervals of $|P|$ influence the similarity measure must be considered according to the individual use case and the similarity measure then adapted accordingly. Another possibility to use data sets with different cardinality and, therefore, work with rectangular matrices within the Hungarian algorithm, is the algorithm presented by Bourgeois and Lassalle [17].

Combining the shifting approach with the Definition 5, we are not only able to realistically compare two different nights of sleep, but also to compare one sleep cycle (cf. Fig. 3) with a whole night to find similar sleep cycles. In other words, the approach is able to find a similar subset within a data set when given only a part of a data set to compare.

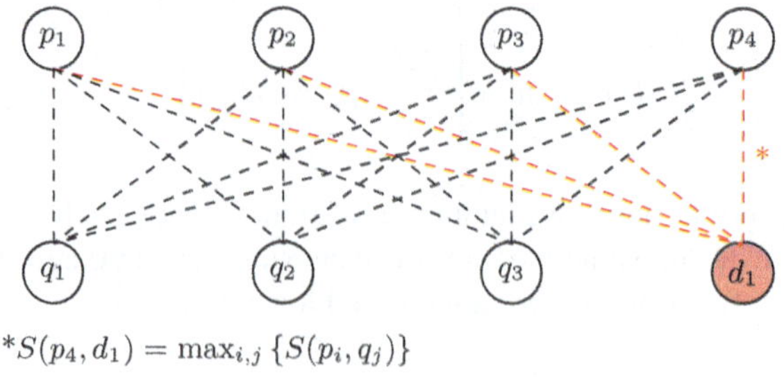

Fig. 7 Bigraph example with one dummy node d_1

5 Discussion and Outlook

The STIDes approach is capable of processing different kinds of similarity views because of the capability to set different weight parameters λ_i according to each specific use case. However, this results in an additional effort in the basic setting of the method since the parameters must be set separately for each application. In addition, the creation of dummy nodes allows a determination of the similarity of two unequal data sets, but the remaining intervals do not yet influence the computation of similarity. With the possibility of applying global changes like time shifts we are able to find similarities between a whole data set and a given subset of another data set. Especially, in this case, the created dummy nodes should not interfere with the similarity calculation.

In the future, we will focus our research on these global changes like time shifts and scaling. Although it was shown that the global minimum within the functions (12) and (13) exists, a statement about the computational complexity could not be made so far. We will also investigate the combined effect of both time shifts and scaling to be able to compare parts of a true-to-scale model with a complete data set. This combined influence can be represented by the structure of the method as a multidimensional function. However, to what extent this affects the complexity of the calculations must also be examined. The possibility to apply different shift and/or scaling factors to different groups of intervals within one data set is also an interesting case, which will be studied in future research. Within the future research, differences in the cardinality of the data sets will again be looked upon to be able to set influence parameters for the similarity measure.

6 Conclusion

In this chapter, a similarity measure depending on the relation of the intervals to each other was introduced. For this purpose, the properties of the intervals, such as the size of the overlap, start and end point distances were defined. From these properties, distance values were derived, which in a weighted sum form the similarity measure of two intervals. This allows to individually weight each interval characteristic an, therefore, is an approach which is adaptable to the considered use case. On the basis of the weighted sum, the STIDes approach was defined, which compares two time interval data sets with one another. For this purpose, the minimum sum of the individual similarities is calculated over all possible interval pairs, which results in the defined similarity measure. An interval pair consists of an interval of each of the two considered time interval data sets. The option to weight each interval property is retained by this approach in the extended similarity measure of two data sets. In order to compensate for a possible cardinality difference between the data sets, dummy nodes were introduced, so that each interval can be assigned one partner from the other set and therefore the STIDes approach can be applied. The desired similarity measure of the possibly modified data sets is determined by the Hungarian algorithm in polynomial time ($O(n^3)$), where n represents the number of intervals in the bigger data set. The introduced methodology for identifying similarities also made it possible to incorporate global changes in the intervals of one data set into the analysis. In this context, it has been shown that the defined functions have a global minimum in order to be able to apply the above-described approach, but the complexity changes with the implementation of global changes are not yet researched. The combination of working with different cardinalities and applying global time shifts to one data set, enables the approach to find the most similar subset of a time interval data sets for a given smaller data set.

Overall, the considered approach provides a versatile method for describing similarities, in which all properties of the intervals are included in the similarity analysis and, moreover, various types of dynamic changes within the data sets can be mapped. Due to the general representation of this methodology, the similarity analysis can be applied to a variety of problems, and thus meets the goal of a general description of similarities between time interval data sets.

References

1. Jayal, A., Badurdeen, F., Dillon, O., Jawahir, I.: Sustainable manufacturing: modeling and optimization challenges at the product, process and system levels. CIRP J. Manuf. Sci. Technol. **2**(3), 144–152 (2010)
2. Tanomaru, J.: Staff scheduling by a genetic algorithm with heuristic operators. In: IEEE International Conference on Systems, Man and Cybernetics. Intelligent Systems for the 21st Century, vol. 3, pp. 1951–1956. IEEE (1995)
3. American Academy of Sleep Medicine: Other: the AASM manual for the scoring of sleep and associated events: rules. AASM, Terminology and Technical Specifications. Westchester (2007)
4. Metzler, D., Dumais, S., Meek, C.: Similarity measures for short segments of text. In: European Conference on Information Retrieval, pp. 16–27. Springer (2007)
5. Yu, G., Li, F., Qin, Y., Bo, X., Wu, Y., Wang, S.: Gosemsim: an r package for measuring semantic similarity among go terms and gene products. Bioinformatics **26**(7), 976–978 (2010)
6. Ogata, H., Fujibuchi, W., Goto, S., Kanehisa, M.: A heuristic graph comparison algorithm and its application to detect functionally related enzyme clusters. Nucleic Acids Res. **28**(20), 4021–4028 (2000)
7. Wang, A., et al.: An industrial strength audio search algorithm. In: Ismir, pp. 7–13. Washington, DC (2003)
8. Kostakis, O., Papapetrou, P., Hollmén, J.: Artemis: assessing the similarity of event-interval sequence. Mach. Learn. Knowl. Discov. Datab. 229–244 (2011)
9. Meisen, P., Keng, D., Meisen, T., Recchioni, M., Jeschke, S.: Similarity search of bounded tidasets within large time interval databases. In: 2015 International Conference on Computational Science and Computational Intelligence (CSCI), pp. 24–29. IEEE (2015)
10. Kruskall, J., Liberman, M.: The symmetric time warping algorithm: from continuous to discrete. Time warps, string edits and macromolecules (1983)
11. Chen, Y.L., Chiang, M.C., Ko, M.T.: Discovering time-interval sequential patterns in sequence databases. Expert Syst. Appl. **25**(3), 343–354 (2003)
12. Sadasivam, R., Duraiswamy, K.: Efficient approach to discover interval-based sequential patterns. J. Comput. Sci. **9**(2), 225–234 (2013)

13. Koncilia, C., Morzy, T., Wrembel, R., Eder, J.: Interval OLAP: analyzing interval data. In: International Conference on Data Warehousing and Knowledge Discovery, pp. 233–244. Springer (2014)
14. Munkres, J.: Algorithms for the assignment and transportation problems. J. Soc. Industr. Appl. Math. **5**(1), 32–38 (1957)
15. Haßler, M., Jeschke, S., Meisen, T.: Similarity analysis of time interval data sets regarding time shifts and rescaling. In: International Work-Conference on Time Series Analysis, vol. 02, pp. 995–1006 (2017)
16. Rechtschaffen, A., Kales, A.: A manual of standardized terminology, techniques, and scoring systems for sleep stages of human subjects (1968)
17. Bourgeois, F., Lassalle, J.C.: An extension of the munkres algorithm for the assignment problem to rectangular matrices. Commun. ACM **14**(12), 802–804 (1971)

Logical Comparison Measures in Classification of Data—Nonmetric Measures

Kalle Saastamoinen

Abstract In this chapter, we will create and use generalized combined comparison measures from t-norms (T) and t-conorms (S) for comparison of data. Norms are combined by the use of generalized mean, where t-norms give minimum and t-conorms give maximum compensation. From this intuitively thinking follows that when these norms are aggregated together, these new comparison measures should be able to find the best possible classification result in between minimum and maximum. We will use classification as our test bench for the suitability of these new comparison measures created. In these classification tasks, we have tested five different types of combined comparison measures (CCM), with t-norms and t-conorms. That were Dombi family, Frank family, Schweizer-Sklar family, Yager family, and Yu family. In classification, we used the following datasets: ionosphere, iris, and wine. We will compare the results achieved with CCM to the ones achieved with pseudo equivalences and show that these new measures tend to give better results.

Keywords Ionos · Iris · Wine · Similarity · Comparison measure · Logical Classification · Data

1 Introduction

Traditionally, measures used for comparison have been metric-based similarities. It is a common belief that measures for comparison should hold true for some properties of metric spaces. This belief originates from the blinkered view that the comparison of objects should always have something to do with distance. This has been questioned in many papers [1–5]. In practice, it seems that properties of distance have little or no

K. Saastamoinen (✉)
Department of Military Technology, National Defence University,
P.O. Box 7, FI-00861 Helsinki, Finland
e-mail: kalle.saastamoinen@mil.fi; normimies@gmail.com

I. Rojas et al. (eds.), *Time Series Analysis and Forecasting*,
Contributions to Statistics, https://doi.org/10.1007/978-3-319-96944-2_12

affect at all on the results that can be achieved from the use of different comparison measures. This becomes empirically clear when one looks at the test results presented in this paper.

Much of the fuzzy set theory's original inspiration and further developments originate from the problems of pattern classification and cluster analysis. Essentially, this is the reason why classification is chosen to be the test bench for many-valued logic-based comparison measures in this chapter. In classification, normally, the question is not whether a given object is or is not a member of a class, but the degree to which the object belongs to the class. This means that most classes in real situations are fuzzy in nature [6]. This fuzzy nature of real-world classification problems may shed some light on the general problem of decision- making [7].

This chapter will suggest a general definition for comparison measure and will show how results presented in article [8] achieved with pseudo equivalences can get better.

The chapter is organized as follows. In the first section, logical comparison measures, combined comparison measure (CCM) (12), and the theory behind them are presented. The second section presents classification schemata and data sets used for testing. The third section presents results achieved and these results are compared to the results achieved before. In the fourth section, conclusions are done and some future directions are given.

2 Logical Comparison Measures

Definition 1 A set function g defined on X, where X is a fuzzy set and has the following properties called as fuzzy measure:

1. $g(\emptyset) = 0, g(X) = 1$
2. If $A, B \in X$ and $A \subseteq B$, then $g(A) \leq g(B)$
3. If $A_n \in B, A_1 \subseteq A_2 \subseteq \cdots \subseteq A_{n-1} \subseteq A_n$, then $\lim_{n\to\infty} g(A_n) = g\left(\lim_{n\to\infty} A_n\right)$

It is suggested here that the comparison measures used, where comparison is done feature by feature and then, these comparisons are aggregated, could actually be any measures which fulfill the following properties:

1. The comparison measure used has a clear logical structure, e.g., it is an Archimedean t-norm or t-conorm (like Frank (2), (7)) or S-equivalence [8].
2. The comparison measure is monotone. This condition ensures that a decrease (or increase) in any values that are to be compared cannot produce an increase (or decrease) in the comparison result.
3. The comparison measure is associative. This guarantees that the final comparison results are independent of the grouping of the arguments and that one can expand these comparisons to more than two arguments.
4. The comparison measure is continuous. This guarantees that one can safely compute with the values that are to be compared.

The idea behind using logical structures instead of, for example, simple distances lie in the fact that logical structures always have some kind of linguistic content inside them. For example, *t*-norms and *t*-conorms can be seen as corresponding to the words "and" and "or", equivalence as corresponding to the expression "if and only if". One can see that just by using these logical measures, it is possible to give some linguistic meaning to the comparison procedure.

Some criteria for comparison measures are suggested here. The following criteria arc almost the same as Lowen gives for aggregation operators [9] and originally, they are presented by Bellman-Giertz [10]. It has also been suggested that not all of these criteria are necessary [11]. One can, however, see that the criteria by Bellman R. and Giertz M. also applies well to the comparison measures presented in this chapter.

1. **Axiomatic strength**. It is suggested here that the operator is better, if the axioms of the operator satisfies are less limiting, this is equivalent to Lowen [9]. It is seen that depending on the choice of the logical structure used, this will fit well with the definition given in (2).
2. **Flexibility**. Through the flexibility, three things are met that are of an empirical fit, adaptability, and compensation. Adaptability comes from the fact that all comparison measures created in this article are parameterized. Compensation property follows from the use of a generalized mean to combine the different values. Empirical fit follows then from the use of logical structures, adaptability, and compensation. Empirical fit can naturally only finally be proven by empirical testing, as is done in this article.
3. **Numerical efficiency**. Some operators such as *min* and *max* are numerically more efficient than, for example, Frank's t-norm and t-conorm. In large problems, this will always be problematic to some degree. However, it is gradually becoming less of problem as computers computing power is constantly increasing.
4. **Range of compensation**. In general, the larger the range of compensation the better the compensatory operator. In some comparison measures presented in this article, the range of compensation has been increased by combining t-norms and t-conorms and in all comparison measures, a generalized mean has been used.
5. **Aggregating behavior of the comparison measure**. Aggregating behavior in the comparison measures presented here, be adjusted by the use of proper mean value in the generalized mean. For example, if a parameter value of 0 is used with a generalized mean, a geometric mean will be obtained, which is to say that one attains the product of the values and subsequently each value "added" normally decreases the resulting aggregate degrees of membership.
6. **Required scale level of membership functions**. Comparison measures presented in this article have very little restrictions concerning scale levels.

T-norms and T-conorms as the Measures for Comparison

In the paper, [12] measures have been defined based on the use of the generalized mean, weights, t-norms, and t-conorms. Below, these results are added to the definition of the combination measure of the t-norm and t-conorm.

Connectives play an important role when trying to model reality by equations. For example, when linguistic interpretations such as "*AND*" or "*OR*" are used for connectives in conjunction and disjunction, quite often this does not require or mean crisp connectives, but that these connectives are only needed to some degree. In such cases, connectives called t-norms or t-conorms may be used. The t-norm gives minimum compensation, while the t-conorm gives maximum compensation. This means that t-norms tend to give more value for the small values, while t-conorms give more value for the big values in the interval in which they are used. In practice, neither of these connectives fit the collected data appropriately. There is still a lot of information that is left in between of these two connectives. An important issue when dealing with t-norms and t-conorms is the question of how to combine them in a meaningful way, since neither of these connectives alone gives a general compensation for the values where they are adapted. For this reason, one should use a measure that somewhat compensates this gap in between the values of these two norms. Article [13] shows how the generalized mean works as the compensative connective between minimum and maximum connectives. The scope of aggregation operators is demonstrated in Fig. 1.

The first researchers to try the compensation of t-norms and t-conorms were Zimmermann and Zysno [14]. They used the weighted geometric mean in order to compensate the gap between fuzzy intersections and unions. When one uses the geometric mean, equal compensation is allocated to the all values, and problems might occur if some of the values combined are relatively very low or high.

Created Comparison Measures From T-norms and T-conorms The following is a brief representation of the algebraic equations that can be created by combining weights into some important t-norms and t-conorms and then, the combining values are given that were achieved by aggregating them with a generalized mean. Archimedean t-norms and t-conorms are a good choice since they are continuous and monotonic [15].

The comparison measure (12) has been tested by combining it here with different kinds of t-norms and t-conorms. It has been tested without weights ω_{ci} and ω_{di}, since the weighting process was too time-consuming with differential evolution. All the comparison measures mentioned in this subchapter have been tested in classification

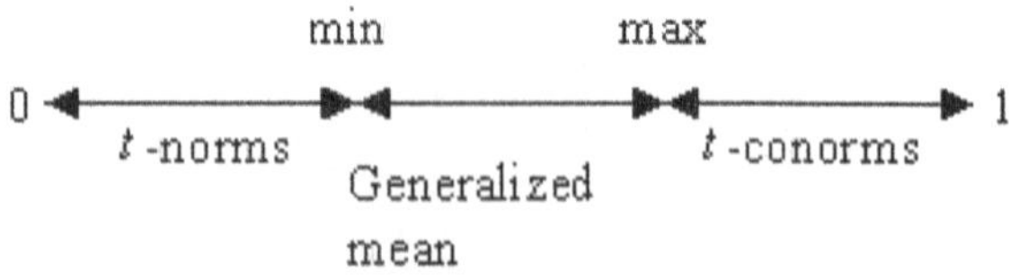

Fig. 1 Compensation of t-norms and t-conorms

tasks. T-norms and t-conorms are tested with weights, where a generalized mean has been used to aggregate and compensate the values.

Parameterized families of t-norms and t-conorms are used here. Families tested in classification were the Dombi family [16], Frank family [17], Schweizer-Sklar family [18], Yager family [19], and Yu family [20]. The Frank and Schweizer-Sklar families of t-norms are also copula families [21] so they have good statistical properties see Fisher [22].

From t-norm and t-conorm families, we have created the following comparison measures.

Definition 2 Measure based on Dombi [16] class of t-norm with generalized mean and weights:

$$T_D \langle f_1(i), f_2(i) \rangle = \left(\sum_{i=1}^{n} \omega_{ci} \left(1 + \left[\left(\frac{1}{f_1(i)} - 1 \right)^p + \left(\frac{1}{f_2(i)} - 1 \right)^p \right]^{\frac{1}{p}} \right)^{-m} \right)^{\frac{1}{m}}, \quad (1)$$

where $p > 0$ and $i = 1, \ldots, n$.

Definition 3 Measure based on Frank [17] class of t-norm with generalized mean and weights:

$$T_F \langle f_1(i), f_2(i) \rangle = \left(\sum_{i=1}^{n} \omega_{ci} \left(\log_p \left[1 + \frac{\left(p^{f_1(i)} - 1\right)\left(p^{f_2(i)} - 1\right)}{p - 1} \right] \right)^m \right)^{\frac{1}{m}}, \quad (2)$$

where $p > 0, p \neq 1$ and $i = 1, \ldots, n$.

Definition 4 Measure based on Schweizer and Sklar [18] class of t-norm with generalized mean and weights:

$$T_{SS} \langle f_1(i), f_2(i) \rangle = \left(\sum_{i=1}^{n} \omega_{ci} \left(\max \left\{ 0, (f_1(i))^p + (f_2(i))^p - 1 \right\} \right)^{\frac{m}{p}} \right)^{\frac{1}{m}}, \quad (3)$$

where $p \neq 0$ and $i = 1, \ldots, n$.

Definition 5 Measure based on Yager [19] class of t-norm with generalized mean and weights:

$$T_Y \langle f_1(i), f_2(i) \rangle = \left(\sum_{i=1}^{n} \omega_{ci} \left(1 - \min \left\{ 1, \left[(1 - f_1(i))^p + (1 - f_2(i))^p \right]^{\frac{1}{p}} \right\} \right)^m \right)^{\frac{1}{m}}, \quad (4)$$

where $p > 0$ and $i = 1, \ldots, n$.

Definition 6 Measure based on Yu [20] class of t-norm with generalized mean and weights:

$$T_{Yu}\langle f_1(i), f_2(i)\rangle = \left(\sum_{i=1}^{n} \omega_{ci} \left(\max\{0, (1+p)(f_1(i)+f_2(i)-1) - p \cdot f_1(i) f_2(i)\}\right)^m\right)^{\frac{1}{m}}, \quad (5)$$

where $p > -1$ and $i = 1, \ldots, n$.

Definition 7 Measure based on Dombi [16] class of t-conorm with generalized mean and weights:

$$S_D\langle f_1(i), f_2(i)\rangle = \left(\sum_{i=1}^{n} \omega_{di} \left(1 + \left[\left(\frac{1}{f_1(i)} - 1\right)^{-p} + \left(\frac{1}{f_2(i)} - 1\right)^{-p}\right]^{-\frac{1}{p}}\right)^{-m}\right)^{\frac{1}{m}}, \quad (6)$$

where $p > 0$ and $i = 1, \ldots, n$.

Definition 8 Measure based on Frank [17] class of t-conorm with generalized mean and weights:

$$S_F\langle f_1(i), f_2(i)\rangle = \left(\sum_{i=1}^{n} \omega_{di} \left(1 - \log_p \left[1 + \frac{\left(p^{1-f_1(i)} - 1\right)\left(p^{1-f_2(i)} - 1\right)}{p-1}\right]\right)^m\right)^{\frac{1}{m}}, \quad (7)$$

where $p > 0, p \neq 1$ and $i = 1, \ldots, n$.

Definition 9 Measure based on Schweizer and Sklar [18] class of t-conorm with generalized mean and weights:

$$S_{SS}\langle f_1(i), f_2(i)\rangle = \left(\sum_{i=1}^{n} \omega_{di} \left(1 - \left(\max\left\{0, (f_1(i))^p + (f_2(i))^p - 1\right\}\right)^{\frac{1}{p}}\right)^m\right)^{\frac{1}{m}}, \quad (8)$$

where $p \neq 0$ and $i = 1, \ldots, n$.

Definition 10 Measure based on Yager [19] class of t-conorm with generalized mean and weights:

$$S_Y\langle f_1(i), f_2(i)\rangle = \left(\sum_{i=1}^{n} \omega_{di} \left(\min\left\{1, \left[(f_1(i))^p + (f_2(i))^p\right]^{\frac{1}{p}}\right\}\right)^m\right)^{\frac{1}{m}}, \quad (9)$$

where $p > 0$ and $i = 1, \ldots, n$.

Definition 11 Measure based on Yu [20] class of t-conorm with generalized mean and weights:

$$S_{Yu}\langle f_1(i), f_2(i)\rangle = \left(\sum_{i=1}^{n} \omega_{di} \left(\min\{1, f_1(i) + f_2(i) + p \cdot f_1(i) f_2(i)\}\right)^m\right)^{\frac{1}{m}}, \quad (10)$$

where $p > -1$ and $i = 1, \ldots, n$.

Definition 12 Combined comparison measure (CCM) based on the t-norm and t-conorm with a generalized mean and weights [30]:

$$C\langle f_1, f_2\rangle = \left(\sum_{i=1}^{n}\left(w_i T_i^p \langle f_1(i), f_2(i)\rangle + (1-w_i)\left(S_i^p \langle f_1(i), f_2(i)\rangle\right)\right)^m\right)^{\frac{1}{m}} \quad (11)$$

where $i = 1, \ldots, n$, p is a parameter combined to the corresponding class of fuzzy intersections T_i and unions S_i and w_i are weights and $i = 1, \ldots, n$.

3 Classification

Many times, there are given a set of data which is already grouped into classes and the problem is then to predict which class each new data belongs to. This is normally referred to as classification problem. The first set of data is referred to as training set, while this new set of data is referred to as test set [23]. Classification is seen as a comparison between training set and test set.

Description of the Similarity Based Classifiers

Objects, each characterized by one feature vector in $[0, 1]^n$, is classified into different classes. The assumption that the vectors belong to $[0, 1]^n$ is not restrictive since the appropriate shift and normalization can be done for any space $[a, b]^n$. The comparison measures can be used to compare objects to classes. Below is the used classifier in the algorithmic form:

SIMILARITY BASED CLASSIFIER

Require: *data*
  scale *data* between [0, 1]
Require: *test*,*learn*[1...*n*],*weights*,*dim*
  for $i = 1$ to n **do**
    $idealvec[i] = IDEAL[learn[i]]$
    $maxcomp[i] = \left(\frac{1}{\dim}\right)^{1/m}\left(\sum_{j=1}^{\dim} weights[j]\left(CCM\left(idealvec[i,j], test[j]\right)\right)^m\right)^{1/m}$
  end for
  $class = \arg\max_i maxcomp[i]$

In the algorithm, the combined comparison measure (CCM) with a generalized mean is used. *IDEAL* is the vector that best characterizes the class i and here the generalized mean vector of the class as an *IDEAL*-operator has been used.

When we choose to use randomized weights (RWs) instead of using differential evolution (DE), we achieve a significant saving in computing time. RW is approximately 150,000 times faster.

Evolutionary algorithm is used because of its diversity and robustness to find weights in classification process, information about evolutionary algorithms, in general, can be found, for example, from [24–27]. Obviously, other optimizers can be used as well. Evolutionary algorithm used here is based on differential evolution [28]. DE is a simple population-based stochastic function minimizer. The objective of DE is to iterate each member of the population and compare its value to the trial member value, and the superior member stays for the next iteration. The evolution strategy defines the way in which a trial member is generated. DE tries to seek weights that will give the maximal similarity compared to the values set by experts. This is done so that DE tries to minimize the value of the objective function with trial member values. The objective function is the total difference between classification defined by experts and the classification defined by similarity used here for all learning data sets. Finally, DE gives the optimal weight values.

The flow of the classification task has been described in the flowchart (Fig. 2). Classification procedure uses part of the data (*learning*) for weight optimization either using differential evolution or randomized weights depending on the choice. After this, rest of the data (*test*) is used for classification and then, this result is saved, now if loop is done, N-times *max*, *min*, and *mean* values are saved and then, this same classification procedure is done from the beginning for the next parameter value p. After all, parameter values p have been done we start from the next mean value m the loop again.

Data Sets

We tested our measures with three different data sets which are available from the [29]. The data sets chosen for the test were: ionosphere, iris, and wine. These sets differ greatly in the magnitude of instances and the number of predictive attribute values.

Ionos: This is radar data, where the targets were free electrons in the ionosphere. Here are two classes: "Good" and "Bad". "Good" radar returns are those showing evidence of some type of structure in the ionosphere. "Bad" returns are those that do not; their signals pass through the ionosphere. The number of instances is 351. The number of attributes is 34 plus the class attribute.

Iris: Perhaps, it is the best-known database to be found in the pattern recognition literature. The number of attributes is 4 plus the class attribute. The data set contains 3 classes of 50 instances each, where each class refers to a type of iris plant.

Wine: The data is the result of a chemical analysis of wines grown in the same region in Italy but derived from three different cultivars. The analysis determined the

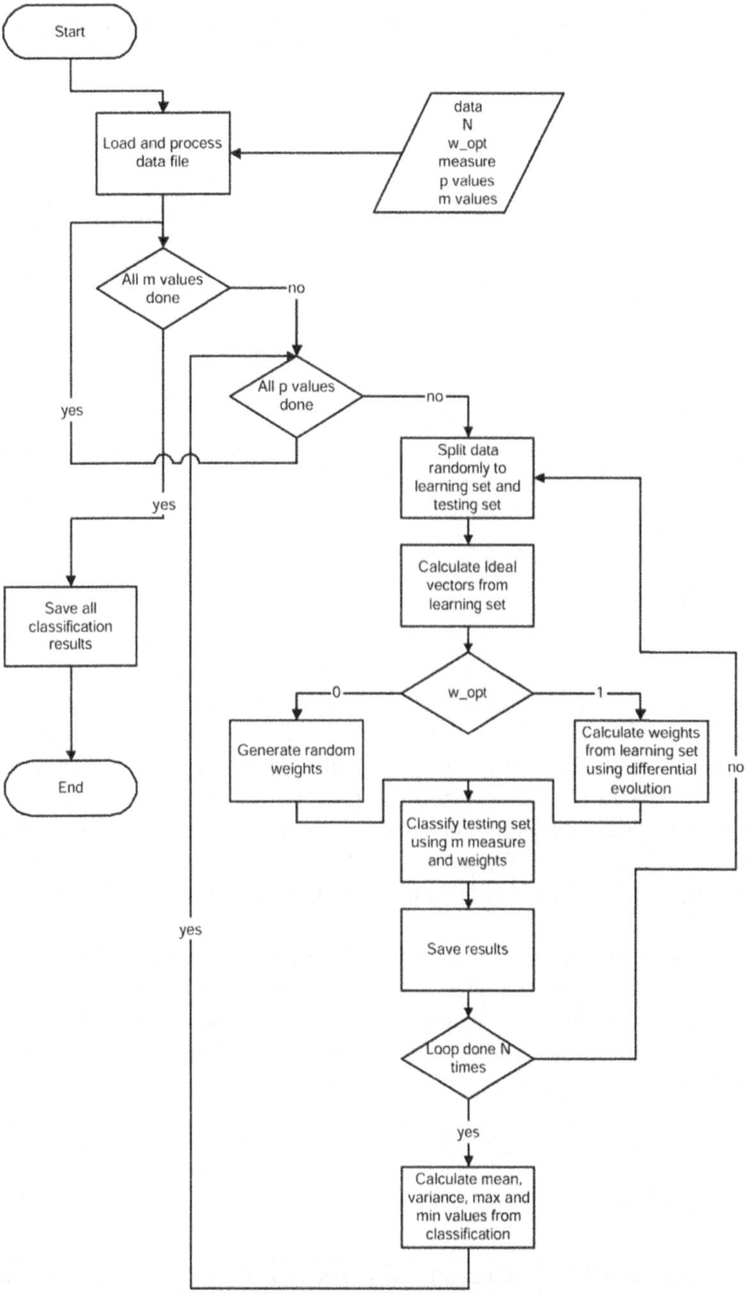

Fig. 2 Simplified flowchart of the classification procedure

Table 1 Mean (Av), Maximum (Max) classification results with different comparison measures (CMs), Optimized (DE), and Randomized (RND) Weights and Variances (VAR) versus previous best results with pseudo equivalences

CM	Iono_{Av}	Iono_{Max}	Iris_{Av}	Iris_{Max}	Wine_{Av}	Wine_{Max}
Frank CCM_{DE} (%)	83.68	94.89	91.68	100	90.47	100
Frank CCM_{RND} (%)	79.86	94.32	71.61	100	87.22	100
Frank CCM_{VAR}	0	0.0075833	0	0.022331	0.00013855	0.022344
Equivalences (%)	80.16	93.75	98.84	100	96.35	100

quantities of 13 constituents found in each of the 3 types of wines. The number of instances are class 1 59, class 2 71, and class 3 48.

4 Results

In our classification tasks, we have tested five different types of combined comparison measures (CCM), with t-norms and t-conorms from Dombi, Frank, Schweizer-Sklar, Yu, and Yager families [30]. In all these tested families, the combination of Frank norms always managed to give the best results. From the Table 1, one can see a comparison between the previous best true positive classification results with many-valued pseudo equivalences [8] versus to the best true positive classification results with CCM presented here. We tested our classifications in both weights that were randomly selected 200 times (RND) and with weights that were optimized 10 times (DE) for each p- and m-value.

Table 1 shows the mean and maximum of the classification results from all combinations of weights and parameters. Results were better using CCM measure Eq. 11 with Frank norms Eq. 2 and 7 with Ionosphere data set than when we used pseudo equivalences for the classification. With Iris and Wine data sets pseudo equivalences gave slightly better results.

5 Conclusions

In classification and the development of expert systems, the typical problem is choosing the right function for comparison. When data has different dependencies, different operators should be used. Usually, the simplest operators are selected, which are not normally the optimal choice. As a solution to this problem, this paper has offered combined comparison measure 12 that combines with generalized mean t-norm and

t-conorm. This measure is on a logically sound basis. It also has been shown that these comparison measures give reasonable results.

Comparison measures introduced in this article consistently give good and stable results in classification, which can be seen from the Table 1. Combined comparison measure (11) based on Frank type of t-norm (2) and t-conorm (7) gave the best classification results, which are the same or better than those attained from the pseudo equivalences. One can also see that the improvements in classification results due to changing to the right comparison measures were quite significant.

From the tested combined comparison measures (11), use of a combination of Frank type t-norm and t-conorm is recommended.

These new comparison measures can be used in, for example, pattern recognition, clustering, expert systems, medical diagnosis systems, decision support systems, fuzzy control, etc. Classification results were not only good but also stable, which makes these comparison measures usable.

References

1. Tversky, A., Krantz, D.H.: The dimensional representation and the metric structure of similarity data. J. Math. Psychol. **7**(3), 572–596 (1970)
2. Santini, S., Jain, R.: Similarity measures. IEEE Trans. Pattern Anal. Mach. Intell. **21**(9), 871–883 (1999)
3. France, R.K.: Weights and Measures: An Axiomatic Model for Similarity Computations. Internal Report, Virginia Tech (1994)
4. De Cock, M., Kerre, E.: Why fuzzy T-equivalence relations do not resolve the Poincar paradox, and related issues. Fuzzy Sets Syst. **133**(2), 181–192 (2003)
5. Tversky, A.: Features of similarity. Psychol. Rev. **84**(4), 327–352 (1977)
6. Zadeh, L.A., Fuzzy sets and their application to pattern classification and clustering analysis. In: Van Ryzin, J. (Ed.) Classification and Clustering. Academic Press, pp. 251–299 (1977)
7. Pal, S.K., Dutta-Majumder, D.K.: Fuzzy Mathematical Approach to Pattern Recognition. Wiley (Halsted), N. Y. (1986)
8. Saastamoinen, K.: Classification of data with similarity classifier. In: International Work-Conference on Time Series, Keynote Speech, ITISE2016 Proceedings
9. Lowen, R.R.: Fuzzy Set Theory: Basic Concepts, Techniques, and Bibliography. Kluwer Academic Publishers, Dordrecht (1996)
10. Bellman, R., Giertz, M.: On the analytic formalism of the theory of fuzzy sets. Inf. Sci. **5**, 149–165 (1973)
11. Kang, T., Chen, G.: Modifications of Bellman-Giertz's theorem. Fuzzy Sets Syst. **94**(3), 349–353 (1998)
12. Saastamoinen, K., Ketola, J.: Using generalized combination measure from Dombi and Yager type of T-norms and T-conorms in classification. In: Proceedings of the ECTI-CON 2005 Conference (2005)
13. Dyckhoff, H., Pedrycz, W.: Generalized means as model of compensative connectives. Fuzzy Sets Syst. **14**, 143–154 (1984)
14. Zimmermann, H.-J., Zysno, P.: Latent connectives in human decision making. Fuzzy Sets Syst. **4**, 37–51 (1980)
15. Bilgiç, T., Türkşen, I.B.: Measurement theoretic justification of connectives in fuzzy set theory. Fuzzy Sets Syst. **76**(3), 289–308 (1995)
16. Dombi, J.: Basic concepts for a theory of evaluation: the aggregative operator. Eur. J. Oper. Res. **10**, 282–293 (1982)

17. Frank, M.J.: On the simultaneous associativity of $F(x, y)$ and $x + y - F(x, y)$. Aeqationes Math. **19**, 194–226 (1979)
18. Schweizer, B., Sklar, A.: Associative functions and abstract semigroups. Publ. Math. Debr. **10**, 69–81 (1963)
19. Yager, R.R.: On a general class of fuzzy connectives. Fuzzy Sets Syst. **4**, 235–242 (1980)
20. Yu, Y.: Triangular norms and TNF-sigma algebras. Fuzzy Sets Syst. **16**, 251–264 (1985)
21. Sklar, A.: Fonctions de $r\acute{e}$partition $\acute{a}$ n dimensions et leurs marges. Publ. Inst. Statist. Univ. Paris **8**, 229–231 (1959)
22. Fisher, N.I.: Copulas. In: Kotz, S., Read, C.B., Banks, D.L. (eds.) Encyclopedia of Statistical Sciences, vol. 1, pp. 159–163. Wiley, New York (1997)
23. Hastie, T., Tibshirani, R.: The Elements of Statistical Learning: Data Mining, Inference, and Prediction. Springer Series in Statistics, Springer, New York (2001)
24. Goldberg, D.E.: Real-coded genetic algorithms, virtual alphabets, and blocking. Technical Report 9001, University of Illinois at Urbana-Champain (1990)
25. Mantel, B., Periaux, J., Sefrioui, M.: Gradient and genetic optimizers for aerodynamic desing. In: ICIAM 95 Conference, Hamburgh (1995)
26. Michalewics, Z.: Genetic Algorithms + Data Structures = Evolution Programs Artificial Intelligence. Springer, New York (1992)
27. Grefenstette, J.J.: Optimization of control parameters for genetic algorithms. IEEE Trans. Syst. Man Cybern. **16**(1), 122–128 (1986)
28. Price, K.V., Storn, R.M., Lampinen, J.A.: Differential Evolution—A Practical Approach to Global Optimization. Springer, Natural Computing Series (2005)
29. UC Irvine Machine Learning Repository. http://archive.ics.uci.edu/ml/. Accessed 15 Jan 2017
30. Saastamoinen, K.: Many valued algebraic structures as measures of comparison, Acta Universitatis Lappeenrantaensis. Ph.D. Thesis (2008)

Part IV
Econometric Models

Asymptotic and Bootstrap Tests for a Change in Autoregression Omitting Variability Estimation

Barbora Peštová and Michal Pešta

Abstract A sequence of time-ordered observations follows an autoregressive model of order one and its parameter is possibly subject to change at most once at some unknown time point. The aim is to test whether such an unknown change has occurred or not. A change-point method presented here rely on a ratio type test statistic based on the maxima of cumulative sums. The main advantage of the developed approach is that the variance of the observations neither has to be known nor estimated. Asymptotic distribution of the test statistic under the no-change null hypothesis is derived. Moreover, we prove the consistency of the test under the alternative. A bootstrap procedure is proposed in the way of a completely data-driven technique without any tuning parameters. The results are illustrated through a simulation study, which demonstrates the computational efficiency of the procedure. A practical application to real data is presented as well.

Keywords Change point · Structural change · Change in autoregression Hypothesis testing · Bootstrap · Ratio type statistic · Variance estimation free test

1 Introduction and Main Goals

The focus lies on *autoregressive time series of order one*, i.e., AR(1) series. We try to detect a possible change of the scalar parameter from a stationary autoregressive model using the ratio type test statistic, which allows us to avoid estimating the unknown nuisance dispersion parameter of the time series.

B. Peštová
Institute of Computer Science, The Czech Academy of Sciences, Prague, Czech Republic
e-mail: pestova@cs.cas.cz

M. Pešta (✉)
Faculty of Mathematics and Physics, Charles University, Prague, Czech Republic
e-mail: michal.pesta@mff.cuni.cz; pesta@karlin.mff.cuni.cz

I. Rojas et al. (eds.), *Time Series Analysis and Forecasting*,
Contributions to Statistics, https://doi.org/10.1007/978-3-319-96944-2_13

The results are inspired by [1], where an autoregressive times series model of order p is taken into account and the whole vector of autoregression parameters is subject to change. The authors proposed to detect such change by computing partial sums of weighted residuals based on the maximum type CUSUM test statistics. The results were consequently extended by the bootstrap approach in [2]. The main disadvantage of these methods is that the *variance estimation is problematic*. To overcome such a dilemma, the ratio type test statistic is utilized in the change-point detection.

The remainder of the paper is structured as follows: Sect. 2 introduces a change-point model for AR(1) series together with stochastic assumptions. The ratio type test statistic for the change-point detection is proposed in Sect. 3. Consequently, the asymptotic behavior of the considered test statistic is derived, which covers the main theoretical contribution. Asymptotic critical values are calculated in Sect. 4 by Monte Carlo simulations. Bootstrap extension of the testing procedure is established in Sect. 5. Section 6 contains a simulation study that illustrates the performance of the asymptotic and bootstrap tests. It numerically emphasizes the advantages and disadvantages of the proposed procedures. A practical application of the developed approach to a stock exchange index is presented in Sect. 7. Proofs are given in the Appendix.

2 Autoregressive Model with Possibly Changed Parameter

Let us consider the following time series model with a possible change in parameter β after an unknown time point τ:

$$Y_t = \beta Y_{t-1} + \delta Y_{t-1}\mathcal{I}\{t > \tau\} + \varepsilon_t, \quad t = 2, \ldots, n, \tag{1}$$

where β and $\delta \neq 0$ are fixed (not depending on n) unknown parameters, $1 < \tau = \tau_n \leq n$ is the unknown change point, and $\varepsilon_2, \ldots, \varepsilon_n$ are independent and identically distributed (iid) random errors satisfying further conditions specified later on. For the sake of convenience, we have already suppressed the index n in the observations $Y_{t,n}$ as well as in the parameter τ_n in the model formulation (1) and whenever possible later on.

We are going to test the *null hypothesis* that the autoregression parameter remained constant for the whole observation period

$$H_0 : \tau = n \tag{2}$$

against the *alternative* that a change of the autoregression parameter occurred at some unknown time point τ prior to the latest observed time n, i.e.,

$$H_1 : \tau < n, \ \delta \neq 0. \tag{3}$$

3 Test Statistic for Change in Autoregression

The ratio type test statistics for the simple change in mean were introduced in [3]. We utilize this idea and propose the following *ratio type test statistic* to detect the change in the autoregression of order one:

$$\mathcal{V}_n = \max_{n\gamma \le k \le n-n\gamma} \frac{\max_{2\le i\le k} \left|\sum_{j=1}^{i-1} Y_j(Y_{j+1} - \widehat{\beta}_{1k}Y_j)\right|}{\max_{k+1\le i\le n-1} \left|\sum_{j=i}^{n-1} Y_j(Y_{j+1} - \widehat{\beta}_{2k}Y_j)\right|}, \tag{4}$$

where $0 < \gamma < 1/2$ is a given constant, $\widehat{\beta}_{1k}$ is an ordinary least squares estimate of the parameter β based on the observations $Y_1, \dots, Y_k$ and $\widehat{\beta}_{2k}$ is an ordinary least squares estimate of β based on the observations $Y_{k+1}, \dots, Y_n$. Being more formal, the estimate $\widehat{\beta}_{1k}$ is obtained when regressing the vector of responses $\mathbf{y}_{1,k} := (Y_2, \dots, Y_k)^\top$ on the vector of covariates $\mathbf{x}_{1,k} := (Y_1, \dots, Y_{k-1})^\top$. Analogously, the estimate $\widehat{\beta}_{2k}$ is obtained when regressing the vector of responses $\mathbf{y}_{k+1,n} := (Y_{k+2}, \dots, Y_n)^\top$ on the vector of regressors $\mathbf{x}_{k+1,n} := (Y_{k+1}, \dots, Y_{n-1})^\top$.

The motivation for constructing the ratio type test statistic $\mathcal{V}_n$ comes from the linear regression setup (so-called normal equations). so-called normal equations. Being more specific, the estimate $\widehat{\beta}_{1k}$ is a solution of

$$\mathbf{x}_{1,k}^\top \left(\mathbf{y}_{1,k} - \mathbf{x}_{1,k}b\right) = 0$$

with respect to $b \in \mathbb{R}$ and the estimate $\widehat{\beta}_{2k}$ is a solution of

$$\mathbf{x}_{k+1,n}^\top \left(\mathbf{y}_{k+1,n} - \mathbf{x}_{k+1,n}b\right) = 0$$

with respect to $b \in \mathbb{R}$. Then, one may define partial sums of the weighted residuals as

$$\mathbf{x}_{1,i}^\top \left(\mathbf{y}_{1,i} - \mathbf{x}_{1,i}\widehat{\beta}_{1k}\right), \quad i = 2, \dots, k$$

and

$$\mathbf{x}_{i,n}^\top \left(\mathbf{y}_{,n} - \mathbf{x}_{i,n}\widehat{\beta}_{2k}\right), \quad i = k+1, \dots, n.$$

Consequently, these partial sums can be used as basis for the maxima of partial sums in the numerator and the denominator of $\mathcal{V}_n$. Note that, this approach—usage of the ratio type test statistics—can be generalized for the change of a vector autoregression parameter of the stationary autoregressive AR(p)-process, when $p \ge 2$, using the notation from [1].

Before deriving asymptotic properties of the ratio type test statistic, we formulate several stochastic assumptions on the time series model (1):

Assumption 1 $\beta \in (-1, 1) \setminus \{0\}$.

Assumption 2 $\beta + \delta \in (-1, 1) \setminus \{0\}$.

Assumption 3 $\{\varepsilon_t, t = 0, \pm 1, \ldots\}$ are iid random variables having $\mathsf{E}\varepsilon_t = 0$, $\mathsf{Var}\,\varepsilon_t = \sigma^2 > 0$, and $\mathsf{E}\varepsilon_t^4 < \infty$ for all t. Observation Y_1 is independent of $\{\varepsilon_2, \varepsilon_3, \ldots\}$.

Assumptions 1, 2, and 3 mean that the time series is a stationary autoregressive sequence of order one (and not an iid sequence) before and even after the possible change point.

The limit behavior of the test statistic under the null hypothesis is characterized by the following theorem.

Theorem 1 (Under null) *Suppose that $Y_1, \ldots, Y_n$ follow model* (1), *assume that Assumptions 1 and 3 hold. Then, under null hypothesis* (2),

$$\mathcal{V}_n \xrightarrow[n\to\infty]{\mathscr{D}} \sup_{\gamma \le t \le 1-\gamma} \frac{\sup_{0 \le u \le t} |\mathcal{W}(u) - u/t\mathcal{W}(t)|}{\sup_{t \le u \le 1} \left|\widetilde{\mathcal{W}}(u) - (1-u)/(1-t)\widetilde{\mathcal{W}}(t)\right|}, \tag{5}$$

where $\{\mathcal{W}(x), x \in [0, 1]\}$ is a standard Wiener process and $\widetilde{\mathcal{W}}(x) = \mathcal{W}(1) - \mathcal{W}(x)$.

The next theorem describes the test statistic's behavior under a fixed alternative.

Theorem 2 (Under alternative) *Suppose that $Y_1, \ldots, Y_n$ follow model* (1), *assume that alternative* (3) *holds for some fixed $\delta \neq 0$, and $\tau = [\zeta n]$ for some $\gamma < \zeta < 1-\gamma$. Then, under Assumptions 1, 2, and 3,*

$$\mathcal{V}_n \xrightarrow[n\to\infty]{\mathsf{P}} \infty.$$

The previous theorem provides *consistency* of the studied test statistic under the given assumptions. The null hypothesis is rejected for large values of the ratio type statistic. Being more formal, we reject H_0 at significance level α if $\mathcal{V}_n > v_{1-\alpha,\gamma}$, where $v_{1-\alpha,\gamma}$ is the $(1-\alpha)$-quantile of the asymptotic distribution (5).

4 Asymptotic Critical Values

The explicit form of the limit distribution (5) is not known. The critical values may be determined by simulations from the limit distribution from Theorem 1. Theorem 2 ensures that we reject the null hypothesis for large values of the test statistic. We tried to simulate the asymptotic distribution (5) by *discretizing* the Wiener process and using the relationship of a random walk to the Wiener process. We considered 1000 as the number of discretization points within [0, 1] interval and the number of simulation runs equals to 100,000. discretization points and simulations were tried as well, but only negligible differences in the critical values were acquired. In Table 1, we present several critical values for $\gamma = 0.1$ and $\gamma = 0.2$.

Note that, the numerator and denominator in the test statistic $\mathcal{V}_n$ can be interchanged and such a modified test statistic can still be used for detection of the change in autoregression (but using different critical values).

Table 1 Simulated critical values corresponding to the asymptotic distribution of the test statistic $\mathcal{V}_n$ under the null hypothesis

$100(1-\alpha)\%$	90%	95%	97.5%	99%
$\gamma = 0.1$	6.298815	7.293031	8.283429	9.589896
$\gamma = 0.2$	4.117010	4.745884	5.368286	6.159252

5 Bootstrap Test Procedure

A possible extension of the proposed methods is *bootstrapping*. Using the bootstrap techniques implemented similarly as in [4] for the change in means, one can obtain critical values in an alternative way compared to the presented asymptotic approach.

A wide range of literature has been published on bootstrapping in the change-point problem, e.g., [2]. Here, we concentrate on a *residual bootstrap*, for which an estimate of the unknown change point in autoregression is prerequisite. One can *estimate* the change point τ in a similar fashion as in [5]:

$$\widehat{\tau} = \arg\max_{2\leq k\leq n} \left| \sum_{j=1}^{k-1} Y_j(Y_{j+1} - \widehat{\beta}_{1n} Y_j) \right| . \tag{6}$$

Since the errors ε_t's are supposed to be independent, we base the errors' estimates on *residuals*

$$\widehat{\varepsilon}_t := \begin{cases} Y_t - \widehat{\beta}_{1\widehat{\tau}} Y_{t-1}, & t = 2, \ldots, \widehat{\tau}, \\ Y_t - \widehat{\beta}_{2\widehat{\tau}} Y_{t-1}, & t = \widehat{\tau}+1, \ldots n. \end{cases} \tag{7}$$

We build up the bootstrap test on the resampling with replacement of the residuals $\{\widehat{\varepsilon}_2, \ldots, \widehat{\varepsilon}_n\}$. This provides bootstrapped residuals $\{\widehat{\varepsilon}_2^*, \ldots, \widehat{\varepsilon}_n^*\}$. Then, the bootstrapped residuals $\widehat{\varepsilon}_t^*$ are *centered* by their conditional expectation $\frac{1}{n-1}\sum_{i=2}^{n}\widehat{\varepsilon}_i$ yielding

$$\widetilde{\varepsilon}_t^* := \widehat{\varepsilon}_t^* - \frac{1}{n-1}\sum_{i=2}^{n}\widehat{\varepsilon}_i . \tag{8}$$

To develop a proper bootstrap version of our test statistics $\mathcal{V}_n$, it is useful to note that, under the null hypothesis H_0, the expressions from the numerator and the denominator of (4) can be alternatively rewritten as

$$\sum_{j=1}^{i-1} Y_j(Y_{j+1} - \widehat{\beta}_{1k} Y_j) = \sum_{j=1}^{i-1} Y_j \varepsilon_{j+1} - \frac{\sum_{j=1}^{i-1} Y_j^2}{\sum_{j=1}^{k-1} Y_j^2} \sum_{j=1}^{k-1} Y_j \varepsilon_{j+1} \tag{9}$$

and

$$\sum_{j=i}^{n-1} Y_j(Y_{j+1} - \widehat{\beta}_{2k} Y_j) = \sum_{j=i}^{n-1} Y_j \varepsilon_{j+1} - \frac{\sum_{j=i}^{n-1} Y_j^2}{\sum_{j=k}^{n-1} Y_j^2} \sum_{j=k}^{n-1} Y_j \varepsilon_{j+1}. \tag{10}$$

Then, the bootstrap test statistic is just a modification of the test statistic $\mathcal{V}_n$, where the original errors ε_t appearing in (9) and (10) are replaced by their bootstrap counterparts $\widetilde{\varepsilon}_t^*$:

$$\mathcal{V}_n^* = \max_{n\gamma \le k \le n-n\gamma} \frac{\max_{2 \le i \le k} \left| \sum_{j=1}^{i-1} Y_j \widetilde{\varepsilon}_{j+1}^* - \frac{\sum_{j=1}^{i-1} Y_j^2}{\sum_{j=1}^{k-1} Y_j^2} \sum_{j=1}^{k-1} Y_j \widetilde{\varepsilon}_{j+1}^* \right|}{\max_{k+1 \le i \le n-1} \left| \sum_{j=i}^{n-1} Y_j \widetilde{\varepsilon}_{j+1}^* - \frac{\sum_{j=i}^{n-1} Y_j^2}{\sum_{j=k}^{n-1} Y_j^2} \sum_{j=k}^{n-1} Y_j \widetilde{\varepsilon}_{j+1}^* \right|}.$$

An *algorithm* for the bootstrap is illustratively shown in Procedure 1.

Procedure 1 Bootstrapping test statistic $\mathcal{V}_n$.

Input: Time series of length n, i.e., sequence of observations $\{Y_1, \ldots, Y_n\}$.

Output: Bootstrap distribution of $\mathcal{V}_n$, i.e., the empirical distribution where probability mass $1/B$ concentrates at each of ${}_{(1)}\mathcal{V}_n^*, \ldots, {}_{(B)}\mathcal{V}_n^*$.

1: estimate the change point by calculating $\widehat{\tau}$ from (6)
2: compute the residuals $\widehat{\varepsilon}_t$ as in (7)
3: **for** $b = 1$ to B **do** // repeat in order to obtain the empirical distribution
4: $\quad \{\widehat{\varepsilon}_2^*, \ldots, \widehat{\varepsilon}_n^*\}$ resampled with replacement from the original residuals $\{\widehat{\varepsilon}_2, \ldots, \widehat{\varepsilon}_n\}$
5: $\quad$ calculate the centered bootstrap residuals $\{\widetilde{\varepsilon}_2^*, \ldots, \widetilde{\varepsilon}_n^*\}$ via (8)
6: $\quad$ compute bootstrap test statistics ${}_{(b)}\mathcal{V}_n^*$
7: **end for**

---

6 Simulation Study

A simulation experiment was performed to study the *finite sample properties* of the asymptotic and bootstrap tests for the change in the AR(1) parameter. In particular, the interest lies in the *empirical size* of the proposed tests under the null hypothesis and in the empirical *rejection rate* (power) under the alternative. Random samples (1000 each time) are generated from the time series change-point model (1). The number of observations is set to $n = 200$, $n = 400$, and $n = 800$ in order to demonstrate the performance of the testing approach in case of different sample sizes. Two

Table 2 Empirical size of the asymptotic (*asym*) and bootstrap (*boot*) tests for change in autoregression under H_0 using $\gamma = 0.1$, considering the significance level α. Innovations (*innov*) are iid having Student t_5 and standard normal $\mathsf{N}(0,1)$ distribution

		α	0.100		0.050		0.025		0.010	
		innov	$\mathsf{N}(0,1)$	t_5	$\mathsf{N}(0,1)$	t_5	$\mathsf{N}(0,1)$	t_5	$\mathsf{N}(0,1)$	t_5
$n = 200$	$\beta = -0.6$	asym	0.258	0.342	0.172	0.266	0.126	0.216	0.088	0.152
		boot	0.231	0.314	0.157	0.243	0.109	0.192	0.078	0.140
	$\beta = 0.2$	asym	0.296	0.400	0.206	0.318	0.158	0.234	0.106	0.176
		boot	0.273	0.361	0.182	0.290	0.142	0.217	0.095	0.158
$n = 400$	$\beta = -0.6$	asym	0.218	0.238	0.136	0.160	0.080	0.108	0.046	0.072
		boot	0.199	0.207	0.118	0.144	0.069	0.097	0.040	0.063
	$\beta = 0.2$	asym	0.186	0.220	0.124	0.152	0.086	0.106	0.044	0.072
		boot	0.159	0.196	0.108	0.137	0.077	0.094	0.039	0.056
$n = 800$	$\beta = -0.6$	asym	0.157	0.193	0.098	0.122	0.059	0.081	0.022	0.048
		boot	0.141	0.172	0.084	0.103	0.051	0.068	0.019	0.043
	$\beta = 0.2$	asym	0.135	0.187	0.078	0.115	0.054	0.073	0.023	0.049
		boot	0.124	0.175	0.066	0.101	0.049	0.067	0.020	0.042

values of the autoregression parameter are taken into consideration, i.e., $\beta = -0.6$ and $\beta = 0.2$ to represent stronger negative dependence and weaker positive dependence. The innovations are obtained as iid random variables from a standard normal $\mathsf{N}(0,1)$ or Student t_5 distribution. Simulation scenarios are produced as all possible combinations of the abovementioned settings. Parameter γ is set to 0.1. The number of bootstrap replications used is 2000.

To assess the theoretical results under H_0 numerically, Table 2 provides the empirical sizes (empirical probabilities of the type I error) of the tests for change in the autoregression parameter, where the significance level is α.

The proportion of rejecting the null hypothesis is getting closer to the theoretical significance level as the number of time series observations increases. Better performance of the tests under the null hypothesis is observed, when the innovations have lighter tails (represented by $\mathsf{N}(0,1)$ distribution). Note that, the test statistics $\mathcal{V}_n$ and $\mathcal{V}_n^*$ are based on the L_2 regression approach. There is no visible direct effect of the value of the autoregression parameter on the empirical rejection rates based on this particular simulation study. Generally, the empirical sizes are higher than they should be, i.e., the tests reject the null hypothesis more often than one would expect. We may see that comparing to the critical values obtained by simulations from the asymptotic distribution, the critical values obtained by bootstrapping are more accurate, because the empirical sizes coming from the bootstrap procedure are closer to the theoretical ones.

Table 3 Empirical power of the asymptotic (*asym*) and bootstrap (*boot*) tests for change in autoregression under H_1 using $\gamma = 0.1$, considering the significance level α and $\delta = 0.5$. Innovations (*innov*) are iid having Student t_5 and standard normal $\mathsf{N}(0, 1)$ distribution

			α	0.100		0.050		0.025		0.010	
			innov	$\mathsf{N}(0, 1)$	t_5	$\mathsf{N}(0, 1)$	t_5	$\mathsf{N}(0, 1)$	t_5	$\mathsf{N}(0, 1)$	t_5
$n = 200$	$\beta = -0.6$	$\tau = \frac{n}{2}$	asym	0.924	0.920	0.888	0.888	0.834	0.848	0.774	0.774
			boot	0.935	0.929	0.899	0.897	0.845	0.859	0.793	0.792
		$\tau = \frac{n}{3}$	asym	0.930	0.896	0.894	0.866	0.838	0.830	0.772	0.766
			boot	0.944	0.910	0.906	0.885	0.851	0.843	0.788	0.780
	$\beta = 0.2$	$\tau = \frac{n}{2}$	asym	0.788	0.828	0.718	0.766	0.640	0.694	0.548	0.602
			boot	0.801	0.839	0.729	0.779	0.653	0.708	0.551	0.615
		$\tau = \frac{n}{3}$	asym	0.774	0.784	0.676	0.698	0.596	0.606	0.478	0.508
			boot	0.787	0.798	0.688	0.711	0.608	0.619	0.492	0.516
$n = 400$	$\beta = -0.6$	$\tau = \frac{n}{2}$	asym	0.984	0.984	0.968	0.958	0.926	0.924	0.856	0.888
			boot	0.990	0.989	0.980	0.969	0.939	0.936	0.871	0.901
		$\tau = \frac{n}{3}$	asym	0.992	0.982	0.972	0.962	0.958	0.944	0.924	0.926
			boot	0.997	0.986	0.981	0.972	0.970	0.965	0.938	0.939
	$\beta = 0.2$	$\tau = \frac{n}{2}$	asym	0.948	0.938	0.906	0.904	0.864	0.852	0.792	0.798
			boot	0.961	0.952	0.920	0.919	0.877	0.865	0.804	0.807
		$\tau = \frac{n}{3}$	asym	0.898	0.892	0.826	0.828	0.752	0.758	0.634	0.642
			boot	0.909	0.903	0.840	0.837	0.769	0.771	0.645	0.665
$n = 800$	$\beta = -0.6$	$\tau = \frac{n}{2}$	asym	0.999	0.999	0.996	0.996	0.992	0.995	0.980	0.981
			boot	0.999	0.999	0.999	0.999	0.997	0.998	0.989	0.992
		$\tau = \frac{n}{3}$	asym	0.999	0.999	0.996	0.998	0.995	0.997	0.987	0.988
			boot	0.999	0.999	0.999	0.999	0.998	0.999	0.994	0.996
	$\beta = 0.2$	$\tau = \frac{n}{2}$	asym	0.996	0.989	0.988	0.978	0.972	0.963	0.938	0.931
			boot	0.999	0.994	0.996	0.989	0.985	0.972	0.951	0.950
		$\tau = \frac{n}{3}$	asym	0.981	0.980	0.960	0.959	0.926	0.929	0.860	0.866
			boot	0.992	0.991	0.981	0.979	0.944	0.946	0.878	0.881

The performance of both testing procedures under H_1 in terms of the empirical rejection rates is shown in Table 3, where the change point is set to $\tau = n/2$ or $\tau = n/3$. Parameter δ is chosen as $\delta = 0.5$.

We may conclude that the power of the tests increases as the number of observations increases, which was expected. The test power drops when switching from a change point located in the middle of the time series to a change point closer to the beginning or the end of the time series. Innovations with heavier tails (i.e., t_5) yield slightly smaller power than innovations with lighter tails. Negative dependence seems to give higher power of the tests based on this simulation study. Generally, the bootstrap outperforms with respect to power the classical asymptotics in all scenarios. So, thc bootstrap extension of the developed asymptotic procedure provides improvement from a numerical and computational point of view.

In contrast to the slightly lower power in case of relatively small sample size and moderate change in the autoregression parameter, one may try to consider a larger change in β from -0.8 to 0.8 in case of $n = 150$. Here, the simulated power, even in case of the asymptotic version of the test, reaches 0.994 (for $\alpha = 0.05$). Hence, for a large change in autoregression, the tests achieve high power. To improve the computational performance of the tests for detecting the change in autoregression, longer time series of observations are a general solution.

7 Application to Stock Exchange Index

As an illustrative example of the proposed technique for detecting of the change in autoregression, we concentrate on the Prague Stock Exchange index called *PX Index* (formerly *PX50*). It is a capitalization-weighted index of major stocks that trade on the Prague Stock Exchange.

The starting exchange day for the Index PX50 was April 5, 1994. We consider a time series consisting of daily PX50 values starting from November 16, 1994 up to September 27, 2001. Only business days were taken into account, providing 1850 observations. The starting date of the observation period was chosen later than the starting day of the exchange, since only weekly (not daily) values of the PX50 records were available at the beginning. Moreover, the market after opening the exchange was not as stable as later on. The last observation date was chosen in order to avoid effects of the attacks on September 11, 2001. The considered time series can be seen in Fig. 1. The PX50 data can also be downloaded from [6].

We denote the original data of the PX50 index as $\{X_t\}_t$. First, we transform the PX50 index by taking into account the differences of logarithms, i.e., $Y_t = \log(X_t/X_{t-1})$. This transformation can be interpreted as considering logarithms of daily returns of the PX50 index. Besides that, using this approach stationary time series before and even after a possible change point are obtained. The transformed index values are shown in Fig. 3.

Let us assume that $Y_1, \dots, Y_n$ follow autoregressive change point model (1). We are going to decide whether the change in the AR(1) parameter occurred or not based on the proposed asymptotic and bootstrap tests. The value of the test statistic $\mathcal{V}_n$ for $\gamma = 0.1$ is 7.321143, which is larger than the 95%-critical value 7.293031 simulated from the limit distribution under the null hypothesis as well as the 95%-critical

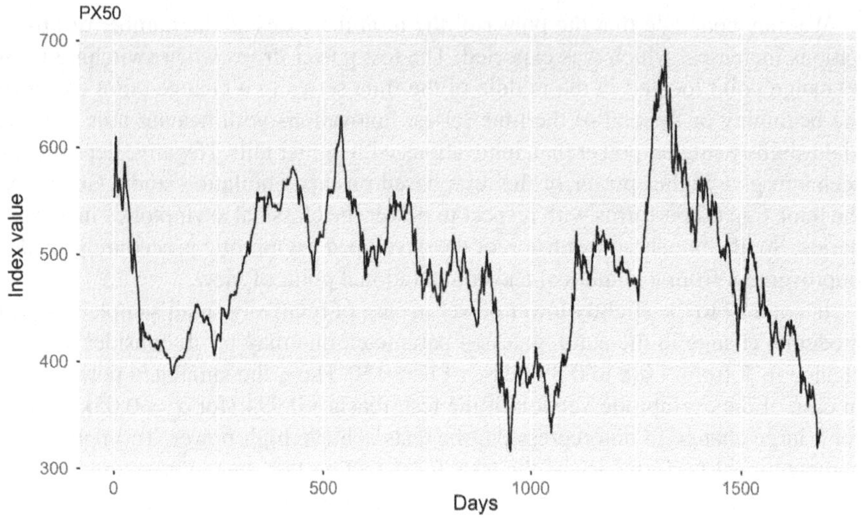

Fig. 1 Daily prague stock exchange index (PX50) values from November 16, 1994 to September 27, 2001

value 6.043474 provided by the bootstrap resampling. Therefore, we reject the null hypothesis of no change in the autoregressive parameter based on both approaches. The progress of the ratio of the test statistic

$$Q_k = \frac{\max_{2\leq i\leq k}\left|\sum_{j=1}^{i-1} Y_j(Y_{j+1} - \widehat{\beta}_{1k}Y_j)\right|}{\max_{k+1\leq i\leq n-1}\left|\sum_{j=i}^{n-1} Y_j(Y_{j+1} - \widehat{\beta}_{2k}Y_j)\right|}, \quad n\gamma \leq k \leq n - n\gamma$$

is depicted in Fig. 2 together with the asymptotic and bootstrap 95%-critical values.

We can *estimate* the unknown change point τ as in (6). This leads to $\widehat{\tau} = 949$, which corresponds to October 7, 1998. The log returns of PX50 together with the depicted estimated change point for the change in autoregression are displayed in Fig. 3.

The explanation of the detected change in autoregression is possibly connected to the Russian financial crisis (also called Ruble crisis) that hit Russia on August 17, 1998. It resulted in the Russian government and the Russian Central Bank devaluing the ruble and defaulting on its debt. In 1998 influenced by Russian financial crisis, the index reached its historical bottom on October 8 with 316 points, which is the first day after the detected change in autoregression of the PX50 log returns.

Moreover, the estimate of the AR(1) parameter before the detected change is $\widehat{\beta}_{1\widehat{\tau}} = 0.3347$ and the estimate of the AR(1) parameter after the detected change is $\widehat{\beta}_{2\widehat{\tau}} = 0.0538$. This coincides with Fig. 3, where the time series seems more 'correlated' before the detected change than after it.

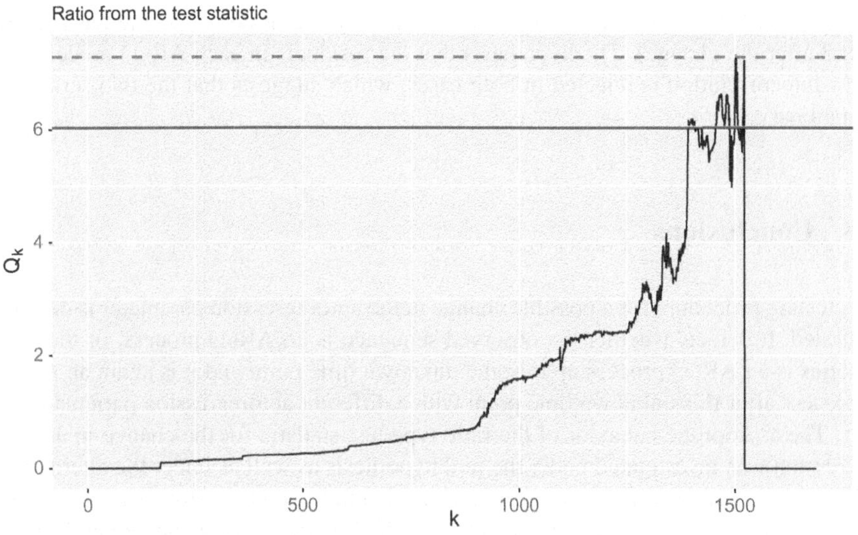

Fig. 2 The values of Q_k for the PX50 index data with $\gamma = 0.1$. The dashed purple horizontal line represents the asymptotic 95%-critical value and the solid blue horizontal line stands for the bootstrap 95%-critical value

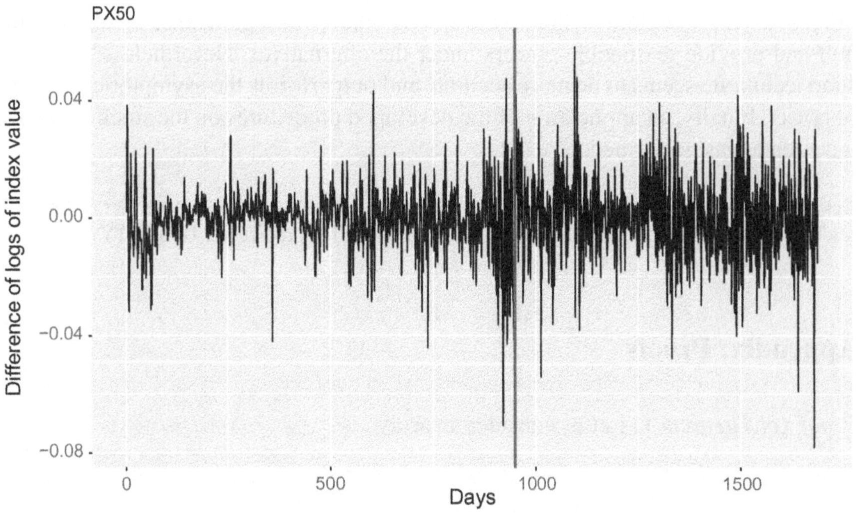

Fig. 3 Log returns of PX50 with the estimated change point $\widehat{\tau}$ depicted by the orange vertical line

Finally, we investigated the eligibility of the model. The ACF (autocorrelation function) and PACF (partial autocorrelation function) plots of the time series before and after the estimated change point is employed. Both ACF plots go to zero at an exponential rate, while both PACF plots become zero immediately after the first lag.

We applied the Ljung–Box test on the residuals of the fitted AR(1) models (before and after the change). The hypothesis that the residuals in each AR(1) model have no autocorrelation is rejected in both cases, which suggests that the two series are stationary.

8 Conclusions

A testing procedure for a possible change in the autoregression parameter is demonstrated. It detects whether the observed sequence is an AR(1) process, or the time series is an AR(1) process up to some unknown time point and it is again an AR(1) process after this unknown time point with a different autoregression parameter.

The asymptotic behavior of the ratio type test statistic for the change in autoregression was investigated under the null hypothesis as well as under the alternative. The theoretical limiting distribution under the null hypothesis provided critical values for the test, which were obtained by simulation. The main advantage of the ratio type statistics in hypotheses testing is that they provide an alternative to the non-ratio type statistics mainly in situations, in which variance estimation is not straightforward. The regression bootstrap method is investigated as an add-on to the asymptotic procedure. The simulations reveal that both presented approaches—based on the traditional asymptotics and on the bootstrapping—keep the significance level under the null and provide reasonable powers under the alternatives. Nevertheless, the bootstrap technique seems to be more accurate and outperforms the asymptotics in terms of power. Finally, an application of the developed procedures on the stock exchange index data was performed.

Acknowledgements Institutional support to Barbora Peštová was provided by RVO:67985807. Michal Pešta was supported by the Czech Science Foundation project No. 18-01781Y.

Appendix: Proofs

Proof (*of Theorem* 1) Let us consider an array

$$U_{n,i} = \frac{\sqrt{1-\beta^2}}{\sigma^2\sqrt{n-1}} Y_{i-1}\varepsilon_i, \quad i = 2, \ldots, n$$

and a filtration $\mathcal{F}_{n,i} = \sigma\{\varepsilon_j,\ j \le i\}$, $i = 2, \ldots, n$ and $n \in \mathbb{N}$. Then, $\{U_{n,i}, \mathcal{F}_{n,i}\}$ is a martingale difference array such that

$$\mathsf{E}U_{n,i}^2 = \frac{1-\beta^2}{\sigma^4(n-1)}\mathsf{E}Y_{i-1}^2\varepsilon_i^2 = \frac{1}{n-1}.$$

Moreover,

$$\sum_{i=2}^{n} U_{n,i}^2 - \sum_{i=2}^{n} \mathsf{E} U_{n,i}^2 = \frac{1-\beta^2}{\sigma^4(n-1)} \sum_{i=2}^{n} (Y_{i-1}^2 \varepsilon_i^2 - \mathsf{E} Y_{i-1}^2 \varepsilon_i^2).$$

Furthermore,

$$\frac{1}{n-1} \sum_{i=2}^{n} (Y_{i-1}^2 \varepsilon_i^2 - \mathsf{E} Y_{i-1}^2 \varepsilon_i^2)$$
$$= \frac{1}{n-1} \sum_{i=2}^{n} [Y_{i-1}^2 (\varepsilon_i^2 - \sigma^2)] + \frac{1}{n-1} \sum_{i=2}^{n} (Y_{i-1}^2 - \mathsf{E} Y_{i-1}^2) \sigma^2.$$

Since $\{Y_{i-1}^2(\varepsilon_i^2 - \sigma^2)\}$ is a martingale difference array again with respect to $\mathcal{F}_{n,i}$, we have under Assumption 3 from the Chebyshev's inequality that

$$\frac{1}{n-1} \sum_{i=2}^{n} [Y_{i-1}^2 (\varepsilon_i^2 - \sigma^2)] \xrightarrow[n\to\infty]{\mathsf{P}} 0.$$

Similarly, as a consequence of Lemma 4.2 in [1],

$$\frac{1}{n-1} \sum_{i=2}^{n} (Y_{i-1}^2 - \mathsf{E} Y_{i-1}^2) \xrightarrow[n\to\infty]{\mathsf{P}} 0.$$

Thus,

$$\sum_{i=2}^{n} U_{n,i}^2 \xrightarrow[n\to\infty]{\mathsf{P}} 1. \tag{11}$$

Next, for any $\epsilon > 0$,

$$\mathsf{P}\left(\max_{2\le i\le n} U_{n,i}^2 > \epsilon \right) \le \sum_{i=2}^{n} \mathsf{P}\left(\frac{1-\beta^2}{\sigma^4(n-1)} Y_{i-1}^2 \varepsilon_i^2 > \epsilon \right)$$
$$\le \frac{(1-\beta^2)^2}{\epsilon^2 \sigma^8 (n-1)^2} \sum_{i=2}^{n} \mathsf{E} Y_{i-1}^4 \mathsf{E} \varepsilon_i^4 \xrightarrow[n\to\infty]{} 0. \tag{12}$$

Additionally,

$$\lim_{n\to\infty} \sum_{i=2}^{[nt]} \mathsf{E} U_{n,i}^2 = \lim_{n\to\infty} \frac{[nt]-1}{n-1} = t \tag{13}$$

for all $t \in [0, 1]$.

According to Theorem 27.14 from [7] for the martingale difference array $\{U_{n,i}, \mathcal{F}_{n,i}\}$, where the assumptions of this theorem are satisfied due to (11), (12), and (13), we get

$$\sum_{i=2}^{[nt]} U_{n,i} \xrightarrow[n\to\infty]{\mathscr{D}[0,1]} \mathcal{W}(t).$$

Therefore,

$$\frac{1}{\sqrt{n-1}}\left(\sum_{i=2}^{[nt]} Y_{i-1}\varepsilon_i, \sum_{i=[nt]+2}^{n} Y_{i-1}\varepsilon_i\right) \xrightarrow[n\to\infty]{\mathscr{D}^2[0,1]} \frac{\sigma^2}{\sqrt{1-\beta^2}}\left(\mathcal{W}(t), \widetilde{\mathcal{W}}(t)\right), \quad (14)$$

where $\widetilde{\mathcal{W}}(t) = \mathcal{W}(1) - \mathcal{W}(t)$.

Let us define

$$\mathbf{Y}_{j,l} = (Y_j, \ldots, Y_l)^\top \quad \text{and} \quad \boldsymbol{\varepsilon}_{j,l} = (\varepsilon_j, \ldots, \varepsilon_l)^\top.$$

Hence, for the expression in the numerator of $\mathcal{V}_n$, it holds

$$\begin{aligned}
&\sum_{j=1}^{i-1} Y_j(Y_{j+1} - \widehat{\beta}_{1k} Y_j) = \mathbf{Y}_{1,i-1}^\top \left(\mathbf{Y}_{2,i} - \mathbf{Y}_{1,i-1}\widehat{\beta}_{1k}\right) \\
&= \mathbf{Y}_{1,i-1}^\top \left(\mathbf{Y}_{1,i-1}\beta + \boldsymbol{\varepsilon}_{2,i} - \mathbf{Y}_{1,i-1}\beta - \mathbf{Y}_{1,i-1}\left(\mathbf{Y}_{1,k-1}^\top \mathbf{Y}_{1,k-1}\right)^{-1} \mathbf{Y}_{1,k-1}^\top \boldsymbol{\varepsilon}_{2,k}\right) \\
&= \mathbf{Y}_{1,i-1}^\top \boldsymbol{\varepsilon}_{2,i} - \mathbf{Y}_{1,i-1}^\top \mathbf{Y}_{1,i-1}\left(\mathbf{Y}_{1,k-1}^\top \mathbf{Y}_{1,k-1}\right)^{-1} \mathbf{Y}_{1,k-1}^\top \boldsymbol{\varepsilon}_{2,k}. \quad (15)
\end{aligned}$$

Similarly, for the expression in the denominator of $\mathcal{V}_n$,

$$\begin{aligned}
&\sum_{j=i}^{n-1} Y_j(Y_{j+1} - \widehat{\beta}_{2k} Y_j) \\
&= \mathbf{Y}_{i,n-1}^\top \boldsymbol{\varepsilon}_{i+1,n} - \mathbf{Y}_{i,n-1}^\top \mathbf{Y}_{i,n-1}\left(\mathbf{Y}_{k+1,n-1}^\top \mathbf{Y}_{k+1,n-1}\right)^{-1} \mathbf{Y}_{k+1,n-1}^\top \boldsymbol{\varepsilon}_{k+2,n}. \quad (16)
\end{aligned}$$

Lemma 4.2 in [1] gives

$$\sup_{\gamma\leq t<1} \frac{1}{[nt]}\left|\sum_{s=1}^{[nt]}(Y_s^2 - \mathsf{E}Y_s^2)\right| = o_{\mathsf{P}}(1) \quad (17)$$

and

$$\sup_{0<t\leq 1-\gamma} \frac{1}{[n(1-t)]}\left|\sum_{s=[nt]+1}^{n-1}(Y_s^2 - \mathsf{E}Y_s^2)\right| = o_{\mathsf{P}}(1), \quad (18)$$

as $n \to \infty$, where $[nt]$ and $[n(1-t)]$ mean truncated number to zero decimal digits. Finally, (14) together with (15), (16), (17), and (18) implies

$$\frac{1}{\sqrt{n-1}}\begin{pmatrix} \sup\limits_{0\le u\le t}\left|\sum_{j=1}^{[nu]-1} Y_j(Y_{j+1}-\widehat{\beta}_{1[nt]}Y_j)\right| \\ \sup\limits_{t\le u\le 1}\left|\sum_{j=[nu]+1}^{n-1} Y_j(Y_{j+1}-\widehat{\beta}_{2[nt]}Y_j)\right| \end{pmatrix}$$

$$\xrightarrow[n\to\infty]{\mathscr{D}^2[\gamma,1-\gamma]} \frac{\sigma^2}{\sqrt{1-\beta^2}}\begin{pmatrix} \sup\limits_{0\le u\le t}\left|\mathcal{W}(u)-u/t\mathcal{W}(t)\right| \\ \sup\limits_{t\le u\le 1}\left|\widetilde{\mathcal{W}}(u)-(1-u)/(1-t)\widetilde{\mathcal{W}}(t)\right| \end{pmatrix}.$$

Then, the assertion of the theorem directly follows. □

Proof (*of Theorem* 2) Let us take $k=\tau+2$, $k=[\xi n]$ for some $\zeta<\xi<1-\gamma$ and $i=\tau+1$. Then,

$$\sum_{j=1}^{\tau} Y_j(Y_{j+1}-\widehat{\beta}_{1(\tau+2)}Y_j)$$
$$=\mathbf{Y}_{1,\tau}^{\top}\boldsymbol{\varepsilon}_{2,\tau+1}-\mathbf{Y}_{1,\tau}^{\top}\mathbf{Y}_{1,\tau}\left(\mathbf{Y}_{1,\tau+1}^{\top}\mathbf{Y}_{1,\tau+1}\right)^{-1}\mathbf{Y}_{1,\tau+1}^{\top}\boldsymbol{\varepsilon}_{2,\tau+2}-\mathbf{Y}_{1,\tau}^{\top}\mathbf{Y}_{1,\tau}\delta.$$

According to the proof of Theorem 1, as $n\to\infty$,

$$\frac{1}{\sqrt{n-1}}\left(\mathbf{Y}_{1,\tau}^{\top}\boldsymbol{\varepsilon}_{2,\tau+1}-\mathbf{Y}_{1,\tau}^{\top}\mathbf{Y}_{1,\tau}\left(\mathbf{Y}_{1,\tau+1}^{\top}\mathbf{Y}_{1,\tau+1}\right)^{-1}\mathbf{Y}_{1,\tau+1}^{\top}\boldsymbol{\varepsilon}_{2,\tau+2}\right)=\mathcal{O}_{\mathsf{P}}(1).$$

Lemma 4.2 from [1] gives

$$\frac{1}{\sqrt{n-1}}\left|\mathbf{Y}_{1,\tau}^{\top}\mathbf{Y}_{1,\tau}\delta\right|\xrightarrow[n\to\infty]{\mathsf{P}}\infty.$$

Now,

$$\frac{1}{\sqrt{n-1}}\max_{2\le i\le k}\left|\sum_{j=1}^{i-1}Y_j(Y_{j+1}-\widehat{\beta}_{1k}Y_j)\right|\xrightarrow[n\to\infty]{\mathsf{P}}\infty.$$

For $\tau<k=[\xi n]$, the denominator in (4) divided by $\sqrt{n-1}$ has the same distribution as under the null hypothesis and it is, therefore, bounded in probability. It follows that the maximum of the ratio has to tend in probability to infinity as well, while $n\to\infty$. □

References

1. Hušková, M., Prášková, Z., Steinebach, J.: On the detection of changes in autoregressive time series I. Asymptotics. J. Stat. Plan. Infer. **137**(4), 1243–1259 (2007)
2. Hušková, M., Kirch, C., Prášková, Z., Steinebach, J.: On the detection of changes in autoregressive time series, II. Resampling procedures. J. Stat. Plan. Infer. **138**(6), 1697–1721 (2008)
3. Horváth, L., Horváth, Z., Hušková, M.: Ratio tests for change point detection. In: Balakrishnan, N., Peña, E.A., Silvapulle, M.J. (eds.) Beyond Parametrics in Interdisciplinary Research: Festschrift in Honor of Professor Pranab K. Sen, vol. 1, pp. 293–304. IMS Collections, Beachwood, Ohio (2009)
4. Peštová, B., Pešta, M.: Testing structural changes in panel data with small fixed panel size and bootstrap. Metrika **78**(6), 665–689 (2015)
5. Peštová, B., Pešta, M.: Change point estimation in panel data without boundary issue. Risks **5**(1), 7 (2017)
6. Prague Stock Exchange: PX Index 2015. https://www.pse.cz/dokument.aspx?k=Burzovni-Indexy. Updated April 30, 2015; Accessed April 30, 2015
7. Davidson, J.: Stochastic Limit Theory: An Introduction for Econometricians. Oxford University Press, New York (1994)

Distance Between VARMA Models and Its Application to Spatial Differences Analysis in the Relationship GDP—Unemployment Growth Rate in Europe

Francesca Di Iorio and Umberto Triacca

Abstract In this paper, a novel distance measure for evaluating the closeness of two vector autoregressive moving average models is presented and its main properties are discussed. The proposed distance is used to investigate the presence of spatial differences in the dynamic link between unemployment rate variation and GDP growth in some European Union countries.

Keywords AR metric · Distance · Unemployment · GDP · VARMA models

1 Introduction

Vector autoreressive moving average (VARMA) models are a class of models that are designed to capture joint movements and dynamic patterns in an array of multiple variables. These models have been applied in various research fields. Their success is mainly based on the fact that they are considered to be data-driven, i.e., the underlying structure in the estimated model is determined by the data. However, there is no canonical way to measure the dissimilarity between two different VARMAs. The need for such a distance measure arises in both clustering and classification of multivariate time series. In this paper, we propose a distance measure for evaluating the closeness of two VARMA models and we use such notion of distance to investigate the presence of spatial differences in the dynamic link between unemployment rate variation and economic growth in some European Union economies.

F. Di Iorio (✉)
Università di Napoli Federico II, Naples, Italy
e-mail: fdiiorio@unina.it

U. Triacca
Università dell'Aquila, L'Aquila, Italy
e-mail: umberto.triacca@ec.univaq.it

I. Rojas et al. (eds.), *Time Series Analysis and Forecasting*,
Contributions to Statistics, https://doi.org/10.1007/978-3-319-96944-2_14

The rest of paper is organized as follows. Section 2 introduces a distance measure between pairs of VARMA models. In Sect. 3, we present the application. Finally, we conclude in Sect. 4.

2 A Distance Measure Between VARMA Models

Let us remind that a k-dimensional process $\left\{\mathbf{y}_t = (y_{1t}, \ldots y_{kt})'; \; t \in \mathbb{Z}\right\}$ is a VARMA(p, q) process if it can be represented as

$$\boldsymbol{\Phi}(L)\mathbf{y}_t = \boldsymbol{\Theta}(L)\mathbf{u}_t \tag{1}$$

where $\left\{\mathbf{u}_t = (u_{1t}, \ldots u_{kt})'; \; t \in \mathbb{Z}\right\}$ is a k-variate white-noise process with zero mean vector and nonsingular covariance matrix $\boldsymbol{\Sigma}_{\mathbf{u}}$. The $(k \times k)$ matrices $\boldsymbol{\Phi}(L)$ and $\boldsymbol{\Theta}(L)$ have finite polynomial elements in lag operator L and are assumed to be of full rank. They can be expressed as $\boldsymbol{\Phi}(L) = \mathbf{I} - \boldsymbol{\Phi}_1 L - \cdots - \boldsymbol{\Phi}_p L^p$ and $\boldsymbol{\Theta}(L) = \mathbf{I} - \boldsymbol{\Theta}_1 L - \cdots - \boldsymbol{\Theta}_q L^q$ where $\mathbf{I}$ is the $(k \times k)$ identity matrix and $\{\boldsymbol{\Phi}_i\}$ and $\{\boldsymbol{\Theta}_i\}$ are $(k \times k)$ matrices of parameters. It is assumed that $\boldsymbol{\Phi}(z)$ and $\boldsymbol{\Theta}(z)$ have no common factors. Process (1) is stationary if the roots of the determinantal equation det $[\boldsymbol{\Phi}(z)] = 0$ are outside the unit circle, and invertible if the roots of determinantal equation det $[\boldsymbol{\Theta}(z)] = 0$ are outside the unit circle.

In this paper, we assume that

$$\det[\boldsymbol{\Phi}(z)] \neq 0, \quad |z| < 1 \;\; \text{for } z \in \mathbb{C} \tag{2}$$

It is important to note that the condition (2) allows for nonstationarity. However, it excludes explosive processes from our consideration.

Univariate Models Implied by a VARMA Model

In this subsection, following [9], we show that a VARMA process implies a particular specification of its individual elements in terms of univariate ARMA processes. We first observe that

$$\text{adj}[\boldsymbol{\Phi}(L)]\,\boldsymbol{\Phi}(L) = \det[\boldsymbol{\Phi}(L)]\,\mathbf{I}$$

where adj $[\boldsymbol{\Phi}(L)]$ is the adjoint of the matrix $\boldsymbol{\Phi}(L)$. Then, premultiplying both sides of (1) by adj $[\boldsymbol{\Phi}(L)]$, we obtain the autoregressive final form,

$$\det[\boldsymbol{\Phi}(L)]\,\mathbf{y}_t = \text{adj}[\boldsymbol{\Phi}(L)]\,\boldsymbol{\Theta}(L)\mathbf{u}_t.$$

Consequently, the marginal model for the ith element of $\mathbf{y}_t$ is given by

$$\det [\boldsymbol{\Phi}(L)]\, y_{it} = \mathbf{B}_i(L)\mathbf{u}_t \tag{3}$$

where $\mathbf{B}_i(L)$ denotes the ith row of the matrix $\mathbf{B}(L) = \text{adj}\, [\boldsymbol{\Phi}(L)]\, \boldsymbol{\Theta}(L)$. As the right-hand side of (3) is the sum of k finite moving averages, it can also be represented as a finite moving average $\theta_i(L)\epsilon_{it}$, where ϵ_{it} is a white noise process, such that

$$\theta_i(L)\epsilon_{it} = \mathbf{B}_i(L)\mathbf{u}_t \tag{4}$$

The coefficients of the polynomial $\theta_i(L)$ are found by equating the autocovariances in the two representations. In particular we obtain a nonlinear system of equations where the unknowns are the $\theta_i(L)$ coefficients and the variance of the ϵ_{it}. The invertibility condition ensures a unique solution. Considering (3) and (4), the univariate ARMA models implied by (1) are given by

$$\det [\boldsymbol{\Phi}(L)]\, y_{it} = \theta_i(L)\epsilon_{it}.$$

Thus, we have $y_{it} \sim \text{ARMA}(p^*, q^*)$ $i = 1, ..., k$, where it is well known that $p^* \leq kp$ and $q^* \leq (k-1)p + q$. We note that the innovations in the different ARMA models are correlated and, in absence of any cancelation, all the univariate models have identical autoregressive parts.

The Proposed Distance

In the previous subsection, we have shown that a set $A_\mathbf{y}$ of k invertible univariate ARMA processes corresponds to every k-variate VARMA process, $\mathbf{y}$. We note that any univariate process $y_i \in A_\mathbf{y}$ has an (possibly infinite order) AR representation,

$$y_{it} = \sum_{l=1}^{\infty} \pi_{il} y_{it-l} + \epsilon_{it}.$$

Given two invertible ARMA processes, x, y, following [4], we consider the quantity

$$d(x, y) = \left[\sum_{l=1}^{\infty} (\pi_{xl} - \pi_{yl})^2 \right]^{\frac{1}{2}}.$$

as a measure of distance between the two invertible ARMA processes. Thus, in the class of the k-variate VARMA processes, V_k, could seem natural to consider the following definition of dissimilarity between two VARMA processes.

Definition 1 Let $\mathbf{x}$ and $\mathbf{y}$ be two VARMA models in V_k; then their distance $\mathcal{D}(\mathbf{x}, \mathbf{y})$ is given by the sum of the distances between the implied ARMA models component by component as follows:

$$\mathcal{D}(\mathbf{x}, \mathbf{y}) = \sum_{i=1}^{k} d(x_i, y_i), \quad \mathbf{x}, \mathbf{y} \in V_k \tag{5}$$

where x_i and y_i ($i = 1, ..., k$) are the univariate invertible ARMA processes implied by k-variate VAR processes $\mathbf{x}$ and $\mathbf{y}$, respectively. ■

Example 1 For illustrative purposes, consider two VARMA(1, 1) processes, $\mathbf{y}$ and $\mathbf{x}$, respectively, defined by the following equations:

$$\begin{bmatrix} 1-0.2L & 0.3L \\ -0.5L & 1-0.7L \end{bmatrix} \begin{bmatrix} y_{1t} \\ y_{2t} \end{bmatrix} = \begin{bmatrix} 1-0.2L & 0.8L \\ 0.3L & 1-0.5L \end{bmatrix} \begin{bmatrix} a_{1t} \\ a_{2t} \end{bmatrix} \tag{6}$$

with covariance error matrix $\Sigma_a = I$ and

$$\begin{bmatrix} 1-0.6L & -0.5L \\ 0.5L & 1+0.3L \end{bmatrix} \begin{bmatrix} x_{1t} \\ x_{2t} \end{bmatrix} = \begin{bmatrix} 1+0.2L & -0.7L \\ -0.8L & 1+0.3L \end{bmatrix} \begin{bmatrix} b_{1t} \\ b_{2t} \end{bmatrix} \tag{7}$$

with covariance error matrix $\Sigma_b = I$. The univariate models implied from (6) are given by

$$\left(1 - 0.9L + 0.29L^2\right) y_{1t} = \epsilon_{1t} - 0.88\epsilon_{1t-1} + 0.04\epsilon_{1t-2} \ \text{ with } \ \sigma^2_{\epsilon_1} = 1.25,$$

$$\left(1 - 0.9L + 0.29L^2\right) y_{2t} = \epsilon_{2t} - 0.51\epsilon_{2t-1} + 0.28\epsilon_{2t-2} \ \text{ with } \ \sigma^2_{\epsilon_2} = 1.79.$$

The univariate models implied from (7) are

$$\left(1 - 0.3L + 0.07L^2\right) x_{1t} = e_{1t} + 0.47e_{1t-1} - 0.32e_{1t-2} \ \text{ with } \ \sigma^2_{e_1} = 1.06,$$

$$\left(1 - 0.3L + 0.07L^2\right) x_{2t} = e_{2t} - 0.29e_{2t-1} - 0.06e_{2t-2} \ \text{ with } \ \sigma^2_{e_2} = 2.71.$$

The distance between these VARMA processes is equal to $\mathcal{D}(\mathbf{y}, \mathbf{x}) = 2.03$. ■

It is important to stress that contrary to the distance $d(x_i, y_i)$ for univariate ARMA model, the distance $\mathcal{D}(\mathbf{x}, \mathbf{y})$ depends on variance matrices of the considered processes.

Example 1-continued. An illustration of this aspect can be see considered for the bivariate process (7) the following covariance matrix:

$$\Sigma_b = E\left(\begin{bmatrix} b_{1t} \\ b_{2t} \end{bmatrix} \begin{bmatrix} b_{1t} \ b_{2t} \end{bmatrix}\right) = \begin{bmatrix} 1 & 1 \\ 1 & 1 \end{bmatrix}$$

In this case, the implied univariate models become

$$\left(1 - 0.3L + 0.07L^2\right) x_{1t} = e_{1t} + 0.30e_{1t-1} - 0.40e_{1t-2}$$

$$\left(1 - 0.3L + 0.07L^2\right) x_{2t} = e_{2t} - 1.41e_{2t-1} + 0.46e_{2t-2}$$

and then the new distance between processes (6) and (7) is $\mathcal{D}(\mathbf{y}, \mathbf{x}) = 4.55$. ■
The next proposition provides some main properties of the distance $\mathcal{D}$.

Proposition *Let V_k be the class of the k-variate VARMA processes. The function $\mathcal{D} : V_k \times V_k \to \mathbb{R}$ defined as*

$$\mathcal{D}(\mathbf{v}_1, \mathbf{v}_2) = \sum_{i=1}^{k} d(v_{1i}, v_{2i}) \quad v_1, v_2 \in V_k,$$

satisfies the following properties:

i. Nonnegativity: $\mathcal{D}(\mathbf{v}_1, \mathbf{v}_2) \geq 0 \;\; \forall\, \mathbf{v}_1, \mathbf{v}_2 \in V_k$;
ii. Symmetry: $\mathcal{D}(\mathbf{v}_1, \mathbf{v}_2) = \mathcal{D}(\mathbf{v}_2, \mathbf{v}_1) \;\; \forall\, \mathbf{v}_1, \mathbf{v}_2 \in V_k$;
iii. Triangularity: $\mathcal{D}(\mathbf{v}_1, \mathbf{v}_2) \leq \mathcal{D}(\mathbf{v}_1, \mathbf{v}_3) + \mathcal{D}(\mathbf{v}_3, \mathbf{v}_2) \;\forall\, \mathbf{v}_1, \mathbf{v}_2, \mathbf{v}_3 \in V_k$;
iv. $v_1 = v_2$ *implies* $\mathcal{D}(\mathbf{v}_1, \mathbf{v}_2) = 0 \;\; \forall\, \mathbf{v}_1, \mathbf{v}_2 \in V_k$.

Proof Evidently, $\mathcal{D}(\mathbf{v}_1, \mathbf{v}_2)$ is a nonnegative function. Further, since $d(v_{1i}, v_{2i}) = d(v_{2i}, v_{1i})$ for $i = 1, ..., k$, we have that

$$\mathcal{D}(\mathbf{v}_1, \mathbf{v}_2) = \sum_{i=1}^{k} d(v_{1i}, v_{2i}) = \sum_{i=1}^{k} d(v_{2i}, v_{1i}) = \mathcal{D}(v_2, v_1)\; \mathbf{v}_1, \mathbf{v}_2 \in V_k.$$

and hence $\mathcal{D}(\mathbf{v}_1, \mathbf{v}_2)$ is a symmetric function. In order to show the triangle inequality, we first note that

$$d(v_{1i}, v_{2i}) \leq d(v_{1i}, v_{3i}) + d(v_{3i}, v_{2i}),$$

where v_{3i} is the ith univariate invertible ARMA component implied by the k-variate VARMA processes $\mathbf{v}_3$. Thus

$$\begin{aligned}\mathcal{D}(\mathbf{v}_1, \mathbf{v}_2) &\leq \sum_{i=1}^{k} [d(v_{1i}, v_{3i}) + d(v_{3i}, v_{2i})] = \sum_{i=1}^{k} d(v_{1i}, v_{3i}) + \sum_{i=1}^{k} d(v_{3i}, v_{2i}) \\ &= \mathcal{D}(\mathbf{v}_1, \mathbf{v}_3) + \mathcal{D}(\mathbf{v}_3, \mathbf{v}_2).\end{aligned}$$

Finally, it is clear that if $\mathbf{v}_1 = \mathbf{v}_2$, we have that $v_{1i} = v_{2i}$ for $i = 1, ..., k$ and hence $d(v_{1i}, v_{2i}) = 0$ for $i = 1, ..., k$. It follows that $\mathcal{D}(\mathbf{v}_1, \mathbf{v}_2) = 0$. ■

Using the proposed distance (5), we can introduce the notion of *norm* of a VARMA proccss.

Definition 2 Let $\mathbf{y}$ be a VARMA model in V_k, defined by the equation

$$\boldsymbol{\Phi}(L)\mathbf{y}_t = \boldsymbol{\Theta}(L)\mathbf{u}_t.$$

The norm of $\mathbf{y}$ is given by $\|\mathbf{y}\| = \mathcal{D}(\mathbf{y}, \mathbf{u})$, where $\mathbf{u} = \{\mathbf{u}_t\}$. ■

Example 2 Consider, for instance, the VARMA process $\mathbf{x}$ given by equation (7). The norm of this process is $\|\mathbf{x}\| = 1.63$. ■

The norm of a VARMA process is defined as the distance between the process and its innovation. We observe that the norm of the k-variate process $\mathbf{v} \in V_k$, depends on the sequences $\{\pi_{1i}\}$, ..., $\{\pi_{ki}\}$. Since these sequences contain all information about the stochastic dependence structure of the process $\mathbf{v}$, we can interpret the norm of a VARMA process like a measure of the stochastic dependence structure of the process. To illustrate this point, we consider the following VARMA process $\mathbf{y}$ given by equation (8):

Example 3

$$\begin{bmatrix} 1-0.1L & 0 \\ 0 & 1+0.1L \end{bmatrix}\begin{bmatrix} y_{1t} \\ y_{2t} \end{bmatrix} = \begin{bmatrix} 1+0.3L & 0.2L \\ 0 & 1+0.2L \end{bmatrix}\begin{bmatrix} a_{1t} \\ a_{2t} \end{bmatrix} \tag{8}$$

with covariance error matrix $\Sigma_a = I$. In process (8), there is less "structure" than in the process (7), in fact its the norm $\|\mathbf{y}\| = 0.40$, is less than to the norm of the process (7) $\|\mathbf{x}\| = 1.63$. ■

We conclude this subsection by presenting an example useful to motivate our distance. In particular, we consider the following VARMA(1, 1) processes given in [7] (p. 323):

Example 4

$$\begin{bmatrix} x_{1t} \\ x_{2t} \end{bmatrix} - \begin{bmatrix} -0.8 & -2 \\ 0 & 0 \end{bmatrix}\begin{bmatrix} x_{1t-1} \\ x_{2t-1} \end{bmatrix} = \begin{bmatrix} a_{1t} \\ a_{2t} \end{bmatrix} - \begin{bmatrix} -0.5 & 0 \\ 0 & 0 \end{bmatrix}\begin{bmatrix} a_{1t-1} \\ a_{2t-1} \end{bmatrix}$$

$$\begin{bmatrix} y_{1t} \\ y_{2t} \end{bmatrix} - \begin{bmatrix} -0.8 & -2+\eta \\ 0 & \omega \end{bmatrix}\begin{bmatrix} y_{1t-1} \\ y_{2t-1} \end{bmatrix} = \begin{bmatrix} a_{1t} \\ a_{2t} \end{bmatrix} - \begin{bmatrix} -0.5 & \eta \\ 0 & \omega \end{bmatrix}\begin{bmatrix} a_{1t-1} \\ a_{2t-1} \end{bmatrix}$$

A simple way to compare them could be through a parametric approach, considering the Euclidean distance between the vectors $[-0.8, -2, 0, 0, -0.5, 0, 0, 0]'$ and $[-0.8, -2+\eta, 0, \omega, -0.5, \eta, 0, \omega]'$, given by

$$\delta = \left[2\eta^2 + 2\omega^2\right]^{1/2}$$

We note that these processes are equal for any nonzero ω and η, but $\delta > 0$. Thus, we reach a wrong conclusion. Instead, using distance (5) we correctly conclude that the distance between these two processes is 0. ■

Two remarks are in order (i) the distance $d(x, y)$ for the univariate ARMA model is a special case of $\mathcal{D}(\mathbf{x}, \mathbf{y})$ for $k = 1$; (ii) given the same set of initial values, if $\mathcal{D}(\mathbf{x}, \mathbf{y}) = 0$, then the corresponding VARMA models $\mathbf{x}$ and $\mathbf{y}$ produce the same forecasts. However, unlike the distance $d(x, y)$, the vice-versa is not true, since the distance (5) depends on the covariance matrices of the error terms. As a consequence, two VARMAs with the same parameters may have distance different from zero if they have different covariance error matrices.

Distance Estimation Procedure

The estimation of the proposed distance is based on usual VARMA model estimates carried on two datasets. It must be underlined that the theoretical results illustrated in the previous sections remain valid even if the estimated VARMA models in the two datasets do not have the same lag order. This is because it is a matter of fact that a smaller order VARMA process can be considered as a restriction of larger one.

In order to obtain an estimate of $\mathcal{D}(\mathbf{x}, \mathbf{y})$ we use the following procedure:

Procedure

1. Estimate on two observed data $\mathbf{x}$ and $\mathbf{y}$ the suitable VARMA(p, q) models, obtaining the estimated parameter vectors $\vartheta_x = (\phi_1, \ldots, \phi_{p_x}; \theta_1, \ldots, \theta_{q_x})$ and $\vartheta_y = (\phi_1, \ldots, \phi_{p_y}; \theta_1, \ldots, \theta_{q_y})$, where possible $p_x \neq p_y$ and $q_x \neq q_y$;
2. Using the estimated parameters ϑ_x and ϑ_y from step 1 obtain the implied ARMA models for the univariate processes x_i and y_i $(i = 1, ..., k)$;
3. Evaluate the AR(∞) representation truncated a some suitable lag $\tilde{p}$ of the ARMA models in step 2 obtained the estimated autoregressive coefficients $\pi_{x_i l}$ and $\pi_{y_i l}$;
4. Using the coefficients $\pi_{x_i l}$ and $\pi_{y_i l}$ of the AR(∞) representation from step 3 evaluate the estimate distance $\hat{d}(x_i, y_i)$ $(i = 1, ..., k)$;
5. Estimate the VARMA distance $\mathcal{D}(\mathbf{x}, \mathbf{y})$ using $\hat{\mathcal{D}}(\mathbf{x}, \mathbf{y}) = \sum_{i=1}^{k} \hat{d}(x_i, y_i)$.

3 Spatial Variability of the Relationship Between Unemployment and GDP

The linkage between the rate of change in GDP and change in unemployment (the Okuns Law) is one of the most studied issue of empirical macroeconomics (see [1, 5, 8], among others).

The aim of this section is to investigate the presence of spatial differences in the dynamic linkage between unemployment (U) and Gross Domestic Product (GDP) in thirteen European Union economies: Belgium, Denmark, Germany, Ireland, Greece, Spain, France, Italy, Netherlands, Austria, Portugal, Finland, and UK. The used quarterly data, from Eurostat database, are the Gross Domestic Product at market prices,

Table 1 Missing data

	Missing data
Be	1998: Q1, Q3, Q4
Dk	1998: Q1, Q3, Q4
Ge	1998, 1999, 2000, 2001, 2002, 2003, 2004: Q1, Q3, Q4
Ei	1998: Q1, Q3, Q4; 1999: Q1
Fr	1998, 1999, 2000, 2001, 2002: Q1, Q3, Q4
Nl	1998, 1999: Q1, Q3, Q4
Au	1998: Q1, Q3, Q4
Uk	1998: Q1, Q3, Q4

chain-linked volumes index with 2010 = 100 seasonally and calendar adjusted (the *namq_10_gdp* Eurostat dataset) and the Total Unemployment rates (the *lfsq_urgan* Eurostat dataset). The sample period is first quarter 1998—fourth quarter of 2016 (1998Q1–2016Q4). There are some missing values for the Unemployment rates at the beginning of the period for some countries, as reported in Table 1.

The missing data are imputated using backcasting, while the Total Unemployment rates are seasonal adjusted using TRAMO-SEATS. Figures 1 and 2 describe the general behavior of the GDP and the seasonally adjusted Unemployment rates. To get more legible graphics, the countries were divided into two graphs with the same scale. The considered economies share more or less the same story before and after the 2009 crisis. We observe a downward trend for Unemployment until the 2008 followed by an increasing path and a upward trend for GDP until 2008 and after a growth slowdown, or a fall or a sharp fall as for Greece. A special case is given by Ireland whose GDP shows from 2015 an incredible increase. This is the result of several corporate actions undertaken by some major multinational company based in the country (see for more details: https://www.oecd.org/std/na/Irish-GDP-up-in-2015-OECD.pdf and http://ec.europa.eu/eurostat/documents/24987/6390465/Irish_GDP_communication.pdf).

Using the Augmented Dickey–Fuller unit root test with constant (asymptotic p-value are reported in Table 2) and with lags selection according to the Bai-Ng test, we verified that all the GDP and Unemployment series in logarithms are I(1), with the only exception of the Unemployment of Belgium. Then, we decided to consider the relationship between the rates of growth of unemployment rate and GDP. The usual preliminary analysis on growth rates shows some of the series have outliers, in particular, in 2008:Q2-2009:Q2, that have been corrected using TRAMO. Lags selection procedure for the VARs, based on BIC criteria, choose just one lag for all countries. The relationship is thus analyzed for all countries through a bivariate VAR(1) model for the variables $\Delta log(U)$ and $\Delta log(GDP)$.

We apply the procedure described above evaluating proposed distance (5) between any pair of countries, setting $\tilde{p} = 15$, obtaining the matrix of distances reported in

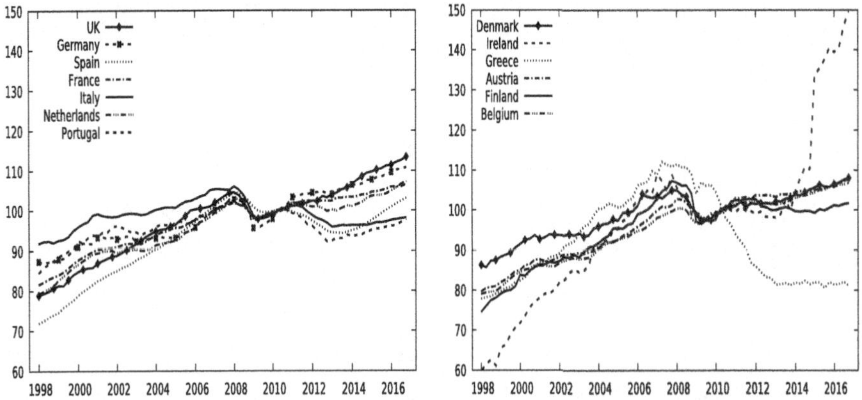

Fig. 1 Gross Domestic Product at market prices, chain-linked volumes index 2010 = 100, seasonal adjusted

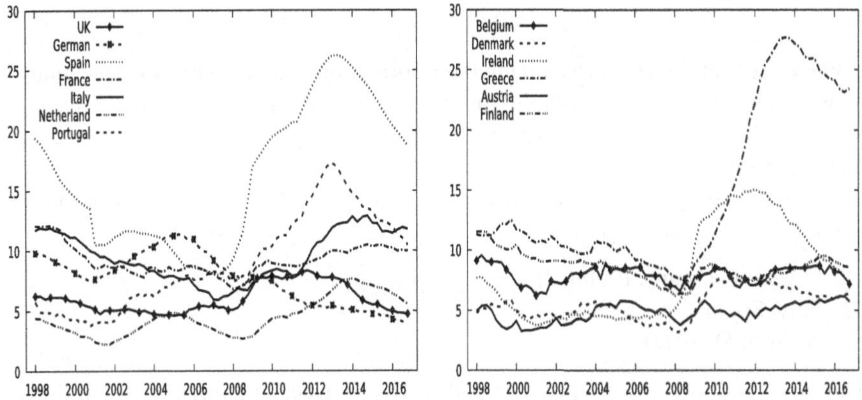

Fig. 2 Total unemployment rates, seasonal adjusted

Table 2 ADF asymptotic p-value

series	Be	Dk	Ge	Ire	Gre	Es	Fra	Ita	NL	Au	Por	Fin	Uk
log(U)	9.e−05***	0.21	0.98	0.47	0.61	0.49	0.17	0.46	0.33	0.35	0.55	0.06	0.42
log(Gdp)	0.14	0.52	0.77	0.84	0.10	0.48	0.17	0.11	0.25	0.31	0.05	0.07	0.58

***Statistical significance at 1%

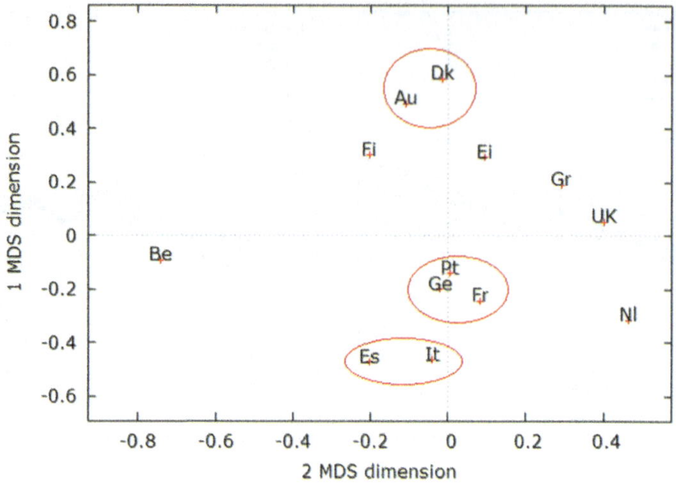

Fig. 3 Two-dimensional scatterplot of countries obtained by multidimensional scaling

Table 3.[1] A useful way of visualizing the information contained in a distance matrix between units is to conduct a MultiDimensional Scaling (MDS) analysis.[2] Given the matrix of distances among VARs presented in Table 3, classical MDS produces the map reported in Fig. 3.

The analysis by MDS of the distance matrix shows that we can identify the following clouds of similarity:

1. France, Germany, Portugal;
2. Italy, Spain;
3. Austria, Denmark

The other countries reveal peculiar paths. This is true, in particular, for Belgium and Netherlands. The relative proximity between Italy and Spain seems to be overwhelmed by the recent dynamics of both the labor market and GDP resulting from the crisis that has hit Europe since 2009. In the same way, the well-known situation in Greece explains its position in the figure. It seems more difficult to explain, however, the position of Portugal near to Germany and France, even if a graphical inspection reveals a similar general dynamic in the GDP growth rate before 2009. Belgium and Netherlands seem to be very far each other; also, in this case, an explanation can be found in the peculiar political situation experimented by Belgium when the political

[1]Full results concerning the estimation of the VAR models are available from the authors on request.

[2]See [3]. From a nontechnical point of view, the purpose of MDS is to provide a visual representation of the pattern of given distances among a set of objects. Given a matrix of distances between various objects, MDS plots the objects on a map such that those objects that are very similar to each other are placed near each other on the map, and those objects that are very different from each other are placed far away from each other on the map. Present elaboration is conducted by the MDS package in Gretl.

Table 3 Distance matrix

	Be	Dk	Ge	Ei	Gr	Es	Fr	It	Nl	Au	Pt	Fi	UK
Be	0.000	0.967	0.682	0.925	1.112	0.707	0.787	0.667	1.216	0.751	0.715	0.569	1.155
Dk		0.000	0.787	0.346	0.411	1.098	0.840	1.050	1.098	0.217	0.727	0.479	0.654
Ge			0.000	0.496	0.463	0.386	0.133	0.285	0.549	0.696	0.062	0.568	0.480
Ei				0.000	0.259	0.840	0.572	0.779	0.794	0.295	0.435	0.370	0.401
Gr					0.000	0.784	0.432	0.644	0.690	0.371	0.424	0.587	0.244
Es						0.000	0.458	0.285	0.885	1.010	0.430	0.903	0.827
Fr							0.000	0.212	0.472	0.752	0.152	0.647	0.370
It								0.000	0.608	0.962	0.344	0.850	0.543
Nl									0.000	1.019	0.531	0.892	0.455
Au										0.000	0.636	0.296	0.567
Pt											0.000	0.508	0.447
Fi												0.000	0.637
Uk													0.000

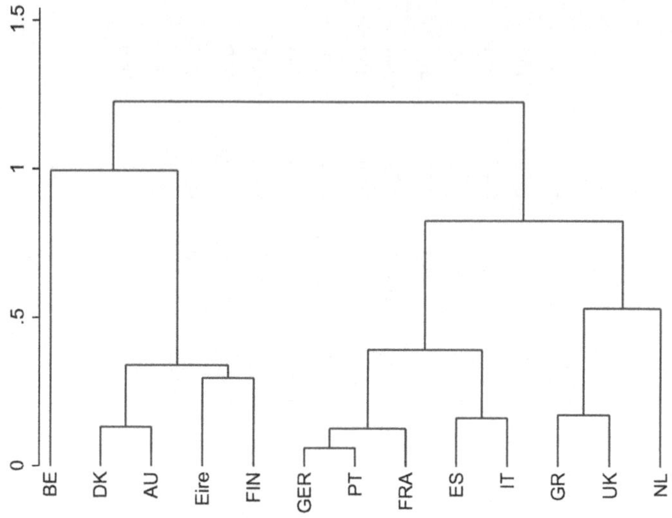

Fig. 4 Cluster analysis on multidimensional scaling distance

parties were not able to form a government for at least 2 years starting from June 2010. The overall conclusion is that despite the ongoing integration within the EU, there are still significant differences among countries regarding the dynamic link between unemployment rate variation and economic growth.

As a robustness check, we have also conducted a cluster analysis to identify homogenous subgroups of implied ARMA models. Results obtained using complete linkage are in Fig. 4 that seems to sustain the naive conclusion based on Fig. 3.

4 Conclusions

There are many circumstances in which is important to compute a distance measure for multivariate time series. In this paper, a novel notion of distance measure between pairs of VARMA models has been introduced and its main properties have been discussed. We have used such notion to investigate the presence of spatial differences in the dynamic linkage between unemployment rate variation and economic growth in 13 European Union economies. The analysis reveals that, despite the ongoing integration within the EU, many countries have special positions of their own.

References

1. Lee, J.: The robustness of Okun's law: evidence from OECD countries. J. Macroecon. 331–356 (2000)
2. Lütkepohl, H.: New Introduction to Multiple Time Series Analysis. Springer, Berlin (2005)
3. Mardia, K.V., Kent, J.T., Bibby, J.M.: Multivariate Analysis. Academic Press (1979)
4. Piccolo, D.: A distance measure for classifying ARIMA models. J. Time Ser. Anal. **11**, 153–164 (1990)
5. Prachowny, M.F.J.: Okuns law: theoretical foundations and revised estimates. Rev. Econ. Stat. **75**(2), 331–336 (1993)
6. Sims, C.: Macroeconomics and reality. Econometrica **48**, 1–48 (1980)
7. Tsay, R.S.: Analysis of Financial Time Series. Wiley (2005)
8. Watts, M., Mitchell, W.: Alleged instability of the Okun's law relationship in Australia: an empirical analysis. Appl. Econ. 1829–1838 (1991)
9. Zellner, A., Palm, F.: Time series analysis and simultaneous equation econometric models. J. Econom. **2**, 17–54 (1974)

Copulas for Modeling the Relationship Between Inflation and the Exchange Rate

Laila Ait Hassou, Fadoua Badaoui, Okou Guei Cyrille, Amine Amar, Abdelhak Zoglat and Elhadj Ezzahid

Abstract Copulas are useful tools for formalizing the dependence structure between variables. They have proven to be very valuable in economics, where the dependence plays a key role. In this chapter, we use copulas to analyze the dependence between inflation and *US/Euro* exchange rates in the Euro area, during different periods. We first explore the dependence between the variables using a nonparametric approach. Then, we select an appropriate parametric copula for each period. Results confirm the sensibility of copulas to macroeconomic fluctuations that occur during the analyzed periods.

Keywords Copulas · Exchange rate · Goodness-of-fit tests · Inflation Nonparametric approaches

L. Ait Hassou · A. Zoglat
Laboratoire de Mathématiques, Statistique et Applications, Faculté des Sciences, Centre de Recherche Mathématiques de Rabat, Université Mohammed V de Rabat, Rabat, Morocco
e-mail: laithassou@gmail.com

A. Zoglat
e-mail: azoglat@gmail.com

F. Badaoui (✉)
Département Statistique, Démographie et Actuariat, Institut National de Statistique et d'Economie Appliquée, Rabat, Morocco
e-mail: fadoua.badaoui@gmail.com

O. Guei Cyrille
Unité de Formation et Recherche (UFR) Environnement, Université Jean Lorougnon Guédé de Daloa, Daloa, Côte d'Ivoire
e-mail: okougueicyrille@gmail.com

A. Amar
Moroccan Agency for Sustainable Energy, Rabat, Morocco
e-mail: amar.abd@gmail.com

E. Ezzahid
Faculty of Law and Economics, Laboratory of Applied Economics, Mohammed V University in Rabat, Rabat, Morocco
e-mail: ezzahidelhadj@yahoo.fr

I. Rojas et al. (eds.), *Time Series Analysis and Forecasting*, Contributions to Statistics, https://doi.org/10.1007/978-3-319-96944-2_15

1 Introduction

Inflation rate is a key macroeconomic indicator. It is defined as a persistent increase in the general level of prices. This increase results in a reduction of the value of money (ILO, IMF, OECD, UNECE, Eurostat, and The World Bank, [1]). One of the highest priorities of governments and central banks is price stability. Low inflation promotes economic growth and full employment. High inflation erodes international competitiveness and reduces the value of saving. To control prices, central banks take into consideration the relationship between inflation and several macroeconomic variables like Gross Domestic Product (GPD), interest rate, and exchange rate.

Economic theory provides conflicting views about the links between inflation and other macroeconomic variables. Empirical research attempts to determine the nature of this relationship depending on countries and periods. Chollete and Ning [2] found a negative dependence between output and prices. Munyeka [3] detected a positive dependence between inflation and real GDP. Fitzgerald and Nicolini [4] documented a negative linear dependence between unemployment and inflation.

The exchange rate is among the macroeconomic variables that are significantly linked to inflation. It directly influences the latter through imported final goods prices, or indirectly via imported intermediate goods prices used in domestic production. Many studies discussed the effect of exchange rate fluctuations on inflation. Arize et al. [5] conducted an empirical investigation to study the links between the variations of exchange and inflation rates in 82 countries. Using a linear regression approach, they showed that exchange rate variability has a statistically significant effect on inflation variability. Naz et al. [6] showed that high exchange rate volatility contributes to a higher exchange rate pass-through to inflation. Xiongtoua and Sriboonchitta [7] used a copulas-based GARCH to analyze the dependence between the exchange and inflation rates in Laos. They found that the two variables have strong nonlinear and positive correlation. Kano [8], Burstein and Gopinath [9], and Engel [10] showed the existence of a relationship between exchange rate and inflation.

This chapter aims to explain the relationship between US/Euro exchange rate and inflation in Euro area during three different periods that cover crisis and noncrisis economic phases. The remaining of this chapter is organized as follows. The second section provides a theoretical background and the methodology. The third section is devoted to the presentation of the data used to illustrate our approach, the results, and a discussion. The fourth section contains a brief conclusion.

2 Methodology and Theoretical Background

Economic data are often multidimensional and require a joint modeling of several variables. The relationship between variables can be modeled using families of bivariate distributions. It is common, for some convenience reasons, to use bivariate distributions with marginals belonging to a same family. A classical example of such a situation is the use of bivariate normal distribution family because it is widely studied in the literature and easy to apply. This approach is not suitable for the general case where there is no condition on the marginals.

The copula approach enables overcoming this difficulty because it allows to link any two marginals to their bivariate distribution. In other words, knowing only the marginal distributions of two random variables, we theoretically can construct their joint distribution. This property of copulas makes them widely popular and very attractive in statistics. Copulas have been successfully applied in risk management (Kole et al. [11]), finance and insurance (Grégoire et al. [12], and Genest et al. [13]), and actuarial science (Frees and Valdez [14]). The books by Joe in [15] and Nelsen [16] are excellent references on copulas theory.

A Brief Aperçu of Copulas

Given two random variables X and Y with continuous marginals F and G, the Sklar theorem (Sklar [17]) states that the joint distribution function $H(.,.)$ of (X, Y) can be written in terms of a unique function $C(F(.), G(.))$:

$$H(x, y) = C(F(x), G(y)). \tag{1}$$

The function $C(.,.)$ is known as the copula of (X, Y). It describes how H is coupled with the marginal functions F and G.

The literature is rich in copula families. For the reader convenience, we present below some of them that are very common and will be used in this work.

Elliptical copulas: They are very appropriate to capture symmetric dependence. Gaussian and t-copulas are famous examples of elliptical copulas. The main advantage of a t-copula is that it can capture tail dependence.

Archimedean copulas: They are adequate in the case of tails dependence. The Clayton, Frank, and Gumbel copulas are Archimedean. The Frank copula is symmetric while the Clayton and Gumbel copulas are asymmetric. As the Clayton copula exhibits greater dependence in the lower tail, the Gumbel copula captures upper tail dependence.

Table 1 provides the expressions of copulas employed in this chapter.

Choosing the Appropriate Copula

To identify the best copula fitting a given set of data, we proceed in two steps. First, we use a nonparametric approach (Chi and K-Plots) to identify the families of candidate copulas. Such tools are available as packages in the statistical software R. Then, to select the most appropriate copulas among those families, we use a semi-parametric

Table 1 Some examples of copulas

Copula name	$C(u, v)$ expression
Gumbel	$\exp[-((-\log(u))^{\theta} + (-\log(v))^{\theta})^{\frac{1}{\theta}}], \theta \geq 1$
Clayton	$\max([u^{-\theta} + v^{-\theta} - 1]^{-\frac{1}{\theta}}, 0), 0 \neq \theta \geq -1$
Frank	$-\frac{1}{\theta}\ln(1 + \frac{(\exp(-\theta u)-1)(\exp(-\theta v)-1)}{\exp(-\theta)-1}), \theta \neq 0$
Gaussian	$\Phi_{\Sigma}(\Phi^{-1}(u), \Phi^{-1}(v)), \lvert\theta\rvert \leq 1$
Student	$t_{\Sigma,\nu}(t_{\nu}^{-1}(u), t_{\nu}^{-1}(v)), \lvert\theta\rvert \leq 1$
Plackett	$\frac{(1+(\theta-1)(u+v))-\sqrt{(1+(\theta-1)(u+v))^2-4uv\theta(\theta-1)}}{2(\theta-1)}$, $1 \neq \theta > 0$
Galambos	$uv\exp(((-\log(u))^{-\theta} + (-\log(v))^{-\theta})^{\frac{-1}{\theta}})$, $\theta > 0$

approach, based on Deheuvels copula and Mean Square Error, and a parametric approach based some Goodness-of-fit tests.

Let X and Y be two random variables, F and G their marginal distributions, and H their joint distribution. Consider $(X_i, Y_i)_{1 \leq i \leq n}$ a random sample from the distribution H. We denote by F_n, G_n, and H_n the empirical distribution functions of the X, Y, and the couple (X, Y), respectively

$$F_n(X_i) = \frac{rank(X_i)}{n}, \quad G_n(Y_i) = \frac{rank(Y_i)}{n}, and \quad H_n(X_i, Y_i) = \frac{rank(X_i, Y_i)}{n}$$

For simplicity, we note hereafter $F_n(X_i)$, $G_n(Y_i)$, and $H_n(X_i, Y_i)$, respectively, by F_i, G_i, and H_i.

The Chi-Plot The Chi-Plot is a graphical tool that helps localizing the dependence between two variables X and Y. Fisher and Switzer [18, 19] present Chi-Plot as a plot of the pairs (λ_i, χ_i), where χ_i is the measure of failure of the bivariate distribution function to factorize into a product of marginal distribution functions at the sample argument (X_i, Y_i), and λ_i is the distance from (X_i, Y_i) to the bivariate median. The χ_i and λ_i are defined by

$$\lambda_i = 4\, sign(\widetilde{F}_i\widetilde{G}_i)\max(\widetilde{F}_i^{\,2}, \widetilde{G}_i^{\,2}) \quad and \quad \chi_i = \frac{H_i - F_i * G_i}{\sqrt{F_i(1-F_i)G_i(1-G_i)}}, \tag{2}$$

where $\widetilde{F}_i = F_i - 0.5 \quad and \quad \widetilde{G}_i = G_i - 0.5$.

To help assess random variation in the observed values of χ, Fisher and Switzer [18, 19] built confidence intervals of the form $\pm c_p\sqrt{n}$. They also gave approximate values of the c_p's for different values of $p \in [0, 1]$. In the case of independence, we expect that $p \times 100\%$ of the pairs (λ_i, χ_i) will be uniformly scattered inside the interval $[-c_p\sqrt{n}, c_p\sqrt{n}\,]$. In the case of a positive dependence, the pairs points go scattered above the band, and conversely for the case of negative dependence.

The Kendall (K-Plot) Similarly to the concept of QQ-plot, used for assessing graphically the adequacy of a fitting distribution, Genest and Boies [20] propose the K-Plot a visual tool for assessing dependence in a bivariate random sample The K-Plot is a rank-based procedure for the detection of dependence. The procedure consists in representing the pairs $(W_{i,n}, H_i)$ for $i \in [1, n]$, where $W_{i,n}$ is the expectation of the ith order statistic of a n random sample size. K_0 is a conditional distribution, issued from H, under independence of X and Y. The form of the bivariate distribution K_0 is given by

$$K_0(w) = P(UV \leq w) = w - w\log(w) \tag{3}$$

where U and V are independent uniform random variables on the interval [0, 1], and $W_{i,n}$ is given by

$$W_{i,n} = \frac{n(n-1)!}{(i-1)!(n-i)!}\int_0^1 w\{K_0(w)\}^{(i-1)}\{1 - K_0(w)\}^{(n-i)}\mathrm{d}K_0(w) \tag{4}$$

The closer is the K-Plot to the 45° line, the weaker is the association between the random variables.

Deheuvels Empirical Copula and Mean Squared Error (MSE) The Chi-Plot and K-Plot can provide a wide family of candidate copulas that fit the data. To reduce such a family, we resort to a semi-parametric tool. This approach compares a candidate copula C_{θ_j} to the empirical copula C_n, introduced by Deheuvels [21], using the mean squared error

$$SME_n(C_{\theta_j}) = \frac{1}{n}\sum_{i=1}^{n}\left[C_n(F_i, G_i) - C_{\theta_j}(F_i, G_i)\right]^2,$$

where C_n is defined by H_n (as Sklar theorem states), and $\theta_j = \theta_0 + j * \Delta$ for $1 \leq j \leq m$ with Δ denoting the step variation. The best copula C_{θ_j} to fit the sample is the one that minimizes $SME_n(C_{\theta_j})$.

To confirm the nonparametric approach choice, we perform some nonparametric goodness-of-fit tests.

Goodness-of-Fit Tests In this section, we present the Cramer–von Mises and Kolmogorov–Smirnov tests that we used to select the appropriate copula to fit the data. We give below their expressions based on Kendall's transformation as proposed by Genest et al. [22].

$$S_n = \int_0^1 |\mathbb{K}_n(t)|^2 \partial K_{\theta n}(t), \textit{ and } \quad T_n = \sup_{0\leq t\leq 1} |\mathbb{K}_n(t)|,$$

where $\mathbb{K}_n(t) = \sqrt{n}(K_n(t) - K_c(t))$ with $K_n(t)$ denoting, respectively, the empirical Kendall distribution, and $K_C(t) = P(C(U, V) \leq t)$ denoting the Kendall distribution corresponding to copula $C(u, v)$.

Forecasting

Once the copula $C(.,.)$ modeling the dependence between variables X and Y is identified, forecasting becomes a straightforward operation. For instance, using the conditional copula C_X, we can forecast Y given X. The conditional copula C_X is defined as

$$P[Y \leq y|X = x] = P[V \leq v|U = u] = \frac{\partial C(u, v)}{\partial u} := C_X(F(x), G(Y)), \quad (5)$$

where $V = G(Y)$, $v = G(y)$, $U = F(X)$, and $u = F(x)$.

For $\tau \in [0, 1]$, the τ-th conditional quantile function given $X_t = x$, denoted by $Q_X(\tau|x)$, is defined by

$$Q_X(\tau|x) = G^{-1}(C_X^{-1}(F(x), \tau)), \quad (6)$$

where φ^{-1} denotes the inverse function of a function φ.

3 Results and Discussion

The proposed methodology is applied to monthly inflation and $US/Euro$ exchange rate in the Euro area over different periods. These data are available at the European Central Bank (https://www.ecb.europa.eu). We apply the methodology to model data over different periods (see Fig. 1) to pinpoint the sensitivity of copulas to macroeconomic changes.

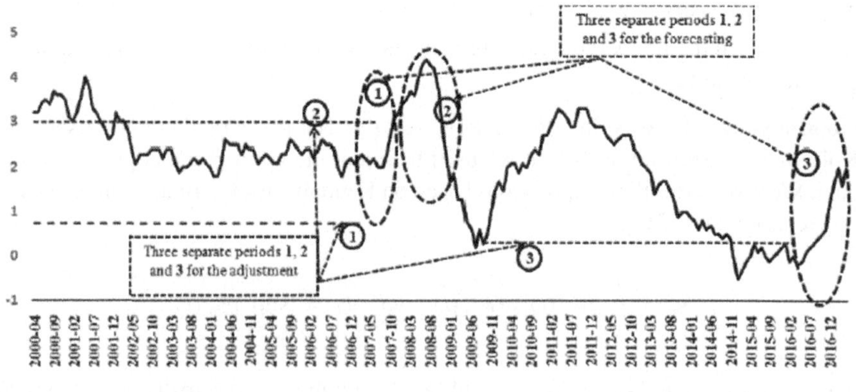

Fig. 1 The evolution of inflation in the Euro area from Apr. 2000 to Dec. 2016

Fig. 2 The exchange rate and inflation evolution during sixteen years

Data Description

The first period begins in April 2000 and ends in December 2007. Observations during 2007, not involved in building the model, are used to check its forecasting validity. A slightly different model is obtained when these data are added to the learning sample. This new model proved to be inadequate in forecasting 2008 observations. This could be due to the fact that the year 2008 was a period of economic crisis. The third period ranges from September 2009 to June 2016. The data from September 2009 to March 2016 account for a learning sample used to choose the appropriate copula. The retained model is used to forecast observations of the second quarter of 2016. The evolution of *US/Euro* exchange and inflation rates in the Euro area is shown in Fig. 2. This graphic shows clearly that these rates are correlated.

Fitting the Data

Financial time series are not immediately suitable for copula modeling because they are serially correlated [23]. We need first to eliminate autocorrelation and seasonality. This can be achieved using the Box-Jenkins approach for modeling time series (Box and Jenkins [24]). In this work, we use this approach and find that our data can be modeled by an AR or a SARIMA process. The models adjusted to X_t and Y_t provide correlated residuals, ε_{X_t} and ε_{Y_t}, which are not autocorrelated. The key idea to model the correlation, between X_t and Y_t, is to determine the best copula that models the dependence between ε_{X_t} and ε_{Y_t}.

Using some goodness-of-fit criteria and statistical tests (AIC, BIC, and significance of estimated parameters), we retain the models in Table 2 to extract the unpredictable component of the time series.

Table 2 Residuals deduced from adjusting different time series models

Series	Adjusted models	Residuals
The inflation (2000–2007)	SARIMA (0, 1, 0)(0, 0, 1)	$\varepsilon_{Inf,period1}$
The *US/Euro* exchange rate (2000–2007)	SARIMA (0, 1, 1)(0, 0, 0)	$\varepsilon_{Exch,period1}$
The inflation (2000–2008)	SARIMA (1, 1, 0)(0, 0, 1)	$\varepsilon_{Inf,period2}$
The *US/Euro* exchange rate (2000–2008)	SARIMA (0, 1, 1)(0, 0, 0)	$\varepsilon_{Exch,period2}$
The inflation (2009–2016)	Exponential smoothing[a]	$\varepsilon_{Inf,period3}$
The *US/Euro* exchange rate (2009–2016)	SARIMA (1, 1, 0)(0, 0, 0)	$\varepsilon_{Exch,period3}$

[a]For details on exponential smoothing models, see Holt [25]

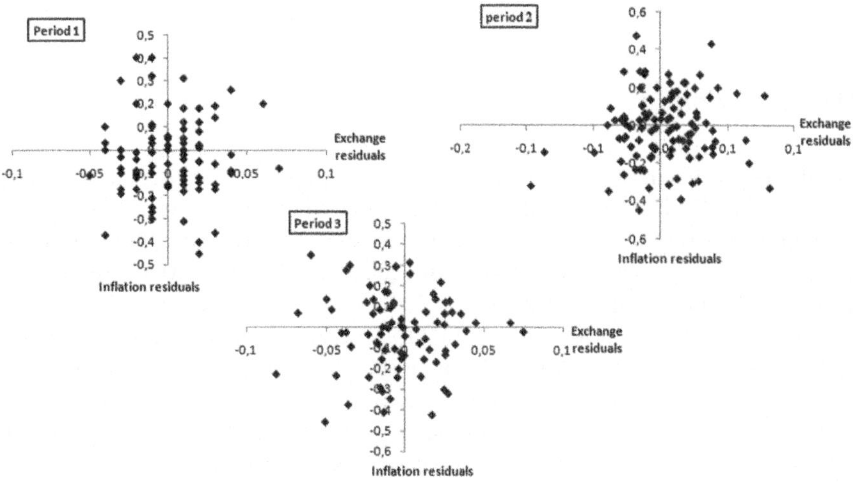

Fig. 3 Plots of inflation residuals against exchange rate residuals

Applying the Kolmogorov–Smirnov goodness-of-fit test, we conclude that the residuals are normally distributed. The plots of inflation residuals against the exchange rate residuals (Fig. 3) do not show any particular pattern. To graphically explore the dependence structure between the residuals, we use the Chi-Plot and K-Plot approach expecting that the plots guide our choice of the appropriate copula families.

On the K-Plot in Fig. 4a, the dots show a clear departure from the diagonal line. This is an evidence of dependence between the two variables. This dependence is confirmed by the Chi-Plot (Fig. 4b) including the whole sample. This graphic is an indicator of a correlation existence on the tails.

Constructing Chi-Plots for data in the tails (Fig. 4c, d), we conclude that there is no correlation in the lower tails, while the upper tails are strongly related.

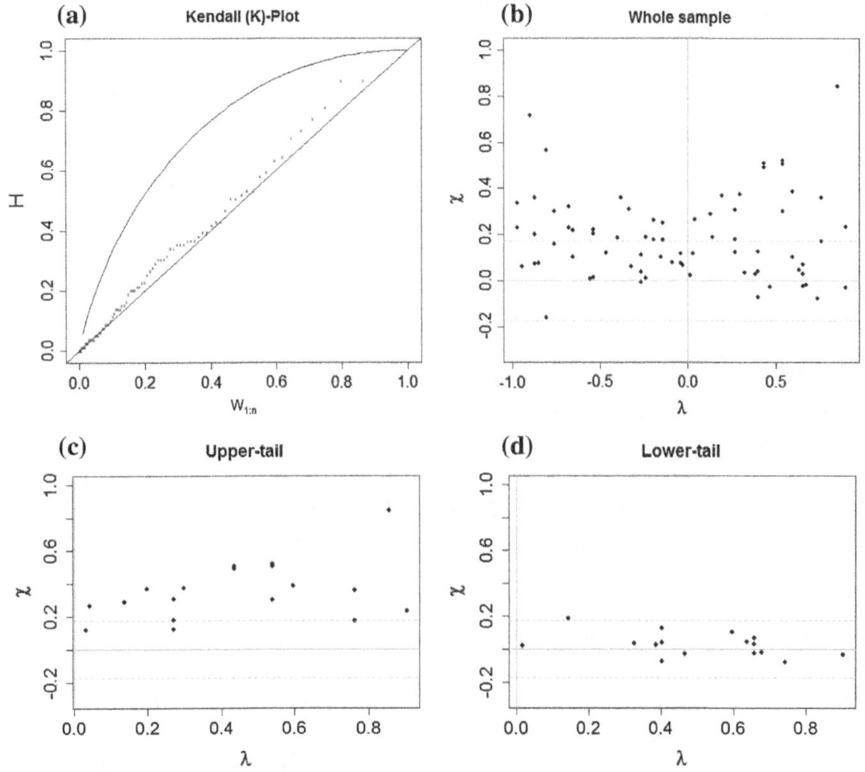

Fig. 4 Chi-Plot and K-Plot for the first period

Table 3 The retained copulas from the empirical investigation for the first period

Period	Copula	Value of θ	SME value
2000–2006	t-Copula	0.02	0.00194476
	Normal	0.02	0.0019655
	Frank	0.1	0.000198502
	Plackett	1.04	0.0019857
	Clayton	0.01	0.000199487
	Gumbel	1	0.000201036
	Galambos	1	0.003321712

These results suggest that the appropriate underlying copula should belong to the Archimedean family. In Tables 3 and 4, we present, respectively, the results of our empirical investigation for the 2000–2006 and 2009–2016 periods.

The copulas in Table 3 are tested using the Cramer–von Mises goodness-of-fit test. The results are shown in Table 5.

Table 4 The retained copulas from the empirical investigation for the third period

Period	Copula	Value of θ	SME value
2009–2016	t-Copula	0.14	0.0002496044
	Normal	0.14	0.0002496305
	Frank	0.72	0.0002510564
	Plackett	1.42	0.0002513465
	Clayton	0.21	0.0002222103
	Gumbel	1.1	0.0002812121
	Galambos	1	0.00248208

Table 5 The final retained copulas, based on a GoF tests

Period	Copulas family	Fitted copula	P-value	Value of θ
2000–2006	Archimedean	Clayton	0.6848	−0.066658
	Plackett	Plackett	0.5759	0.85718
	Elliptical Copulas	t-Copula	0.9615	−0.054132
		Normal	0.9805	−0.054132

Table 6 The final retained copulas for the third period, based on a GoF tests

Period	Copulas family	Fitted copula	P-value	Value of θ
2009–2016	Archimedean	Clayton	0.9595	0.017868
	Plackett	Plackett	0.9486	1.0394
	Elliptical Copulas	t-Copula	0.7248	0.013909
		Normal	0.6958	0.013909

During the 2000–2006 period, the most appropriate copula, describing the dependence between inflation and exchange rate, was a Normal copula. Other families of copulas, namely, the Archimedean and the Plackett copulas, could be used to describe the studied dependence structure.

Based on the same methodology, the Elliptical copulas are retained to model the dependence between inflation and *US/Euro* exchange rate during the 2008 period (a t-copula with a p-value of 0.93 and a Normal copula with a p-value of 0.96).

During the third period (2009–2016), an Archimedean copula (a Clayton copula with a p-value of 0.95 and a Plackett copula with a p-value of 0.94, as showed in Table 6) is retained to capture the link between the two variables. Elliptical copulas give also a high p-value but its p-value is lower than the p-value of the two previous copulas. Archimedean copulas have the great advantage to capture a wide range of dependence. Examples of Archimedean copulas are the product copula which corresponds to independence of the examined variables, the Clayton copula which is used in the case of strong left tail dependence, the Gumbel copula which is employed in the case of highly correlated variables at high values but less correlated at low

Table 7 Predicted values based on the retained copulas

Observed values	t-Copula	Normal	Clayton	Plackett
−0.19	−0.085	−0.092	−0.096	−0.096
−0.09	−0.033	−0.033	−0.037	−0.037
−0.07	−0.01	−0.008	−0.011	−0.011
−0.05	−0.046	−0.049	−0.052	−0.052
0.02	0.001	0.005	0.002	0.002
0	−0.047	−0.05	−0.053	−0.053

values, and the Frank copula which is applied when a tail dependence is weak. Unlike the Elliptical copulas, the Archimedean copulas are not derived from multivariate distribution functions using Sklar's theorem.

We propose to retain the Elliptical and Archimedean copulas for forecasting. This step allows identifying the best model that captures the structure of the dependence between inflation and *US/Euro* exchange rate.

Our results show that during the first period, characterized by a stability, copulas perform well in terms of forecasting (Table 7).

In the opposite, for crisis and instable periods, the forecasting seems to be imprecise. This is due to the volatility of the major macroeconomic aggregates. To overcome this problem, we suggest to model the updated series using dynamic copulas.

4 Conclusion

The study of dependence between economic variables is of prime importance for decision makers. Copulas Techniques are powerful tools to explore this dependence. In this chapter, we attempted to explore the correlation structure between inflation and *US/Euro* exchange rate during the pre-crisis, crisis, and post-crisis periods.

Four main results are to be highlighted. First, during the 2000–2006 period, the most appropriate copula that describes the dependence between inflation and exchange rate, was a Normal copula. Second, the Elliptical copulas are the most suitable to model the dependence between inflation and *US/Euro* exchange rate during the 2008 period. Third, for the 2009–2016 period, an Archimedean copula (a Clayton copula with a p-value of 0.95 and a Plackett copula with a p-value of 0.94, as showed in Table 6) is retained to capture the link between the two variables. Four, it appears that during the stable pre-crisis period, copulas performs well in terms of forecasting. These results provide evidence about the accuracy to use different copulas to model the links between economic variables when different periods are analysed.

References

1. International Labour Office (ILO), International Monetary Fund (IMF), Organization for Economic Co-operation and Development (OECD), Statistical Office of the European Communities (Eurosat), United Nations (UN) and the World Bank: Consumer price index manual. Theory and practice. International Labour Office. Geneva (2004). ISBN, 92–2–113699–X
2. Chollete, L., Ning, C.: The Dependence structure of macroeconomic variables in the US. Working Papers Series 0204, University of Ryerson (2009)
3. Munyeka, W.: An in-depth look at economic growth and employment in post-apartheid South Africa: analysis and policy implications. J. Educ. Soc. Res. **4**(3) MCSER, Rome-Italy (2014)
4. Fitzgerald, T.J., Nicolini, J.P.: Is there a stable relationship between unemployment and future inflation? Evidence from U.S. Cities. Working Paper 713, Federal Reserve Bank of Minneapolis Research Department (2014)
5. Arize, A.C., Malindretos, J., Nippani, S.: Variations in exchange rates and inflation in 82 countries: an empirical investigation. North Am. J. Econ. Fin. **15**, 227–247 (2004)
6. Naz, F., Mohsin, A., Zaman, K.: Exchange rate pass-through in to inflation: new insights in to the cointegration relationship from Pakistan. J. Econ. Model. **29**, 2205–2221 (2012)
7. Xiongtoua, T., Sriboonchitta, S.: Analysis of volatility and dependence between exchange rate and inflation rate in Lao people's democratic republic using copula-based GARCH approach. Model. Depend. Econ. **251**, 201–214 (2014)
8. Kano, T.: Exchange rates and fundamentals: a general equilibrium exploration, Hitotsubashi Institute for Advanced Study, Discussion Paper HIAS–E–19 (2016)
9. Burstein, A., Gopinath, G.: International prices and exchange rates. In: Handbook of International Economics, vol. 4, pp. 391–451. North Holland/Elsevier, London (2014)
10. Engel, C.: Exchange rates and interest parity. In: Handbook of International Economics, vol. 4, pp. 453–522. North Holland/Elsvier, London (2014)
11. Kole, E. et al.: Selecting copulas for risk management. J. Bank. Fin. https://doi.org/10.1016/j.jbankfin.2006.09.010.(2007)
12. Grégoire, V., Genest, C., Gendron, M.: Using copulas to model price dependence in energy markets. Energy Risk **5**, 58–64 (2008)
13. Genest, C., Gendron, M., Bourdeau-Brien, M.: The advent of copulas in finance. Eur. J. Fin. **15**(7/8), 609–618 (2009)
14. Frees, E.W., Valdez, E.A.: Under standing relationships using copulas. North Am. Actu. J. **2**, 1–25 (1998)
15. Joe, H.: Multivariate models and dependence concepts. In: Monographs on Statistics and Applied Probability, vol. 73. Chapman Hall, London (1997)
16. Nelsen, R.B.: An Introduction to Copulas, 2nd edn. Springer, New York (2006)
17. Sklar, A.: Fonctions de répartition á n dimensions et leurs marges. Publications de l'Institut Statistique de l'université de Paris, vol. 8, pp. 229–231. Paris (1959)
18. Fisher, N.I., Switzer, P.: Chi-plots of assessing of dependence. Biometrika **72**, 253–265 (1985)
19. Fisher, N.I., Switzer, P.: Graphical assessment of dependence: is a picture worth 100 tests. Am. Stat. **55**(3), 233–239 (2001)
20. Genest, C., Boies, J.C.: Detecting dependence with kendall plots. Am. Stat. **57**(4), 275–284 (2003)
21. Deheuvels, P.: La fonction de dépendance empirique et ses propriétés. Un test non paramétrique d'indépendance. Acad. Roy. Belg. Bull. Cl. Sci. (5) **65**(6), 274–292 (1979)
22. Genest, C., Rémillard, B., Beaudoin, D.: Goodness-of-fit tests for copulas: a review and a power study. Ins. Math. Econ. **44**, 199–214 (2009)
23. Patton, A.J.: A review of copula models for economic time series. J. Multivariate Anal. **110**, 4–18 (2012)
24. Box, G.E.P., Jenkins, G.M.: Time Series Analysis Forecasting and Control. Holden-Day, San Francisco (1976)
25. Holt, C.C.: Forecasting Trends and Seasonal by Exponentially Weighted Averages. Office of Naval Research Memorandum, vol. 52 (1957)

Part V
Energy Time Series Forecasting

Fuel Consumption Estimation for Climbing Phase

JingJie Chen and YongPing Zhang

Abstract Aiming at the problem of the civil aviation carbon emission, the purpose of this chapter is to present a simplified method to estimate aircraft fuel consumption using an adaptive Genetic Algorithm-Back Propagation (GA-BP) Strong prediction network. This chapter gives a brief overview of the modeling approach and describes efforts to validate and analyze the initial results of this project. The parameters of fuel consumption are analyzed by using QAR flight data, two kinds of fuel consumption prediction model are proposed, it is the BP prediction model and the adaptive (it is abbreviated to A) GA-BP (Genetic Algorithm-Back Propagation) Strong prediction model. The crossover and mutation probability of GA-BP Strong prediction model can be adaptive adjustment, and the BP neural network as a weak predictor, after the limited number of iterations, it can realize error optimization adjustment and solve the complicated nonlinear problem. Results of the simulation indicated the two models have obvious advantages in nonlinear prediction, and the prediction accuracy and the degree of fitting are good. The results of this study illustrate that the two neural network with nonlinear transfer functions can accurately represent complex aircraft fuel consumption functions for climb phases of flight, so the two models are feasible in the field of fuel consumption prediction. The methodology can be extended to cruise and descent phases of flight.

Keywords Flight data · Adaptive GA-BP-AdaBoost network · Fuel consumption Prediction

1 Introduction

Air transport industry acts as a catalyst to the economic and social development of a nation. But the development of air transport industry is faced with major issues like high fuel consumption [1, 2]. Furthermore, according to Henderson et al. [3]

J. Chen (✉) · Y. Zhang
Civil Aviation University of China, Tianjin, China
e-mail: jjchen@cauc.edu.cn

I. Rojas et al. (eds.), *Time Series Analysis and Forecasting*,
Contributions to Statistics, https://doi.org/10.1007/978-3-319-96944-2_16

research that aircraft CO_2 emission is proportional to fuel burn. For this challenge, many researchers have studied about different models on aircraft fuel consumption prediction. Thus, to obtain rational forecast results, various prediction models are put forward. Chang [4] presents the fuzzy logic modeling (FLM) technique, which will be employed to establish the reference lift-to-drag (L/D) model. The model is utilized to predict the deficiencies of lift-to-drag ratio through sensitivity analysis to determine the relative contributions from influencing flight variables for the excessive fuel consumption. Zhang et al. [5] propose a representative support vector network-aided fuel consumption model which is developed using data given in the route date and aircraft performance manual, support vector machine is trained to estimate fuel consumption of a certain aircraft. Performance differences between the aircraft types are natural and they inevitably cause problems in the air traffic system, Cavcar [6] present the excess fuel consumption impact of aircraft performance differences in the air traffic environment. For this purpose, both climb and cruise phases of a flight mission are analyzed. Based on the principle of energy conservation, a model was established to estimate the fuel consumption by Collins [7]. Based on the flight process, Mayer [8] presents an exponential relationship model which is to establish the relationship between the fuel flow and the height in the fall and climb phases. Based on the genetic algorithm, the Baklacioglu [9] researches a genetic algorithm-optimized neural network topology which is designed to predict the fuel flow-rate of a transport aircraft using real flight data. This model used to study the change of the fuel with the air speed and altitude at different time. Baklacioglu [10] develops a new aero-propulsive model (APM) which is derived from the flight manual data of a transport aircraft using Genetic Algorithms (GAs) to perform accurate trajectory predictions. The use of GAs enhanced the accuracy of both propulsive and aerodynamic modeling, and successfully predicted the trajectory for the descent phase. Bartel [11] introduces an updated model which is presented that describes the takeoff thrust to within $\pm 1\%$ of that of the reference engines, for flight speeds up to Mach 0.4. This takeoff thrust model has been adapted to account for the impact of bleed air extraction and altitude effects.

However, the mentioned above have some limitations for the study of fuel consumption. Most of them were based on the level of time series with a single flight as the research object to analyze the characteristics of flight and then built a model to predict the fuel consumption at different times in the flight process. It is lack of practical application.

Using neural network, complex nonlinear function can be easily handled, and the network has an advantage that there is no need to reveal the mathematical equation describing the input–output mapping before the network training. The output of the network function is close to the actual output value, so as to achieve a more accurate prediction effect. The genetic algorithm can realize global optimization and optimize the initial state of the network. Back propagation neural network prediction model and a back propagation based on genetic algorithm (GA-BP) neural network prediction model have many successful applications [12, 13]. Chen et al. [14] proposes a neurons method based on BP neural network to achieve the linear approximation of complex nonlinear relationships and got a satisfactory cruise fuel consumption model. Saravanan et al. [15] used Genetic Algorithm (GA) to estimate electricity demand for different scenarios (low, high growth, and trend lines) based on economic indicators of historical data. Although all these reviews provided vital information on fuel consumption prediction models on different scales, accurate fuel consumption prediction remains a challenge for energy management.

Generally, the flight of the aircraft involves five typical flight stages of takeoff, climb, cruise, descent, and landing. The literature [16] pointed out that the fuel consumption of the climb, cruise, and the descent accounts for over 90% of the total fuel consumption. Many scholars have given a variety of prediction models of cruise fuel consumption. These models have achieved high prediction accuracy. The climb is more typical than the descent, so this chapter studies climb phase. For long voyage, the climb has a small proportion of fuel consumption, and the statistical analysis can obtain the accurate prediction results. The literature [17] presents a method based on Bootstrap statistical theory to establish the prediction model of fuel consumption and gets the high prediction accuracy. However, for short voyage, the climb fuel consumption accounts for a large part, and the result of simple statistical analysis is not satisfactory.

Based on the above analysis, this chapter attempts to develop a suitable method to estimate aircraft fuel consumption using neural network approach to deal with relation between energy consumption and its influence factors. Harshad and Hamsa [18] demonstrates that flight data record the main parameters used for the analysis of fuel consumption. In the study, through the analysis of QAR data, we select the main factors that affect the fuel consumption, but only for the climbing condition.

An adaptive GA-BP neural network fuel consumption forecasting model is established, and a strong prediction is joined in the model. It will enable more accurate climb fuel predictions. In this chapter, the fitting degree of the model output, the average value of the relative error and the mean square error are chosen as the benchmark of the feasibility of the model. Furthermore, as an improvement to the existing models is compared with the BP neural network prediction model, and the simulation results are analyzed. It is proved that these two models have good prediction effect.

2 Fuel Consumption Statistical Analysis

This chapter chooses the flight data for 2012 to analyze its fuel consumption. We use k-means clustering to classify the voyages. We define the fuel consumption as X and randomly select k points as the starting center, k is the number of cluster clusters. Then, calculate the distance between each point and center. To find the nearest center for each point and assign it to the cluster that corresponds to that center. After this step, the center of each cluster is updated to the average of all points of this cluster. Re-cluster and update the center until the results remain unchanged. Through the above method, the voyage is divided into long voyage, middle voyage, and short voyage three categories. Of which, more than 2060 km for the long voyage, 1124–2060 km for the middle voyage and less than 1124 km for short voyage.

Table 1 Statistical analysis of kilometers fuel consumption in different voyage

	Long voyage	Middle voyage	Short voyage
N statistic	695	1061	1741
Minimum statistic	2.22	2.11	1.64
Maximum statistic	4.35	7.80	11.06
Mean statistic	3.023	3.770	4.649
Variance statistic	0.184	0.255	0.899

The fuel consumption of the three voyages is statistically analyzed, respectively. The results are shown in Table 1.

Through the analysis it can be found that as the mileage increases, kilometers fuel consumption is gradually increased. Degree of dispersion of the long voyage kilometer fuel consumption is less, and with the reduction of mileage, kilometers fuel consumption the degree of dispersion gradually increased. Relative to the climb and the descent, the calculation of the cruise fuel consumption is simple, and mainly related to its flight distance. The fuel consumption of the climb has nothing to do with the voyage, and its relative fuel consumption is relatively high during the whole flight, so we mainly study the fuel consumption of the climb.

3 Fuel Consumption Influencing Factors

Flight data QAR records the most flight parameters of the aircraft from takeoff to landing, these parameters are closely related to the fuel consumption of the aircraft, which provides a good basis for the analysis of fuel consumption [19]. So, in this chapter, the QAR data is selected as the data set of model training and prediction.

Based on those flight control parameters, a prediction model can be made to estimate fuel loading. There is required to do data filtering. Sometimes, when data is missed out, interpolation is also required if the filtering frequency is fixed. Similarly, in this chapter, the idea of stepwise linear regression is used to screen and eliminate the multiple co-linear variables in the QAR data [20]. We can screen effect parameters of aircraft climb phase. In this study, according to the absolute value of the correlation coefficient size, we take the initial weight of climbing segment, the climbing distance, the rate of climb, and the force of the wind in the nose and the total temperature of the atmosphere as the input of the model five factors. The outputs of the model are relatively easy to determine according to our modeling objective. They are fuel consumption.

After that, the model input data are normalized to the same dimension. We define x_{max} and x_{min} as the maximum and minimum value in the sample and x_k as the sample normalization value and the function can be written as

$$x_k = \frac{x - x_{\min}}{x_{\max} - x_{\min}} \tag{1}$$

4 Improved Model Algorithm

Basic Procedures and Ideas

We propose a genetic optimization BP neural network. The whole idea of network structure is using the improved adaptive genetic algorithm to obtain the optimal individual and used to optimize the initial weights and threshold of the neural network. After training, the BP network learning to get the fuel consumption forecast output, in order to improve the prediction accuracy, we used BP network as a weak predictor, after a finite number of iterations and outputs the result of strong prediction results. Structure diagram as shown in Fig. 1.

GA-BP-AdaBoost Parameters Setup

In this section, we propose a fuel consumption prediction model based on GA-BP-AdaBoost neural network. We take the 280 sets of QAR data from different flights in the same course and take them as the training dataset and test dataset of the fuel consumption model. The input is the starting weight of aircraft, the climbing distance,

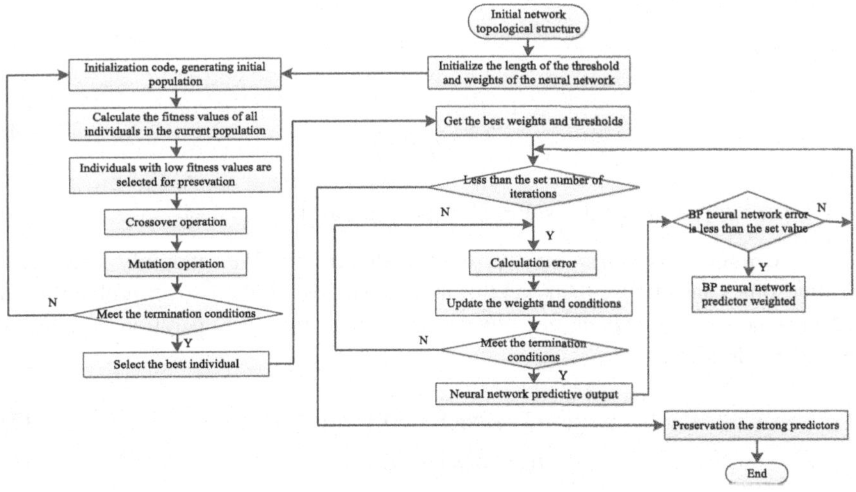

Fig. 1 Based on improved GA-BP strong prediction model

the rate of climb, the wind speed, the total temperature of the atmosphere, set X as input vector, $X = [X_1, X_2, X_3, X_4, X_5]$, Aircraft fuel flow as the output, set Y as the output vector. Genetic algorithm for optimizing network weight needs to design the main parameters.

Step 1. An initial population is generated. A population of W individuals were randomly generated. W is chosen 30.

Step 2. Coding for neural network. In order to get the high precision weight and threshold value, the real number coding method is adopted. Individual coding length is L. There L is equal to $N * M + M + M * K + K$, and the input node number N takes 5, the number of hidden nodes M takes 10, the number of output nodes K takes 1.

Step 3. Fitness Function. Training overall error should be as small as possible, so genetic algorithm uses the minimum objective function value as the fitness function. We calculate the sum of the absolute value of fuel consumption error in the training sample and chromosome adaptive value is

$$f = \sum_{i=0}^{n} abs(y_i - o_i) \tag{2}$$

$n = 200$ is the training samples number. o_i is the expected output of BP network. y_i is the forecasted output of BP network.

Step 4. Selecting operation. The selection operation is used to determine the recombination or crossover parent individuals and the number of offspring individuals generating by the candidate population. We use the following formula to calculate the chromosome fitness and selection probability. The chromosome adaptive value is F_i and the selective probability is P_i.

$$F_i = \frac{1}{f_i} \tag{3}$$

$$P_i = \frac{F_i}{\sum_{i=0}^{P} F_i} \tag{4}$$

P is the reciprocal of fitness. $P = 30$ is population size.

Step 5. Crossover Operator and Crossover Probability. We randomly select two chromosomes and choose their weights and thresholds with crossover probability to form two new individuals a_{ki}, a_{1j}. We use arithmetic crossover operation to generate two new individuals:

$$a_{kj} = a_{ki}(1 - a) + a_{1j}a \tag{5}$$

$$a_{1j} = a_{1i}(1 - a) + a_{kj}a \tag{6}$$

a is a parameter, $a \in (0, 1)$. The crossover probability usually takes the value of the [0.5, 0.98]. The crossover probability of this study is obtained by the following formula:

$$P_c = m + \frac{m(f_{av} + f_i)}{10(f_i - f_{\min})}, f_i \geq f_{av} \quad (7)$$

$$P_c = m - \frac{m(f_{av} + f_i)}{10(f_i - f_{\min})}, f_i < f_{av} \quad (8)$$

$m = 0.7, f_i$ is the smaller value of the fitness value of the two selected individuals, f_{av} is the average value of all fitness values, f_{min} is the minimum value in all fitness values.

Step 6. Mutation Operator and Mutation Probability. We use arithmetic mutation operation to generate two new individuals

$$a_{ij} = a_{ij} + (a_{ij} - a_{\max}) * f(g), r \geq 0.5 \quad (9)$$

$$a_{ij} = (a_{ij} + a_{ij} - a_{\max}) * f(g), r < 0.5 \quad (10)$$

$$f(g) = r_2\left(1 - \frac{g}{G_{\max}}\right) \quad (11)$$

Mutation point is a_{ji} ($a_{ji} \in [a_{\min}, a_{\max}]$). r_2($r_2 \in [0, 1]$) is a random number, g is the number of iteration and G_{max} is the max number of evolution.

Mutation probability P_m usually value between [0.001, 0.05]. It can be represented by an adaptive mutation probability formula.

$$P_m = m_1 + \frac{m_1(f_i^{'} + f_{av})}{10(f_i^{'} - f_{\min})}, f_i^{'} \geq f_{av} \quad (12)$$

$$P_m = m_1 - \frac{m_1(f_i^{'} + f_{av})}{10(f_i^{'} - f_{\min})}, f_i^{'} < f_{av} \quad (13)$$

Here, $m_1 = 0.04$. $f_i^{'}$ is the value of the smaller adaptation for the current two variants.

Step 7. Initialization of BP weights and thresholds. The obtained optimal individual is decomposed into the initialization weight and threshold value of the BP network. In this chapter, we use three layers of BP neural network, the input neurons are 5, and the hidden neurons are 10.

Step 8. Training model. After optimized, the composite model of BP and AdaBoost is established and began to start training.

In step 5 and 6, the formula mentioned, we can know that the crossover probability and mutation probability follow the change of fitness value, which can avoid the divergence of GA algorithm, fall into local minimum and speed up the convergence. In step 8, we can obtain the optimization of initial weights and thresholds. Then, begin to train the BP network, if the predicted output error does not meet the conditions, the iterative adjustment of program can reduce the output error, and prediction model precision is improved, and the practical application of the prediction is enhanced.

5 Model Validations

Model Feasibility Evaluation Criteria

In this chapter, mean relative error (MRE), sum of relative error absolute values, and mean square error of relative error (MSE) were introduced as metrics of the modeling accuracy. Goodness of fit (R) is used to quantify model accuracy. For a data set of n measured outputs o_i and predicted outputs y_i, Relative error, MRE, and MSE is calculated as

$$erro = \frac{o_i - y_i}{y_i} \tag{14}$$

$$MRE = \frac{1}{N}\sum\nolimits_{i=0}^{N} \frac{o_i - y_i}{y_i} \tag{15}$$

$$MSE = \frac{\sum\nolimits_{i=0}^{N}\left(\frac{o_i - y_i}{y_i}\right)^2}{N} \tag{16}$$

$$R = \frac{\sum\nolimits_{i=0}^{N}(o_i - \delta_1)^2}{\sum\nolimits_{i=0}^{N}(y_i - \delta_1)^2} \tag{17}$$

Model Testing and Evaluation

We use Matlab 2016b software platform to build model. It includes the BP neural network model and the GA-BP-AdaBoost model. There are 200 training samples and 80 of the samples are verification samples. The parameters of the BP neural network model are set as follows: the number of training is 30, the training target is 0.002, and the learning rate is 0.1. The parameters of the GA-BP-AdaBoost neural network model are set as follows: the population size is 30, and the evolutionary algebra is 30.

Neural network training index of output value are as follows: the mean relative error value is 0.0035 and the sum of absolute value of relative error is 2.0225, the relative error of standard deviation is 0.0260, the goodness of fit is 0.9801, the mean square error of the relative error is 6.8235e−04. According to evaluation index, the training model is satisfying, and it can be employed to output prediction.

Figure 2 is the contrast diagram of the two models of the prediction of the output and the error. Figure 3 is the neural network curve fitting, which can show the validity of the prediction function. With the different prediction samples 40, 60, 80, Table 1 gives the relative mean error, mean square error, and fitting degree of two kinds of prediction models.

Factors affecting fuel consumption is not limited to the factors considered in this chapter, also affected by other factors, such as weather conditions, the same flight with different routes and airports and other factors such as route congestion. And considering the security requirements of actual flight, the plane will usually carry 45 min, alternate or return flight fuel, which may lead to "Fuel oil consumption" happen. So the prediction value and the expected value of the fuel consumption are allowed to error exist.

It is concluded that adaptive GA-BP-AdaBoost prediction model has a small increase in the prediction accuracy and nonlinear fitting ability, fault tolerance capability. Moreover, when the training data set is less, the prediction precision and dynamic quality of the model are still kept. By analyzing the index value of the simulation and evaluation model, and considering the existence of the actual error, conclusion can be drawn that the results indicate that these two methods used to forecast fuel consumption is feasible, effective, and convenient for practical applications.

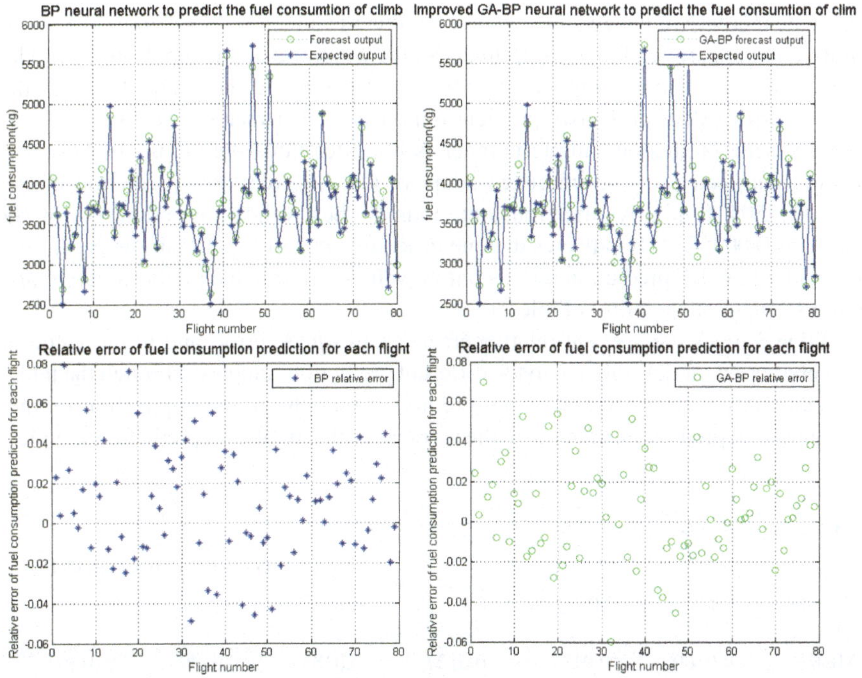

Fig. 2 Output and relative error of model prediction

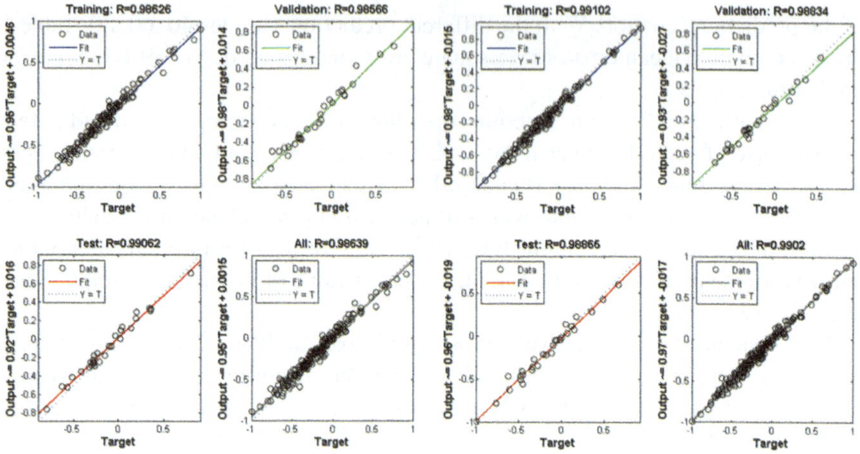

Fig. 3 Analysis of neural network fitting curve

Model Testing and Evaluation

With the increase of the genetic algebra, the fitness value is reduced after 20 generations, and the network fitness is optimal. So the weights and thresholds of the model are optimized which can improve the prediction accuracy of the model. In Fig. 3, the relative error values of the two prediction models are between [−0.1 0.1].

When the prediction sample number is 80, after three iterations, the best validation parameters of the improved adaptive GA-BP-AdaBoost prediction model is 0.0025504. Similarly, after four iterations the best validation parameters of BP model is 0.0035213. It indicates that mean square error of GA-BP strong prediction is smaller than BP prediction. So, its output values closer to the predicted target and has a stronger adaptability (Table 2).

Table 3 gives the average relative error, mean square error, and fitting value of two prediction models relative to the different prediction samples. Through the table, we can analyze the accuracy of the prediction models. In the application of fuel consumption prediction, along with the increase of the number of prediction samples,

Table 2 Model index value

	BP model			Improved GA-BP-AdaBoost model		
Sample number	40	60	80	40	60	80
MRE	0.0100	0.0177	0.0127	0.0185	0.0149	0.0119
SRE	1.0125	1.5365	1.8855	1.1026	1.3550	1.8830
MSE	0.0011	0.0011	8.80e−04	0.0011	7.67e−04	0.0011
R	0.9558	0.9041	0.9043	0.9347	0.9347	0.9633

Table 3 Model comparison

Forecast sample number 80	BP model	FUZZY model	SVR model	Improved GA-BP-AdaBoost model
MRE	0.0177	0.0065	0.0194	0.0119
R	0.9043	0.9049	0.9070	0.9633

the two-forecast model can be effective convergence and the predictive accuracy of adaptive GA-BP-AdaBoost prediction model increases slightly which is converted to the number level of the actual project can achieve a more significant improvement.

The fuzzy logic model of fuel consumption proposed by Chang R C and the SVR model of fuel consumption proposed by HF Zhang are using the same data as the above experimental data to forecast, the prediction sample number is 80. Results in the following Table 3. Analyzing the data in the Table 2, it is found that the Improved GA-BP-AdaBoost model than other model has a good capability of nonlinear fitting, and the average relative error is small than SVR model and BP model. However, the average relative error of fuzzy logic model proposed by Chang R C is small than the average relative error of Improved GA-BP-AdaBoost model, but its sum of relative error is 2.4319, higher than the Improved GA-BP-AdaBoost model error sum 0.5489.it shows that fuzzy logic model is at the cost of model performance to reduce the output error, so that dynamic stability of fuzzy logic model become lower.

Factors affecting fuel consumption is not limited to the factors considered in this chapter, also affected by other factors, such as weather conditions, the same flight with different routes, and airports and other factors such as route congestion. And considering the security requirements of actual flight, the plane will usually carry 45 min alternate or return flight fuel, which may lead to "Fuel oil consumption" happen. So, the prediction value and the expected value of the fuel consumption are allowed to error exist.

It is concluded that adaptive GA-BP-AdaBoost prediction model has a small increase in the prediction accuracy and nonlinear fitting ability, fault tolerance capability. Moreover, when the training data set is less, the prediction precision and dynamic quality of the model are still kept. By analyzing the index value of the simulation and evaluation model, and considering the existence of the actual error, conclusion can be drawn that the results indicate that these two methods, used to forecast fuel consumption, are feasible, effective, and convenient for practical applications.

In the premise of not affect flight safety, it is in order to effectively reduce the fuel consumption, as much as possible to improve the utilization rate of fuel resources. We use the model proposed in this chapter to analyze single factor influence on fuel consumption when defining other influence factors.

Through the prediction of fuel consumption, we can get the single factor variable interval. It can provide a reference for the flight plan, so that flight to achieve the best fuel saving flight state.

6 Conclusion

Accurate forecasts of fuel consumption are vital when demand grows faster, it can guide civil aviation energy policies' effective implementation and reduce flight carbon emissions. Energy consumption forecast is a complex problem due to interactive factors. In this study, in order to reduce the amount of carbon emissions from flights and improve the fuel utilization rate, the fuel consumption forecast model based on GA-BP-AdaBoost is presented. The model is verified by flight climbing stage.

By comparing the BP prediction model, the proposed model in the prediction accuracy, nonlinear fitting ability, and fault-tolerant ability are increased. And with small training data set, the model can still maintain prediction accuracy and dynamic performance index. Analyzing the Simulation and considering the actual causes of errors, it can be concluded that the two models have practical applications. Also, through the analysis of QAR data, we can aim at the different stages of the same voyage flight to establishment two kinds of prediction models are presented in the chapter, they used to predict aircraft in various stages of the fuel consumption. Similarly, in the case of network training as well, if the forecast result of test set is found to have a large deviation from the actual value and then analyzed the error, we can even determine in this flight if there is a fault. Therefore, it is necessary for us to study the prediction model of fuel oil.

In future work, the enhancement of the model presented here is the extension to estimate thrust associated with a fuel burn flight condition parameter such as Thrust Specific Fuel Consumption (TSFC). Preliminary results obtained indicate that TSFC can also be easily characterized using genetic algorithm and TSFC can be regarded as time series. Maybe it is a good research direction to build fuel prediction model related to trajectory prediction of transport aircraft based on time series.

Conflict of Interests The authors declare that there is no conflict of interests regarding the publication of this paper.

References

1. Fan, Y., Xia, Y.: Exploring energy consumption and demand in China. Energy **40**, 23–30 (2012)
2. Yuan, C., Liu, S., Fang, Z., Xie, N.: The relation between Chinese economic development and energy consumption in the different periods. Energy Policy **38**, 5189–5198 (2010)
3. Henderson, R.P., Martins, J.R.R.A., Perez, R.E.: Aircraft conceptual design for optimal environmental performance. Aeronaut. J. **116**, 1–22 (2012)
4. Chang, R.C.: Examination of excessive fuel consumption for transport jet aircraft based on fuzzy-logic models of flight data. Fuzzy Sets Syst. **269**, 115–134 (2015)
5. Zhang, H.F., Wang, X.H., Chen, X.F.: Support vector with ROC optimization method based fuel consumption modeling for civil aircraft. Proc. Eng. **99**, 296–300 (2015)
6. Aydan, C.: Impact of aircraft performance differences on fuel consumption of aircraft in air of management environment. Aircr. Eng. Aerosp. Technol. **76**, 502–515 (2004)
7. Collins, B.P.: Estimation of aircraft fuel consumption. J. AIRCR **19**, 969–975 (1982)

8. Mayer, R.H.: Change-oriented aircraft fuel burn and emissions assessment methodologies. In: ICNS 2012: Bridging CNS and ATM—Conference Proceedings, pp. N5-1–N5-15 (2012)
9. Baklacioglu, T.: Modeling the fuel flow-rate of transport aircraft during flight phases using genetic algorithm-optimized neural networks. Aerosp. Sci. Technol. **49**, 52–62 (2016)
10. Baklacioglu, T.: Aero-propulsive modelling for climb and descent trajectory prediction of transport aircraft using genetic algorithm. Aeronaut. J. **118**, 65–79 (2014)
11. Bartel, M., Young T.M.: Simplified thrust and fuel consumption models for modern two-shaft turbofan engines. J. Aircraft **45**, 1450–1456 (2008)
12. Baklacioglu, T.: Fuel flow-rate modelling of transport aircraft for the climb flight using genetic algorithms. Aeronaut. J. **199**, 173–183 (2015)
13. Yoshiaki, U., Keiichi, H., Ryo, K.: A modified real-coded genetic algorithm con-sidering with fitness-based variability. Int. J. Innovat. Comput. Informat. Control **10**, 1509–1518 (2014)
14. Chen, J.J., Yan, Y., Liu, J.X.: Fuel consumption estimation of cruise phase based on BP net-work and influence structure analysis. Comput. Measure. Control **06**, 2135–2138 (2015)
15. Saravanan, S., Karman, S., Amosedinakaran, S., Thangaraj, C.: India's electricity demand estimation using Genetic Algorithm. In: International Conference on Circuits Power and Computing Technologies, pp. 97–101 (2014)
16. Hu, X.N.: The aircraft climb segment fuel consumption analysis model based on PCA-BP. Technol. Innovat. Appl. **19**, 88–89 (2017)
17. Chen, J.J., Li, M.: Prediction of cruise fuel consumption based on bootstrap for small samples. Measure. Control Technol. **34**, 26–29 (2015)
18. Khadilkar, H., Balakrishnan, H.: Estimation of aircraft taxi fuel burn using flight data. Trans. Environ. **17**, 532–537 (2012)
19. Yang, H., Zhao, C.L.: QAR data fault detection research based on the FP—tree. Comput. Appl. Softw. **10**, 41–44 (2014)
20. Liu, L.X.: Selection of independent variables and the stepwise regression method in the linear regression model. Statist. Decis. **21**, 80–82 (2015)

Time Series Optimization for Energy Prediction in Wi-Fi Infrastructures

David Rodriguez-Lozano, Juan A. Gomez-Pulido and Arturo Duran-Dominguez

Abstract Access points play an important role in Wi-Fi networks and can provide us with useful information about the energy consumption according to the users' behavior. If we predict the energy consumption in a determined access point, we can make easier the maintenance plans for the network infrastructure making the most adequate decisions about the placement of new devices or reinforcement of existing ones, for example. In this work, we propose an energy prediction methodology based on system identification, where the energy measured in the access points is represented as time series. The prediction results were reasonably good for an experimental environment consisting of ten access points in an academic building, modeling the energy patterns along some weeks. Moreover, we found an optimization problem where the main parameters of the identification model can be adjusted in order to provide results more accurate. Given the computational effort required for searching in depth the optimal values, we applied a genetic algorithm, which provided better results in less time with regard to a direct search method.

Keywords Wi-Fi networks · Access point · Energy consumption · Time series System identification · Prediction · Optimization · Genetic algorithm

1 Introduction

The Access Point (AP) is a device that supports the data traffic and the session requests in a Wi-Fi infrastructure. The energy use in the AP comes mainly from the demand for network access by users, among other factors. The energy levels in the APs are optimization objectives in many research works [1], and they give us

D. Rodriguez-Lozano · J. A. Gomez-Pulido (✉) · A. Duran-Dominguez
University of Extremadura, 10003 Cáceres, Spain
e-mail: jangomez@unex.es

D. Rodriguez-Lozano
e-mail: drlozano@unex.es

A. Duran-Dominguez
e-mail: arduran@unex.es

I. Rojas et al. (eds.), *Time Series Analysis and Forecasting*,
Contributions to Statistics, https://doi.org/10.1007/978-3-319-96944-2_17

useful information about the users' behavior. From this knowledge, we can plan the maintenance tasks of the network infrastructure. In this sense, we should know the energy impact of the maintenance work before doing it, predicting the energy in the APs according to the past energy patterns.

We predict the energy in the APs from time series modeling, using a three-step methodology. First, we collect data on the energy from the network usage during a time period where the users had a regular activity. Next, we build as many time series as access points the network has, where each series draws the daily energy level in the corresponding AP. Next, we model the time series applying system identification. Finally, the models obtained are used to predict the next energy data. The prediction could be better if we improve the identification by using metaheuristics that optimize their main parameters.

We have not found works about predicting energy in the APs using time series, but others focus with regards to predict the density of neighboring APs and the corresponding data traffic [2], or the data quota [3], for example. Other aspects of the wireless networks such as users' location [4], data traffic [5], mobility [6, 7] application workloads were studied for prediction purposes.

2 System Identification of Time Series

A Time Series (TS) is a signal $y(k)$ sampled by a period T that describes the behavior of a dynamic system. System Identification (SI) [8] tries to find a parametric mathematical model of the TS from the measures of $y(k)$.

The parametric polynomial description is usual in SI. The ARMAX (Moving-Average Auto-Regressive) model [9] is a well-known option to model a discrete system. If q is the delay unit, and q^{-d} the time delay $(k-d)$, ARMAX is described by $A(q)y(k) = 0$, where $A(q) = 1 + a_1 q^{-1} + \cdots + a_{na} q^{-na}$, being na the model size. In a polynomial description, the ARMAX model is given by (1).

$$y(k) + a_1 y(k-1) + \cdots + a_{na} y(k-na) = 0 \tag{1}$$

The identification of the TS consists in determining the values of a_i from the observation of the NM samples of the signal y. From this model, we can calculate the estimated signal $y_e(k)$ (2), where θ is a one-column matrix containing the a_i parameters and φ is a one-column matrix with observed values. Then, $y_e(k)$ is compared with the real signal $y(k)$ in order to determine the error done.

$$y_e(k) = [-a_1 y(k-1) - \cdots - a_{na} y(k-na)] = \varphi^T(k)\theta \tag{2}$$

There are two possibilities to perform the identification (obtain θ): batch or recursive. We choose the recursive parametric estimation, which estimates and updates θ along the time, in a way that, for each time k, we obtain an ARMAX model.

Obviously, the more past samples we have, the more accurate model we obtain, because we have more information about the behavior of the system.

There are several algorithms for the recursive identification: Kalman Filter, RLS (Recursive Least Squares) and LMS (Least Mean Squares). We choose RLS because of its goodness and accuracy, and because it can be used in environments of dynamic nature where the TS is processed in real time. This algorithm considers the forgetting factor λ, whose value is usually chosen in the interval 0.97 to 0.995 [9]. On the other hand, the accumulated error (3) can be a good measure of the identification accuracy, and it mainly depends on the model size and the forgetting factor, for a particular TS.

$$F(na, \lambda) = \sum_{k=1}^{k=NM} |y(k) - y_e(k)| \quad (3)$$

3 Energy Prediction in Access Points

The prediction of the future behavior of the TS is possible if it is previously identified in order to know its behavior by a mathematical description. The recursive estimation of the ARMAX model allows us to obtain this description. From this approach, the prediction can improve when the identification advances in the time, because we suppose models are more accurate.

Prediction Approach

In order to perform the prediction, we choose the time k_s from which we model the system based on the past behavior. In practice, k_s will be the last known value of the TS. Let us suppose y is the real TS, y_e is the estimated value from the ARMAX model, and y_s is the predicted TS. We have real values of y until $k = k_s$, so the last estimated value will be $y_e(k_s + 1)$, since the estimated value in $k_s + 1$ is calculated from the model built with the real values up to previous time, k_s. From $k_s + 1$, we predict by RLS assuming $y_s(k) = y_e(k)$. Therefore, $y_s(k_s + 1) = y_e(k_s + 1)$, and we apply RLS successively.

Figure 1 shows an example of TS predicted with this approach. The top plot shows the identification of the full TS ($na = 3$ and $\lambda = 0.98$), where y_e is the TS estimated from the model generated with all the data in the TS, and $NM = 39$ is the number of samples. Now, let us suppose we only know the TS up to $k_s = 20$. From this time, we generate the predicted signal y_s, calculated in this way: In the next time $(k_s + 1)$ to the last known value $y(k_s)$, the predicted value y_s is y_e is estimated according to the model ARMAX-RLS built considering the previous known y values. Then, in $k_s + 2$, we perform the identification taking the real value of the TS as y_s, instead

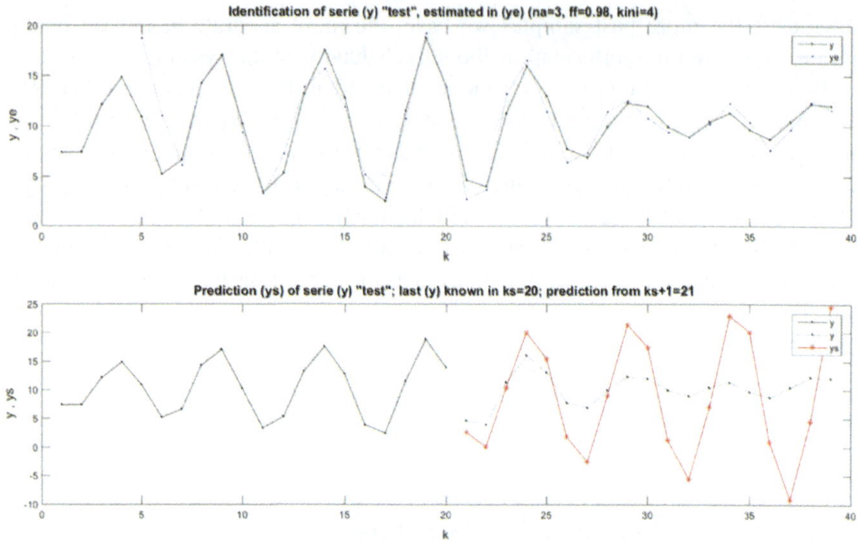

Fig. 1 Example of identification (top) and prediction from $k_s = 20$ (down)

of y (we suppose we do not know it already). Therefore, we calculate the estimated value $y_e(k_s + 2)$, next it is assigned again to the predicted signal y_s at this time, and so on.

We can see in the down plot of the figure the predicted y_s (the dotted plot is the real y which has not been taken into account for the prediction; it is shown only for comparison purpose). Obviously, the more predicted values, the worse prediction we have, because the predicted signal is built with the previous predicted values, instead of the previous real ones.

Energy Data

We have collected data from a library building of the University of Extremadura (UEX), Spain. This Wi-Fi network is composed of 10 access points accessed by 2,907 users along 73 days: 10 full weeks labeled from #1 to #10, and the first 3 days next to the last week; nevertheless, we only consider 5 of these 10 weeks, as we explain later. The energy data in each AP were collected each 60 s in the 12 h daily period from 9 to 21 h, in order to filter the energy data closer to the real users' behavior, since the usual activity of the network users happens in that period in the library building.

We built 10 time series from the collected data, one for each AP, showing the total daily energy due to the users' activity in the period. We want to predict the energy in the APs when we do not have more data. The prediction is more accurate in the next

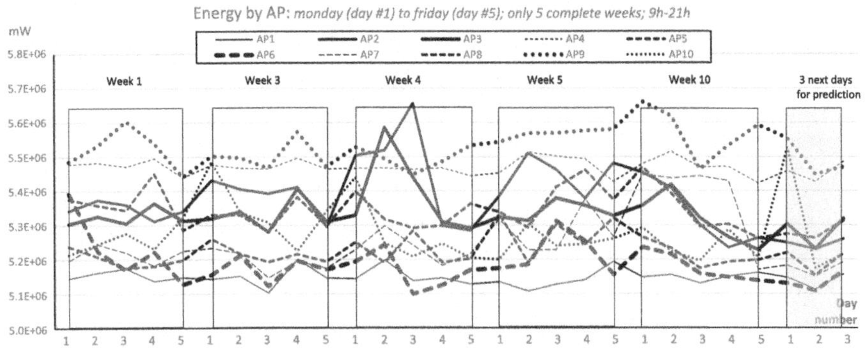

Fig. 2 Ten time series corresponding with the energy in each AP, sampled daily considering only working days: from Monday (day 1) to Friday (day 5). Each series is composed of 25 samples. The 3 last days considered for comparison purposes with the predicted days are also shown

day to the last known day, and it gets worse when the predicted day moves away. The three last days (71–73) are not part of the TS, but they are left for comparison purposes with the 3 first predicted days.

We are interested in analyzing the users' behavior through the energy patterns. For this purpose, we remove all the weekends, since the library is closed then. Moreover, we delete those weeks where one of their days does not reflect a usual behaviour (holiday, network down, etc). Taking into account these constraints, we only consider 5 weeks (25 days), as they are shown in Fig. 2.

Prediction Results

Figure 3 shows the prediction of the time series. Dotted and continuous lines are real (y) and predicted (y_s) series, respectively.

The value of na determines the initial time k_{ini} from which RLS builds the ARMAX model. The more na is, the more value for k_{ini}. Therefore, as the time series has a low number of samples ($NM = 25$), we should consider a low value for na; otherwise, the identification would start later, the recursive calculations would consider few data and, consequently, the identification would be worse.

The predicted values were compared with the real ones using the variable v (variation) defined as $v(\%) = 100 \times (y_s - y)/y$. The predicted data and the absolute values of their corresponding variations are shown graphically in Fig. 4. In general, the predictions are good; as regards they do not move away too much from the real values (less than 2% in almost all the cases). Only 2 of the 30 predictions have variations of 3 and 4%. Besides, the variations for the first predicted day ($k_s + 1$) are under 1% in 8 of the 10 access points.

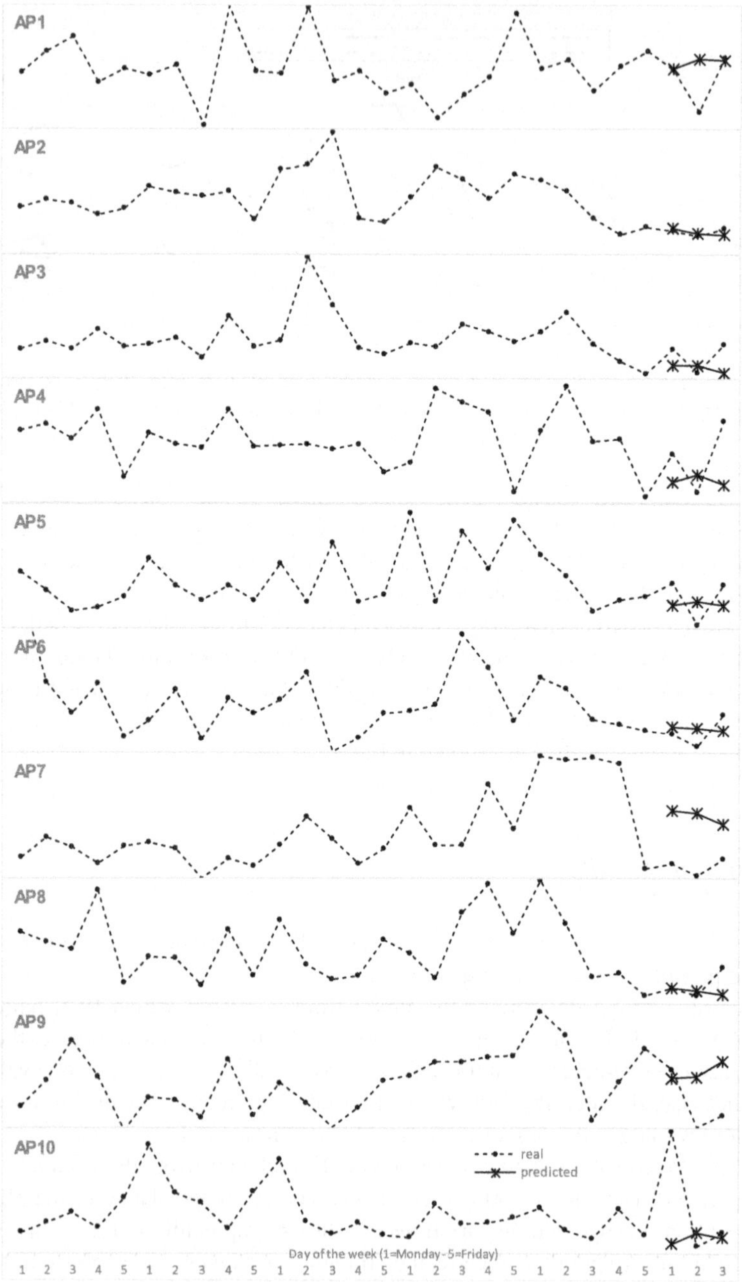

Fig. 3 Predicted energy (continuous line) in each AP in the next 3 days to the last considered day (sample number 25). The known energy data (dotted line) correspond with the working days of the 5 considered weeks, from Monday (1) to Friday (5). The real energies of the 3 last days are not considered in the time series, but they are useful for calculating the prediction performance. Energy is displayed in mW

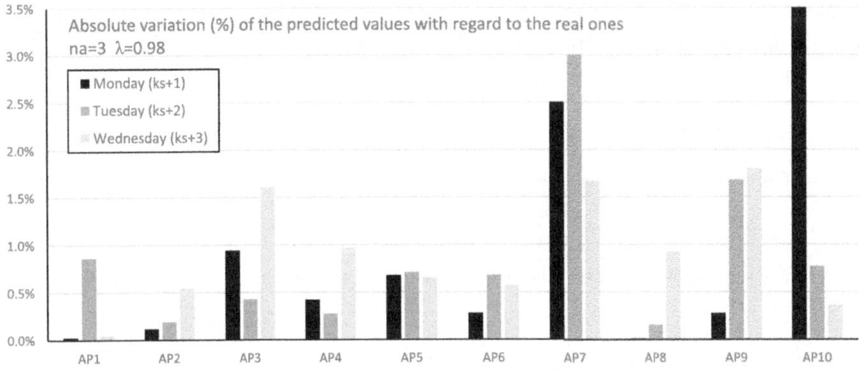

Fig. 4 Absolute variation of the predicted energy with regard to the real energy, in each access point, for the next 3 days to the last known day (k_s): Monday ($k_s + 1$), Tuesday ($k_s + 2$), and Wednesday ($k_s + 3$)

The variation can be positive or negative, showing the trend to increase or decrease the own time series (according to the day in the week, the trend in the network use will be greater or smaller). We can see how the prediction gets worse when it is far from the last known value ($k_s = 25$). This is observable in 6 of the 10 access points; in the other 4 APs, the prediction is different, depending on the behavior and variability of the own time series.

Finally, we remember that the size of the time series and its behavior influences on the prediction results. One the one hand, the more weeks the TS has, the more accurate the prediction will be, because we have more samples of the same day. On the other hand, the variability of the TS (a very different behavior between consecutive days or for the same day in each week) implies a worse identification. Ideally, the prediction will be better as more stable be the behavior of the time series. The worse results of our work can be blamed to the access points that show high variability in its behavior (for example, AP7).

4 Parameter Tuning for Improving the Accuracy

The parameters na (model size) and λ (forgetting factor) have a strong influence on the identification accuracy. In this section, we compare a direct search method and a metaheuristic for founding the optimal pair.

Direct Search

We should establish an upper limit to na if we want to process enough samples of the TS, because RLS only can process from $k_{ini} = na + 1$. The upper limit is NM-MM-1,

Table 1 Optimal pairs (na, λ) and the corresponding identification errors, found after two experiments of direct search with 74,505 (1) and 745,005 (2) identifications

	AP1	AP2	AP3	AP4	AP5	AP6	AP7	AP8	AP9	AP10
na_1	3	3	3	3	3	3	3	3	3	3
λ_1	3.24	3.24	3.24	3.24	3.24	3.24	3.24	3.24	3.24	3.24
F_1	5.65E+6	8.41E+6	1.34E+7	9.42E+6	1.03E+7	6.89E+6	8.35E+6	8.09E+6	8.15E+6	1.26E+7
na_2	3	3	3	3	3	3	3	3	3	3
λ_2	4.32	4.32	4.32	4.32	4.32	4.32	4.32	4.32	4.32	4.32
F_2	5.62E+6	8.98E+6	7.72E+6	9.79E+6	7.79E+6	6.62E+6	7.19E+6	1.31E+7	8.44E+6	2.65E+7

where NM is the number of samples and MM the minimum number of samples that we want to process. As our series has 28 samples, we think that MM = 20 is a reasonable minimum value, because the more behavior information of the series be processed, the more accuracy the prediction will have. On the other hand, the lower limit should be at least 3 because of the ARMAX model is polynomial. Consequently, we establish the limits $3 \leq na \leq 7$. With regard to the forgetting factor, we establish a wide range: $0.1 \leq \lambda \leq 15.0$.

The direct search considers values of the pair (na, λ) generated iteratively by incrementing them using the intervals h_{na} and h_λ for na and λ respectively. The interval $h_{na} = 1$ because the model size is integer, so we consider 5 possible values of na. However, h_λ depends on the desirable accuracy for the direct search. For this purpose, we have performed two experiments with different computational efforts: $h_\lambda = 0.001$ (experiment #1) and $h_\lambda = 0.0001$ (experiment #2). This means to generate $1 + (h_{\lambda u} - h_{\lambda l})/h_\lambda$ values of λ, where λu and λl are the upper and lower limits of λ, respectively. This way, experiment #1 implies to generate 14,901 values for λ and 74,505 identifications, whereas experiment #2 implies to generate 149,001 values for λ and 745,005 identifications, for each TS.

Table 1 shows the optimal pairs and their corresponding minimum identification errors, found after performing both the experiments. We conclude that, for each experiment, the minimum error was found always for the same pair in all the TS: the best model size was $na = 3$, whereas λ oscillates between 3.24 and 5.

Figure 5 shows an example of the direct search. The plot on the left displays the identification errors for the pairs (na, λ) generated in experiment #1 for AP6. This plot does not show clearly the possible minimum peaks, so we have applied a correcting function that highlights the maximum and minimum peaks, maintaining the proportion among all the points. This way, the plot on the side shows a high number of minimum peaks near themselves, where we could sense the existence of a global minimum. This fact is similar in the other series.

We think that the high number of minimum peaks suggests that the optimum pair could be different for each TS; even more, the identification error is very sensitive to the pair setting. Therefore, we should perform a deeper search. Nevertheless, a direct search could waste a high computational effort performing identifications in wide areas where the minimum errors are not present. Besides, larger time series

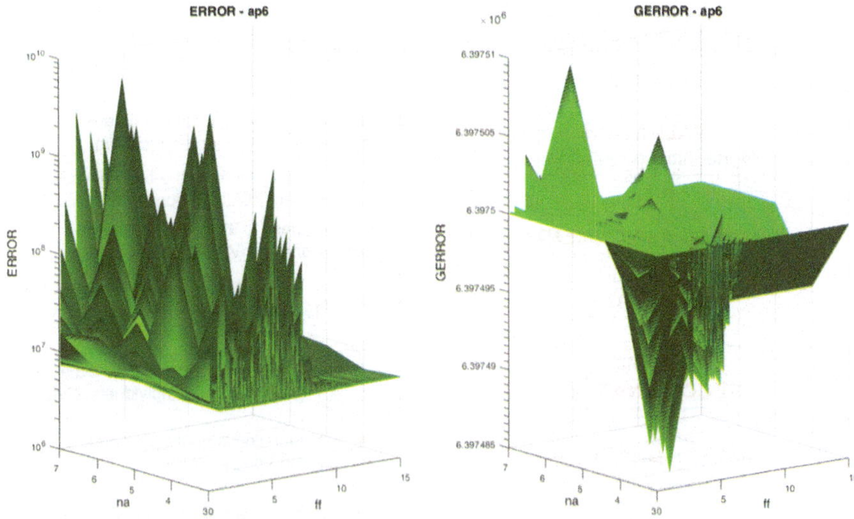

Fig. 5 Error measure in the identification of AP6, for 74,505 values of the pair (na, λ). Plot on the right shows the same plot on the left after applying a correcting function that highlights the maximum and minimum peaks, maintaining the proportion among all the points, in order to show a better display

imply higher computation time for the identification. All these factors move us to consider alternatives that are more efficient instead of direct search. In this sense, metaheuristics are well-proven solutions in optimization problems.

Metaheuristics for Efficient Search

Metaheuristics [10] are approximate algorithms based on heuristic search. They explore efficiently the space of solutions intensifying the search in the nearness of a promising solution, and can be based on trajectory or population. Among population-based metaheuristics, Evolutionary Algorithms (EAs) [11] are inspired in evolution rules at individual level.

Genetic Algorithms (GA) [12] are one of the most known EA. They are stochastic search methods with many successful applications. In order to understand their behavior, we explain some definitions:

1. The objective function $f(X)$ is the function to optimize, and it depends on NVAR decision variables x_i. In our case, we have NVAR = 2 decision variables, $x_1 = na$ and $x_2 = \lambda$, and $f(X)$ is the accumulated error (3) in the TS identification: $f(x_1, x_2) = F(na, \lambda)$. Therefore, each evaluation of the objective function implies to perform a full identification.

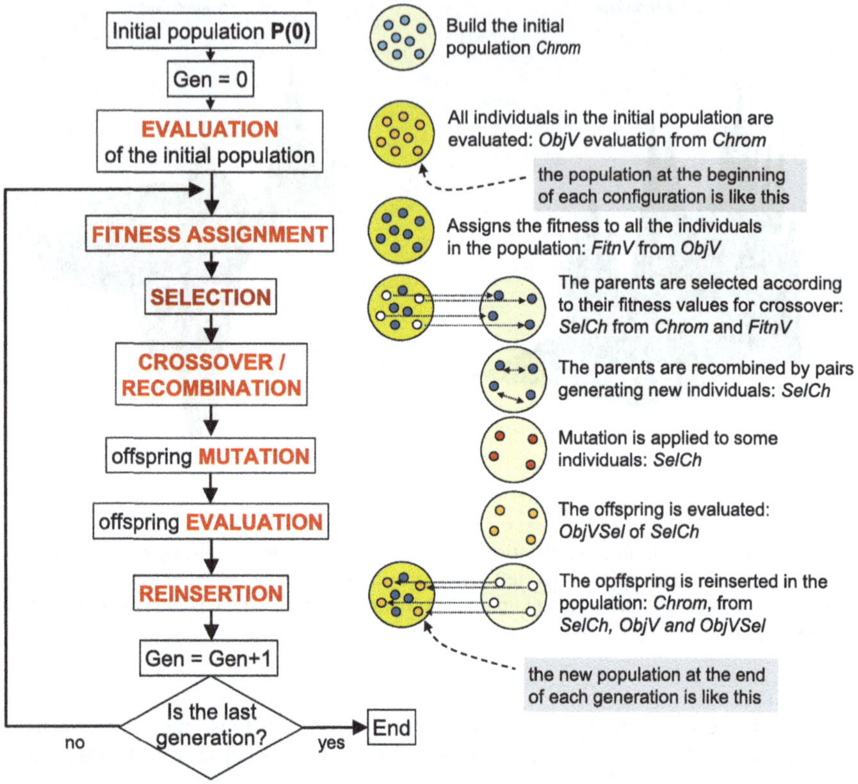

Fig. 6 Scheme of the genetic algorithm used in this work

2. An individual X is a solution of the optimization problem, or a set of values x_i. In our case, a solution is a pair (na, λ).
3. A population is the set of NIND individuals that evolves along generations.
4. The phenotype P_i of individual X_i is defined by their decision variables $P_i = (x_{i,1}, ..., lx_{i,NVAR})$, and the genotype G_i of individual X_i its phenotype is coded according to a determined alphabet. In our case, we work with real numbers, so the genotype has the same representation. The chromosome G of the population is the set of genotypes of all their individuals.
5. Fitness function transforms the objective function into a measure of relative fitness. In our case, the fitness of an individual is its objective value with regard to the entire population.

Figure 6 shows how the GA works:

1. We start generating an initial population: $Chrom$.
2. Next, all the individuals in the population are evaluated: $ObjV$.
3. Assignment phase: In the first iteration (generation) of the GA, a fitness value is assigned to each individual: $FitnV$.

Table 2 Optimal pairs (na, λ) and the corresponding identification errors, found after 10 runs of the GA for each time series, according to the parameter limits

	AP1	AP2	AP3	AP4	AP5	AP6	AP7	AP8	AP9	AP10
na	3	3	4	5	5	4	3	7	7	6
λ	4.87	3.70	5.60	2.23	3.42	2.58	1.70	3.22	1.50	2.19
F	5.66E+6	7.54E+6	6.81E+6	5.94E+6	6.22E+6	6.20E+6	6.92E+6	6.26E+6	6.73E+6	7.13E+6
#run	8	5	7	8	7	2	1	5	7	9

4. Selection phase: the best individuals (parents) are chosen for crossover according to their fitness values: $SelCh$.
5. Recombination phase: The parents are crossed to generate new individuals (offspring), updating $SelCh$.
6. Mutation phase: Particular mutations are applied to some individuals of the offspring, updating $SelCh$.
7. Evaluation phase: The offspring ins evaluated: $ObjVSel$.
8. Reinsertion phase: The offspring individuals are reinstated in $Chrom$.
9. The algorithm goes to the assignment phase again. These phases are performed iteratively along many generations, up to a stop criterion is reached. This criterion can be a number of generations, computation time, etc.

After many experiments, we have tuned the main parameters of the GA as follows: population of 500 individuals, 50 generations as stop criterion, selection rate of 0.9, crossover probability of 0.8, selection by universal stochastic sampling, and crossover by fixed point.

Table 2 shows the optimal solutions found after 10 runs of the GA for each TS, observing the limits established in direct search for na and λ. We note that GA performs less identifications than direct search. Thus, one run of GA performs {evaluation of initial population (500) + 50 generations × [selection rate (0.9) × population size (500)]} = 23,000 identifications, whereas experiments #1 and #2 of direct search implied 74,505 and 745,005 identifications, respectively. Thus, if we compare Tables 1 and 2, we check how GA provides minimum errors in all TS, except for AP1 where the error is practically the same. Consequently, the GA improves the accuracy of the identification, and consequently the prediction, with lesser computational effort with regard to a simple direct search.

5 Conclusions

We have applied the time series analysis to the daily energy consumption in the access points of a Wi-Fi network, in order to predict the values for the next days. This knowledge can be useful for the maintenance of the network infrastructure. We have collected data from a real wireless environment consisting of ten access points in an academic building, accessed by thousands of students along several weeks. The time series, one for each access point, were modeled according to an auto-regressive formulation with recursive estimation and predicted for the three next days. The prediction results were compared with known energy values in order to analyze the method accuracy.

The obtained results are good enough, since the differences between prediction and real values are under 2% in almost all the cases. Nevertheless, the prediction can be improved using optimal values for the prediction model and the forgetting factor in the recursive estimation, since both parameters have a strong influence in the error done. Pursuing this goal, we have checked how a proposal based on genetic algorithms provides better optimal values than direct search methods, even with low computational efforts.

As future research works, we propose to add other parameters different than energy consumption to be predicted for maintenance purpose like number of connected users, network sessions, and data traffic, since these data are easily accessible from the AP itself. In addition, more realistic networks should be considered. For example, time series should be larger, in order to analyze more data with regard to the users' behavior; thus, we may consider a wider window of days in order to give more representation to the different parts of the year (holidays, beginning of the academic year, examination period, etc.).

Acknowledgements This work was partially funded by the Government of Extremadura under the project IB16002, and by the AEI (State Research Agency, Spain) and the ERDF (European Regional Development Fund, EU) under the contract TIN2016-76259-P (PROTEIN project).

References

1. Konstantinidis, A., Yang, K.: Multi-objective K-connected Deployment and Power Assignment in WSNs using a problem-specific constrained evolutionary algorithm based on decomposition. Comput. Commun. **34**, 83–98 (2010)
2. Zhang, H., Chu, X., Guo, W., Wang, S.: Coexistence of Wi-Fi and heterogeneous small cell networks sharing unlicensed spectrum. IEEE Commun. Mag. **53**, 158–164 (2015)
3. Shen, Y., Jiang, C., Quek, T., Zhang, H., Ren, Y.: Pricing equilibrium for data redistribution market in wireless networks with matching methodology. In: 2015 IEEE International Conference on Communications (ICC), pp. 3051–3056 (2015)
4. Scellato, S., Musolesi, M., Mascolo, C., Latora, V., Campbell, A.: NextPlace: A Spatio-temporal Prediction Framework for Pervasive Systems. In: 2011 Proceedings of the 9th International Conference on Pervasive Computing, pp. 152–169 (2011)
5. Hamamoto, R., Takano, C., Obata, H., Ishida, K.: Improvement of throughput prediction method for access point in multi-rate WLANs considering media access control and frame collision. In: 2015 Third International Symposium on Computing and Networking (CANDAR), pp. 227–233 (2015)
6. François, J., Leduc, G.: AP and MN-centric mobility prediction: a comparative study based on wireless traces. In: Proceedings of the 6th International IFIP-TC6 Networking Conference, pp. 322–332 (2007)
7. Gmach, D., Rolia, J., Cherkasova, L., Kemper, A.: Workload analysis and demand prediction of enterprise data center applications. In: Proceedings of the 2007 IEEE 10th International Symposium on Workload Characterization, pp. 171–180 (2007)

8. Soderstrom, T., Stoica, P.: System Identification. Prentice-Hall Int, London (1989)
9. Ljung, L.: System Identification, Theory for the User. Prentice-Hall, Englewood Cliffs, N.J. (1999)
10. Gendreau, M., Potvin, J. (eds.): Handbook of Metaheuristics. Springer (2010)
11. Fogel, G., Corne, D.: Evolutionary computation in bioinformatics. Morgan Kaufmann Publishers (2003)
12. Reeves, C., Rowe, J.: Genetic Algorithms. Principles and perspectives. A guide to GA Theory. Kluwer Academic Publisher, EE.UU (2003)

11. [illegible]
12. [illegible]

An Econometric Analysis of the Merit-Order Effect in Electricity Spot Price: The Germany Case

François Benhmad and Jacques Percebois

Abstract In this paper, we carry out an econometric analysis for Germany, as a country with high penetration of renewable energy sources (RES), in order to investigate impact of wind energy and photovoltaic feed-in on electricity spot price level, the so-called merit-order effect. We have used an ARMA-X-GARCH-X modeling where wind generation and photovoltaic are considered as exogenous variables included in the mean and the variance equation, in order to assess the joint impact of RES on the electricity spot price level as well as on spot price volatility in Germany. Our main empirical findings suggest that wind power and photovoltaic feed-in decreases electricity spot price. However, their impact on electricity spot prices volatility is quite different. Indeed, the solar photovoltaic power has a lowering on impact electricity price volatility, whereas the wind feed-in exacerbates it.

Keywords RES · Electricity spot prices · Merit-order effect · Volatility

1 Introduction

Renewable energy is a key component of the EU energy strategy. It started with the adoption of the 1997 White paper and has been driven by the need to decarbonize the energy sector and address growing dependency on fossil fuel imports from politically unstable regions outside the EU. In 2009, the EU released the First Climate and Energy Package, with 2020 targets (compared to 1990 levels): 20% GHG emissions reduction, 20% renewable energy share in primary energy mix, and 20% energy

F. Benhmad (✉)
Montpellier University, Site Richter, Avenue Raymond Dugrand, CS79606, 34960 Montpellier Cedex 2, France
e-mail: francois.benhmad@umontpellier.fr

J. Percebois
Art-Dev, Montpellier University, Site Richter, Avenue Raymond Dugrand, CS79606, 34960 Montpellier Cedex 2, France
e-mail: jacques.percebois@univ-montp1.fr

I. Rojas et al. (eds.), *Time Series Analysis and Forecasting*,
Contributions to Statistics, https://doi.org/10.1007/978-3-319-96944-2_18

efficiency improvement. The Second Climate and Energy Package with targets for 2030 released in 2014 comprised the objectives submitted to the COP21 in 2015: 40% GHG emissions reductions, 27% renewable energy share in primary energy mix, and 27% energy efficiency improvement.

Various RES supporting schemes are operating in Europe, mainly feed-in tariffs, fixed premiums, and green certificate systems. The German Renewable Energy Act, "Erneuerbare-Energien-Gesetz" (EEG), a well-known support scheme, has provided a favorable feed-in tariff (FIT) for a variety of renewable energy sources (RES) since the year 2000. It also gives priority to electric power in-feed from RES overpower in-feed from conventional power plants, i.e., fossil- and nuclear-fuel thermal and already existing hydro-based power plants. Thus, all renewable sources combined made up to 30 per cent of gross electricity production in 2016 and are Germany's second most important source of electricity generation after coal [3].

One of the central empirical findings in the literature on renewable energy is that an increase in RES generation would put a downward pressure on the spot electricity market price by displacing the conventional power plants with higher marginal cost.

In this paper, we carry out an econometric analysis in order to investigate the impact of the RES on electricity prices, the so-called, merit-order effect using a data sample of daily electricity spot prices in Germany for the 2012–2016 period.

There are two main contributions of this study to the literature. First, in contrast to the previous studies in Germany, we take into account the joint impact of wind feed-in and solar photovoltaic on electricity price with a more recent dataset allowing us to assess the learning curve of new technologies integration in the energy mix of Germany.

Second, an ARMA-X-GARCH-X modeling is used with wind and photovoltaic power generation as exogenous variables included in the mean and the variance equation. The goal is to assess the joint impact of intermittent renewable electricity generation on the electricity spot price level as well as on spot price volatility in Germany.

Our main findings suggest that intermittent wind feed-in and solar photovoltaic power generation not only decrease the spot electricity price in Germany but also have an impact on its price volatility. However, photovoltaic has a downward impact, whereas the wind feed-in has an opposite impact—upward—on electricity spot price volatility.

The so-called merit-order effect has gained increasing attention in the literature both on a theoretical basis and an empirical one. Indeed, Jensen and Skytte [11] point out that RES generation enters at the base of the merit-order function, thus shifting the supply curve to the right and crowding the most expensive marginal plants out from the market, with a reduction of the wholesale clearing electricity price.

Several papers have carried out empirical analysis on the impact of RES in electricity markets, finding evidence of the merit-order effect. Indeed, one of the central empirical findings in the literature on renewable energy sources (RES) is that an increase in intermittent sources generation would put a downward pressure on the spot electricity market price by displacing high fuel-cost marginal generation. RES installations, although they are very capital-intensive, have almost zero marginal generation cost and thus are certainly dispatched to meet demand. More expensive conventional power plants are crowded out, and the electricity price declines.

It is worth noting that several authors have explored this topic. For Germany, Bode and Groscurth [4] find that renewable power generation lowers the electricity price. Neubarth et al. [15] show that the daily average value of the market spot price decreases by 1 €/MWh per additional 1,000 MW wind capacity. Sensfuss et al. [17] show that in 2006, renewables reduced the average market price by 7.83 €/MWh. Nicolosi and Fürsch [16] confirm that in the short run, wind power feed-in reduces prices, whereas in the long run, wind power affects conventional capacity, which could eventually be substituted. For Denmark, Munksgaard and Morthorst [14] conclude that if there is little or no wind (<400 MW), prices can increase up to around 80 €/MWh (600 DKK/MWh), while with strong wind (>1500 MW) spot prices can be brought down to around 34 €/MWh (250 DKK/MWh). Jonsson et al. [12] show that the average spot price is considerably lower at times where wind power production has been predicted to be large. Sáenz de Miera et al. [19] found that wind power generation in Spain would have led to a drop in the wholesale price amounting to 7.08 €/MWh in 2005, 4.75 €/MWh in 2006, and 12.44 €/MWh during the first half of 2007.

Gelabert el al. [10] find that an increase in renewable electricity production by 1 GWh reduces the daily average of the Spanish electricity price by 2 €/MWh. Wurzburg et al. [21] find that additional RES generation by 1 GWh reduces the daily average price by roughly 1 €/MWh in German and Austrian integrated markets. Woo et al. [20] carry out an empirical analysis for the Texas electricity price market and showed a strong negative effect of wind power generation on Texas balancing electricity prices. Ketterer [13] also examined wind power in German electricity markets and found that an additional RES generation by 1GWh led to a reduction of daily spot price by approximatively 1 €/MWh.

Benhmad and Percebois [1, 2] also explored German electricity markets for a more recent dataset and found similar results consisting of a reduction of daily spot price by approximatively 1 €/MWh for each an additional GWh of wind feed-in.

The paper is organized as follows. Section 2 provides an overview of the merit-order effect. In Sect. 3, we carry out an empirical analysis and discuss the main findings. Section 4 provides some concluding remarks.

2 The Merit-Order Effect

In order to supply electricity, different power generation technologies compete with each other according to their availability of supply and their marginal cost of production (fossil fuels such as coal or natural gas, nuclear power, renewable energy sources like hydroelectric generators, wind or solar energy).

The electricity market operates according to day-ahead bidding. Indeed, the transmission system operators basically receive the bids from all power producers for the quantity and cost for each hour of the next day and then assigns the dispatch based on the lowest cost producer until demand is met. All producers who dispatch get the marginal price of the last producer that dispatched. As a result even if the last producer only produced theoretically one kWh, then that is the price of the system. This conventional approach consists of ranking the power plants of the system in ascending order of their marginal cost of generation. This approach is called the merit order.

Traditionally, the hydroelectric power plants are the first to be dispatched on the grid. They are followed, respectively, by nuclear plants, coal-fired, and/or combined-cycle gas turbines (CCGT), and then open cycle gas turbine (OCGT) plants and oil-fired units with the highest fuel costs.

Although power plants with the highest marginal cost correspond to the oil-fired gas turbines, gas plants are usually the marginal producers and as a result the cost of gas is very relevant to the wholesale pricing setting of electricity. But, due to EU ETS price weaknesses, carbon prices have plunged to record low prices making it more expensive to burn gas than coal. Moreover, The U.S. coal surpluses export due to shale gas revolution has lowered coal prices in Europe, whereas oil-indexation of gas contracts and geopolitical concerns have made natural gas more expensive. Therefore, the price competitiveness of more polluting coal-fired plants allows them to be dispatched before the gas turbine and to be the key of electricity price setting.

However, a pricing based on marginal costs could never allow RES to recover their fixed costs. Indeed, the photovoltaic (PV) and wind power plants have a high average cost and their load factor is too low due to intermittency. Therefore, subsidizing renewable energy sources by feed-in tariff scheme allowing their average costs to be recovered corresponds to a support mechanism outside the market. By granting an economic return above the market price, these supporting schemes have promoted RES development in several European electricity markets.

As the renewable energy sources (RES) have priority for grid access at zero marginal cost, i.e., have the privilege of priority dispatch, electricity from RES participating in the auction process at zero marginal cost replaces every other energy source with higher marginal cost. The decoupling of spot market prices and RES in-feed due to FIT support scheme results in lower average equilibrium price level on the spot market. This downward pressure on wholesale electricity prices is the so-called merit-order effect [18].

Indeed, during full and peak times, the marginal power plant is logically a combined-cycle gas-fired plant. However, as they have no fuel costs, RES has a zero marginal cost. Thus, electricity from RES makes the coal-fired plant becoming the marginal plant. The electricity market price is thus lower than it would be if there was no RES power in-feed. Lowering electricity spot prices causes a serious distortion to the electricity market.

Indeed, if the wind or solar power plants were not remunerated according to feed-in tariffs scheme, they could never be profitable because the spot market price at full and peak periods would not allow them to recover their fixed costs.

Furthermore, the insufficient dispatching of the flexible gas-fired plants jeopardizes their profitability as they cannot be operated profitably because peak spot prices are too often below their marginal operation costs. Thus, the RES, by lowering equi-

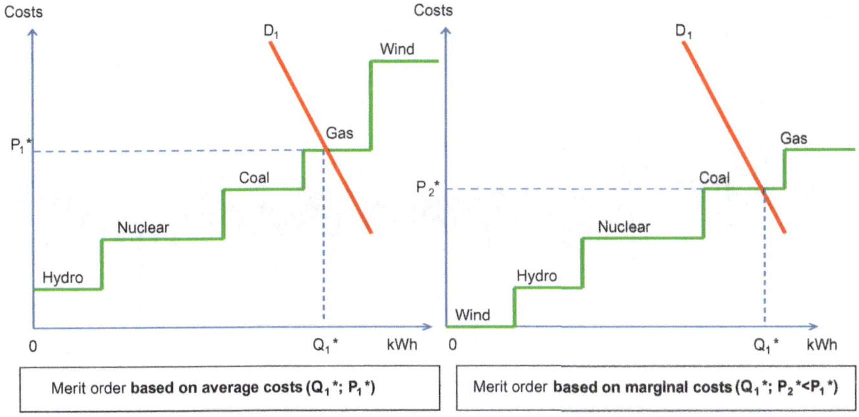

Fig. 1 Merit order based on average and marginal costs

librium spot price level, will squeeze peak load power plants out of the market due to their comparatively higher variable costs. Figure 1 shows the merit-order curve based, respectively, on average and on marginal costs.

3 Empirical Evidence

Data

The analysis is based on time series data of the German power system as provided by the platform of the European Energy Exchange (EEX). The spot market is a day-ahead market, and the spot price is an hourly contract with physical delivery on the next day. The Phelix day base is then calculated as the average, weighted price over these hourly contracts. The sample data covers the period going from January 1, 2012 to December 31, 2016, summing up to 1827 observations.

Figure 2 provides a plot of the data for the whole period. It is easy to see that the data exhibits the typical features of electricity prices and contains several periods of extreme volatility, price spikes, and shows a mean-reverting behavior.

The descriptive statistics of German electricity spot prices summarized in Table 1 show that values of sample mean are close to 34.97 and a standard deviation of 11.74.

The sample kurtosis (6.86) is higher than 3, the kurtosis of a normal distribution, implying that price distribution exhibits fat tails. Furthermore, negative skewness indicates a greater probability of large falls in electricity price than large increases. By the Jarque–Bera statistic, the null hypothesis of normal distributions is also rejected.

For the RES generation, we use daily forecasts of wind power and photovoltaic generation for the full period as illustrated in Figs. 3 and 4.

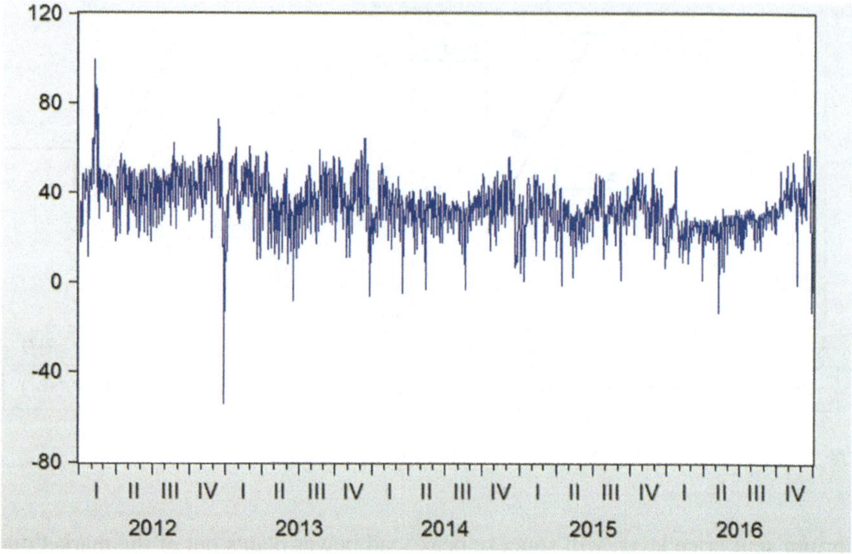

Fig. 2 Daily EEX day-ahead spot prices (€/MWh)

Table 1 Descriptive statistics of German electricity spot prices

Observations	1827
Mean	34.97
Std. dev.	11.74
Skewness	−0.33
Kurtosis	6.86
Jarque-Bera	1171.37
Prob.	0.0000

These forecasts are made by the four German transmission system operators (TSO).[1]

The descriptive statistics (Table 2) show that the wind power and photovoltaic forecasts fed into the grid have, respectively, a daily mean of 6817 and 3651 MWh per day but a high variability.

[1]The data are available in 15 min format. For this study, 15 min MW data are averaged for each hour and again averaged to MWh per day. There is four transmission system operators (TSO) in Germany and one TSO in Austria: *Amprion GmbH, TenneT TSO GmbH, 50 Hz Transmission GmbH, EnBW Transportnetze*, and *APG-Austrian Power Grid AG*.

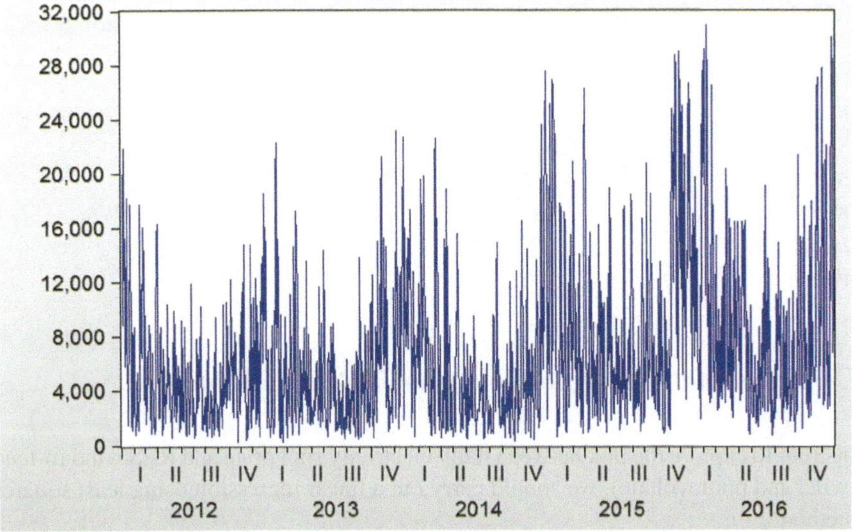

Fig. 3 Wind power feed-in (2012–2016)

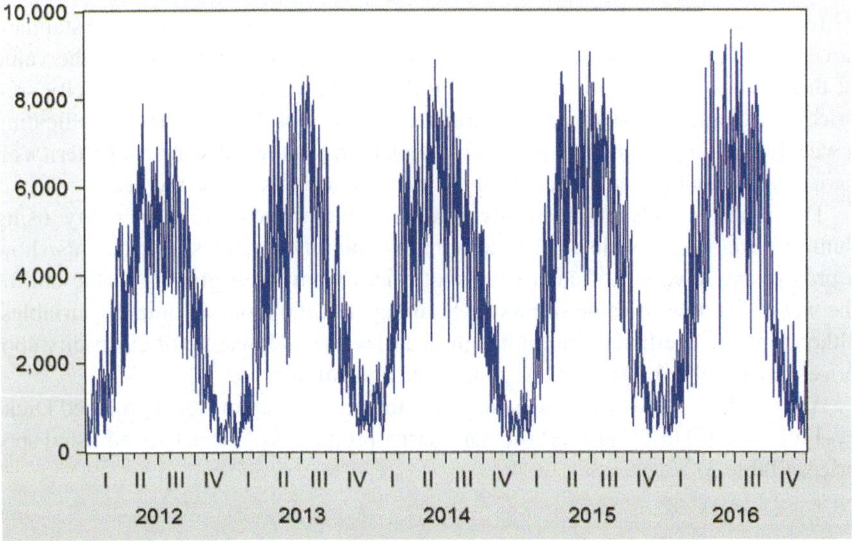

Fig. 4 Photovoltaic power generation (2012–2016)

Table 2 Descriptive statistics of wind feed-in and photovoltaic

	Wind	PV
Observations	1827	1827
Mean	6817.63	3651
Std. dev.	5652.75	2380
Skewness	1.54	0.28
Kurtosis	5.32	1.91
Jarque-Bera	1135.58	113.14
Prob.	0.0000	0.0000

Empirical Methodology: ARMA-X-GARCH-X Model

In order to explore the link between daily electricity spot price and RES wind in-feed (wind and photovoltaic), we should carry out a linear regression using least squares method.

As electricity spot prices deviate from the normal distribution due to more frequent large outliers, outliers should first be removed before conducting the regression analysis.

In line with the literature, we remove values that exceed three times the standard deviation of the original price series. The outliers are then replaced with the value of three times the standard deviation. Furthermore, the analysis of electricity spot prices correlogram shows a strong autocorrelation in lags 7, 14, 21, and 28 indicating a weekly seasonality. Indeed, electricity demand has a typical seasonal pattern as it varies throughout the day and during the week, as well as across the year.

Therefore, models of electricity prices should incorporate seasonality using dummy variables. For the weekly seasonality, dummy variables' coefficients show a progressive lowering of electricity spot prices from the beginning to the end of the week. The lowest value occurs on Saturday. For the monthly dummy variables, although some coefficients are not significant, we see a lowering of electricity spot prices during March, April, May, June, July, and August.

After outliers removal and seasonal adjustment, we carry out an augmented Dickey–Fuller (ADF) test [7] to test for stationarity properties of electricity adjusted spot prices (Table 3).

Table 3 ADF unit root test on adjusted electricity spot prices

		t-statistic	Prob.
Augmented Dickey–Fuller	Test statistic	−8.588311	0.0000
Test critical values	1% level	−3.433739	
	5% level	−2.862924	
	10% level	−2.567554	

Table 4 ADF unit root test on WIND_SA

		t-statistic	Prob.
Augmented Dickey–Fuller	Test statistic	−19.173116	0.0000
Test critical values	1% level	−3.433739	
	5% level	−2.862924	
	10% level	−2.567554	

Table 5 ADF unit root test on PV_SA

		t-statistic	Prob.
Augmented Dickey–Fuller	Test statistic	−23.18669	0.0000
Test critical values	1% level	−3.433739	
	5% level	−2.862924	
	10% level	−2.567554	

The ADF t-statistic is −8.85, whereas the 5% critical value is −2.86. The null hypothesis of a unit root is rejected, spot electricity prices are then stationary. As electricity is not storable, the price tends to spike and then reverts (mean-reverting behavior) as soon as the divergence of supply and demand is resolved [9].

For the wind power, the variable shows seasonal dynamics which could be accounted for using dummy variables. The deseasonalized time series called (wind_sa) is then tested using the ADF test which reveals their stationary behavior (the ADF t-statistic is −19.17, whereas the 5% critical value is −2.86) (Table 4).

For the photovoltaic power, the variable shows seasonal dynamics which could be accounted for using dummy variables. The deseasonalized time series called (pv_sa) is then tested using the ADF test which reveals their stationary behavior (the ADF t-statistic is −23.18, whereas the 5% critical value is −2.86) (Table 5).

Even after removing out seasonality and outliers, electricity spot prices still present high-order serial correlation in its structure which could be filtered out by an autoregressive moving average (ARMA) filter [6]. Therefore, the impact of wind in-feed and photovoltaic on electricity prices is explored according to the following ARMA-X model where the wind feed-in and photovoltaic power considered as exogenous variables X:

$$(spot\_sa)_t = \alpha_0 + \sum_{i=1}^{p} \alpha_1 (spot\_sa)_{t-i} + \sum_{j=1}^{q} \beta_j \varepsilon_{t-j} + \delta wind\_sa_t + \lambda pv\_sa_t + \upsilon_t$$

The selection of autoregressive lag p could depend on AIC minimization, and q is assumed to be 0. According to the Akaike information criterion, the best choice was lag p = 7 which corresponds to a weekly seasonality.[2]

[2] *The results of 7 autoregressive terms, not reported here, are available upon request.*

Table 6 Wind and photovoltaic feed-in impact on electricity prices and volatility

Dependant variable: electricity spot prices		
Sample: 1.1.2012 31.12.2016		
	(A)	(B)
Mean equation		
Constant	−0.46 (0.65)	0.00 (0.99)
Wind	**−0.00099 (0.00)**	**−0.0010 (0.00)**
PV		**−0.00098 (0.00)**
Variance equation		
Constant	3.59 (0.00)	3.44(0.00)
Alpha	0.31 (0.00)	0.31(0.00)
Beta	0.56 (0.00)	0.56 (0.00)
Wind	**0.00016 (0.00)**	**0.00016 (0.00)**
PV		**−0.0004 (0.00)**
Adj. R-squared	0.7467	0.7584
AIC	5.7713	5.6973
BIC	5.8116	5.7426

Note AIC and BIC stand, respectively, for Akaike and Bayesian information criterion, p-values are in parentheses

Table 7 ARCH heteroskedasticity test on regression residuals

Heteroskedasticity test: ARCH			
F-statistic	120.83	Prob. F(1,1816)	0.0000
Obs*R-squared	113.41	Prob. Chi-Square(1)	0.0000

The estimation results reported in Table 6 (Column A) reveal a negative impact of wind power on the electricity price in Germany. Indeed, for each additional GWh of wind feed-in, the electricity price decreases by 1 €/MWh at the spot market. Therefore, and given the average wind electricity generation during 2012–2016, the merit-order effect roughly corresponds to an average price decrease, in absolute terms, of approximately 7 €/MWh.

The residuals of linear regression should then be homoskedastic according to least squares estimator hypothesis. Therefore, an ARCH-effect test following the procedure of Engle [8] is carried out on residuals data (see, Table 7).

We conclude that the time series of residuals is heteroskedastic and the parsimonious GARCH(1, 1) specification [5] could be used to take into account the time-varying volatility feature of spot electricity spot prices.

As our goal consists in exploring the joint impact of wind in-feed on spot electricity price level and also on price volatility dynamics, the wind feed-in should be taken into account as an exogenous variable in the mean equation as well as in the variance equation. Therefore, our empirical analysis is based on ARMA(p, q)-X-GARCH(1, 1)-X modeling where the exogenous variable X represents the wind in-feed. The empirical results based on AR(7)-X-GARCH(1, 1)-X model are reported in Table 6 (Column A).

The model parameters are positive and statistically significant at the 1% level. We can conclude that the introduction of wind electricity in Germany has not only reduced the electricity spot prices (−0.001) but also induced an increase of their volatility (positive sign +0.00016 at the conditional variance equation).

Indeed, wind in-feed, due to the merit-order effect, not only reduces the electricity spot price level making them sometimes negative but induces an increase in electricity price volatility, exacerbating risks in electricity markets.

The estimation results also reported in Table 6 (Column B) reveal not only a negative impact of wind power on the electricity price in Germany but also a negative impact of solar photovoltaic generation on electricity prices of the same magnitude. Indeed, for each additional GWh of photovoltaic feed-in, the electricity price decreases approximatively by 1 €/MWh at the spot market. Therefore, and given the average photovoltaic electricity generation during 2012–2016, the merit-order effect induced by photovoltaic roughly corresponds to an average price decrease, in absolute terms, of approximately 3.65 €/MWh.

To explore the joint impact of wind in-feed and photovoltaic on spot electricity price level and on price volatility dynamics, the two variables should be taken into account as exogenous variable in the mean equation as well as in the variance equation. The empirical results based on AR(7)-X-GARCH(1, 1)-X model are reported in Table 6 (Column B). The model parameters are positive and statistically significant at the 1% level.

We can conclude that the introduction of wind electricity in Germany has not only reduced the electricity spot prices (−0.001), but also induced an increase of their volatility (positive sign +0.00016 at the conditional variance equation). However, photovoltaic electricity not only reduced electricity spot prices (−0.001) but also induced a downward pressure of their volatility (negative sign −0.0004 at the conditional variance equation). Therefore, we conclude that wind and solar photovoltaics have the same effect on electricity spot prices (downward effect) of the same magnitude approximatively, but have the opposite impact on its volatility dynamics.

Thus, the full model containing wind feed-in and photovoltaic electricity generation has a clear superiority on the model based only upon wind feed-in as shown by high level of adjusted R-squared and information criterions (see Table 6).

Indeed, the upward effect on electricity prices volatility induced by highly intermittent wind feed-in is largely offset by the photovoltaic downward effect. Thus, the mixture of installed electricity generation capacities consisting of wind and solar photovoltaic allows German electricity market volatility to be less higher than it would be if Germany had only installed wind generation capacities.

4 Conclusion

The feed-in tariffs support scheme, consisting of buying intermittent electricity at a fixed price off-market considerably higher than the spot market price, has clearly induced a huge market penetration of RES in Germany.

The fact that this intermittent electricity has statutory priority on the grid and at the same time participates in spot market auctions at a zero marginal cost can have negative effects on the functioning of the spot market as it leads to a downward trend in the equilibrium price: the so-called merit-order effect. Indeed, each additional GWh wind (and RES in general) production of electricity will have a crowding effect on higher marginal cost power plants.

The purpose of the paper consists in quantifying the merit-order effect of wind feed-in and photovoltaics in Germany during the 2012–2016 period. One of the major findings is that the day-ahead electricity spot price fell by 1 €/MWh for each additional GWh, respectively, for the two renewable energy sources. Moreover, the wind electricity generation has an increasing effect on the spot prices volatility which is largely offset by photovoltaics with their strong downward impact on volatility.

However, although the volatility is controlled by a mixture of installed capacities of RES, the merit-order effect remains a big challenge for Germany. This negative effect of RES could significantly be limited by the interconnections of between Germany and neighboring countries especially France, allowing it to export its surplus wind power. Therefore, the development of the renewable energy sources should be accompanied by a market coupling in order to address their challenges to European electricity system.

References

1. Benhmad, F., Percebois, J.: Wind power feed-in impact on electricity prices in Germany 2009–2013. Eur. J. Comp. Econ. **13**(1) (2016)
2. Benhmad, F., Percebois, J.: On the impact of wind feed-in and interconnections on electricity price in Germany. Energy Stud. Rev. **23**(1) (2017)
3. BMWi.: Development of renewable energy sources in Germany 2016. Federal Ministry for Economic Affairs and Energy. https://www.erneuerbare-energien.de (2016)
4. Bode, S., Groscurth, H.M.: The effect of the German renewable energy act (EEG) on the electricity price. HWWA Discussion Paper, No. 348 (2006)
5. Bollerslev, T.: Generalized autoregressive conditional heteroskedasticity. J. Econ. **31**, 307–327 (1986)
6. Box, G.E.P., Jenkins, G.M.: Time Series Analysis Forecasting and Control. Holden-Day, San Francisco (1970)
7. Dickey, D.A., Fuller, W.A.: Likelihood ratio statistics for autoregressive time series with a unit root. Econometrica **49**, 1057–1072 (1981)
8. Engle, R.: Autoregressive conditional heteroscedasticity with estimates of the variance of United Kingdom inflation. Econometrica **50**, 987–1007 (1982)
9. Escribano, A., Ignacio, P.J., Villaplana, P.: Modeling electricity prices: international evidence. Oxford Bull. Econ. Stat. **73**, 622–650 (2011)

10. Gelabert, L., Labandeira, X., Linares, P.: An ex-post analysis of the effect of renewable and cogeneration on Spanish electricity prices. Energy Econ. **33**, S59–S65 (2011)
11. Jensen, S.G., Skytte, K.: Interactions between the power and green certificate markets. Energy Policy **30**, 425–435 (2002)
12. Jonsson, T., Pinson, P., Madsen, H.: On the market impacts of wind energy forecasts. Energy Econ. **32**, 313–320 (2010)
13. Ketterer, J.C.: The impact of wind power generation on the electricity price in Germany. Energy Econ. **44**, 270–280 (2014)
14. Munksgaard, J., Morthorst, P.E.: Wind power in the Danish liberalised power market Policy measures, price impact and investor incentives. Energy Policy **36**, 3940–3947 (2008)
15. Neubarth, J., Woll, O., Weber, C., Gerecht, M.: Influence of wind electricity generation on spot prices. Energiewirtschaftliche **56**, 42–45 (2006)
16. Nicolosi, M., Fürsch, M.: The impact of an increasing share of RES-E on the conventional power market—the example of Germany. Zeitschrift für Energiewirtschaft **33**, 246–254 (2009)
17. Sensfuß, F., Ragwitz, M., Genoese, M.: The merit-order effect: A detailed analysis of the price effect of renewable electricity generation on spot market prices in Germany. Energy Policy **36**, 3086–3094 (2008)
18. Sioshansi, F.: Evolution of Global Electricity Markets. Elsevier (2013)
19. Sáenz de Miera, G., del Rio Gonzalez, P., Vizcaino, I.: Analysing the impact of renewable electricity support schemes on power prices: The case of wind electricity in Spain. Energy Policy **36**, 3345–3359 (2008)
20. Woo, C.-K., Horowitz, I., Moore, J., Pacheco, A.: The impact of wind generation on the electricity spot-market price level and variance: the Texas experience. Energy Policy **39**, 3939–3944 (2011)
21. Wurzburg, K., Labandeira, X., Linares P.: Renewable generation and electricity prices: taking stock and new evidence for Germany and Austria. Energy Econ. **40**, 159–171

Part VI
Forecasting in Real Problems

The Analysis of Variability of Short Data Sets Based on Mahalanobis Distance Calculation and Surrogate Time Series Testing

Teimuraz Matcharashvili, Natalia Zhukova, Tamaz Chelidze, Evgeni Baratashvili, Tamar Matcharashvili and Manana Janiashvili

Abstract In this work, we present convenient for short time series approach which is based on the multivariate Mahalanobis distance calculation, combined with the surrogate time series testing. In order to test the ability of this approach to differentiate changes which could occur in complex processes, we analyzed data sets of different origins. We used seismological, meteorological, physiological, and economic data sets. Exactly, we analyzed data sets of inter earthquake times (IET), inter earthquake distances (IED), and differences in consecutive magnitudes (DM) compiled from southern Californian earthquake catalogue, data sets of yearly number of warmer and colder days derived from maximal air temperature data bases in Tbilisi, Georgia, arterial systolic, and diastolic blood pressure time series of healthy persons, as well as components of Index of Economic Freedom (IEF) and exchange rate time series of three southern Caucasian countries. It was shown that used approach, even in the case of relatively short time series, may effectively be used to quantify dynamical changes occurred in different natural complex processes.

T. Matcharashvili (✉) · N. Zhukova · T. Chelidze
M. Nodia Institute of Geophysics, 1. Alexidze str., Tbilisi, Georgia
e-mail: matcharashvili@gtu.ge

N. Zhukova
e-mail: natalia27cpp@gmail.com

T. Chelidze
e-mail: tamaz.chelidze@gmail.com

E. Baratashvili · T. Matcharashvili
Georgian Technical University, 77 Kostava ave, Tbilisi, Georgia
e-mail: barata49@mail.ru

T. Matcharashvili
e-mail: tamar.matcharashvili@gmail.com

M. Janiashvili
Institute of Clinical Cardiology, 4 Liubliana str., Tbilisi, Georgia
e-mail: manana.janiashvili@gmail.com

I. Rojas et al. (eds.), *Time Series Analysis and Forecasting*,
Contributions to Statistics, https://doi.org/10.1007/978-3-319-96944-2_19

Keywords Short time series · Mahalanobis distance · Surrogate time series

1 Introduction

According to ancient Pythagoreans "All things are numbers". Nowadays, in an era of digital technologies, this truth about numbers, through which we explore the world around us, has stopped being an abstract philosophical statement and became a part of everyday life. Not always, but often we manage to do so that these numbers (commonly measured, or calculated somehow) would be represented in the form of equally spaced in time—evenly sampled time series, though often they can be available to us only in the form of unevenly sampled sequences of data. To get these data sets is not an easy job, so far as we deal with structures and processes of nature, which presently often are named as complex. Complexity is a frequently used, though still poorly defined, concept [1, 2]. At present, it looks almost hopeless endeavor to try to find the accepted conventional, clear-cut and single-valued explanation of the term "complexity" in the plenty of available textbooks or regular scientific articles. At the same time, it is clear that complexity is related with the crosscutting hierarchical organization, multitude of out of equilibrium dispersed mutual interactions, etc. All these cause the fact that structurally or functionally the whole in such systems is not equal to the sum of its parts, what in varying degree is characteristic for all nonlinear systems.

Examples of complexity can be found in very different areas, such as atmosphere (climate and weather change), geophysics (tides, earthquakes, volcanoes, magnetic field variations), social systems (crowd behavior), medicine and biology (rhythms, physiological cycles, epidemics), economy (financial market behavior, exchange rates), engineering (friction, fracturing), communication (electronic networks, internet packet dynamics), etc.

Anyway, in spite of unavoidable complexity of natural world, often we succeed to identifying and quantifying a signal, which characterize interesting for us structures or processes [2]. In such cases, we deal with the varying time (at least in time) usually discrete, evenly or unevenly sampled data sequences. Multitude of examples of such data sets from different complex processes can be listed from diverse fields of research and practical activity.

Presently, complex time series analysis is in the focus of interdisciplinary research interests. Many conceptual solutions and interesting results related with the quantification of complex processes can be listed in different fields [2, 3]. At the same time, the problems arise when available for researchers, usually, not stationary data sets do not fulfill necessary for correct complex data analysis, strong requirements imposed on data sets such as length, quality, etc. [2, 3].

In such cases, we are forced to combine different approaches in order to have an understanding on general features of complex process based on existed imperfect data sets. In this work, we describe our approach which is based on the combination of multivariate analysis and surrogate testing, of the relatively short data sets

[2–4]. Exactly, we used multivariate Mahalanobis distance calculation to evaluate changes that occurred in the analyzed process as assessed by two or three of its main characteristics. After, in order to test whether observed changes are indeed related with the internal dynamical structure of process, we compared original data sequence with surrogate time series with the distorted dynamical structure. We also present the results of analyses carried out on relatively short complex time series of different origin.

Thus, the objective of this research can be reduced to a problem of classification, having in mind the separation of states of the analyzed complex processes assessed by selected two–three characteristics.

Because of the fact that data sets by means of which we try to understand dynamical features of natural complex systems rarely fulfill all high standards of the contemporary complex time series analysis, the approaches like that presented in this work will have of great importance for different scientific and practical applications.

2 Data and Methods of Analysis

In this research, four different types of data sequences have been used to demonstrate how in the case when only short data sets are available the multivariate analysis can be used to assess dynamical features of complex natural processes.

We started from the seismic data sets. We analyzed inter earthquake times, inter earthquake distances, and differences in consecutive earthquake magnitudes compiled from the southern Californian earthquake catalogue, at M4.7 representative threshold, as presented by Kagan and colleagues [5]. Exactly, we used part of this catalogue from 1929 to 1952.

Next data set, we used, was maximum daily temperatures measured in Tbilisi, Georgia from 1915 to 2013. This data set fulfilled all standards of data quality control and homogeneity. From these time series, we compiled a daily max temperature anomalies time series consisting of deviations of daily max air temperatures from the long-term average for the same day of the year. From the daily max air temperature anomalies data sets, we compiled sequences of yearly number of days when anomalies of daily max temperatures significantly deviate from the 99-year mean of anomalies for that days. We considered separately sequences of number of days, when anomalies of max daily temperatures were significantly larger (warmer days-WD) or were significantly lower (colder days-CD) than 99-year mean values. These (WD and CD) sequences strongly characterize features of the warm and cold tails of the distribution, and thus are more relevant to characterize aspects of the anomalous changes in max daily temperatures that occurred as a result of temperature increase [6].

After we proceeded to physiological data sets. We considered arterial blood pressure time series from 70 volunteers. The age of participants varied from 30 to 50 years. All subjects gave informed consent to participate in the study. Participants of the study were not given medicines for 2–3 days preceding the examination. The monitoring of blood pressure was carried out from 12.00 to 12.00 AM of the next day, taking into consideration the physiological regime of participants of the study. We analyzed systolic (SBP) and diastolic (DBP) arterial blood pressures. These data have been obtained from 24 h ambulatory monitoring recordings at 15 min sampling time.

Used in this research economical data sets represented components of Index of Economic Freedom (IEF) as well as daily increments of currency exchange data for three southern Caucasian Countries, Azerbaijan, Georgia, and Armenia (2013–2016).

The main goal of our research was akin to a general classification problem, i.e., a problem, regarding changes that occurred in certain process (no matter seismic, climatic, physiological, or economical). For this purpose, in the present research, the Mahalanobis distance (MD) calculation was applied [4] which is a popular method of distinguishing multivariable data groups by using a univariate distance measure that is defined by several performance parameters. Mahalanobis distance can be calculated as follows:

$$D^2 = (\bar{x}_1 - \bar{x}_2)^T S^{-1} (\bar{x}_1 - \bar{x}_2) \tag{1}$$

where $\bar{x}_1$ and $\bar{x}_2$ are sample means from sample sets of sizes n_1 and n_2, The "T" superscript denotes the transpose operator. S is pooled covariance matrix

$$S = \frac{((n_1 - 1)S_1 + (n_2 - 1)S_2)}{n_1 + n_2 - 2} \tag{2}$$

where Si are covariance matrices of corresponding groups.

Generally, the two conditions or states of systems are more similar (more probable to belong to the same class or group), if their MD value is smaller. After computation of MD values, in order to assess the significance of separability between compared groups, Hotelling's two-sample T2 statistics was used, which was then converted into F value and assessed by F-test. F-value was calculated as follows:

$$F = \frac{n_1 n_2}{n_1 + n_2} \frac{n_1 + n_2 - p - 1}{(n_1 + n_2 - 2)p} D^2 \tag{3}$$

where p and $(n_1 + n_2 - p - 1)$ are degrees of freedom. Then, calculated F values were compared with a critical value, Fc, which, for certain degrees of freedom values, can be easily found in different statistical textbooks. When calculated $F > Fc$, then null hypothesis, that there is no separation between considered groups, can be rejected, and thus statistically significant difference between groups is established at a specified probability.

Usually, MD calculation is preferable for multivariate comparison for two main reasons. First, it reduces a multivariate system to a univariate system, and second, MD is sensitive to inter-variable changes in a multivariate system [4].

For the purposes of our analysis, we needed to assess the influence of long- and short-range temporal correlations in the considered data sets on the results of the separability analysis. For this, it was necessary, these results to be compared with the results, obtained for surrogate data sets where the original dynamical structure has been intentionally distorted in a certain manner [2, 3]. In present research, however, we restrict our analysis only by the shuffled surrogates, in which dynamical structure of original process is completely destroyed. Results of comparison with other surrogates in which original time structure is just partly destroyed will be presented in future. Thus, here, we accomplished an analysis on original as well as corresponding randomized (shuffled) data sets. For each of the four, above mentioned, original data sets, we compiled 50 of such randomized sequences. It is logical that these data sets were much closer to a normal distribution, and thus comparison with such data sets was important also for the correctness of the interpretation of results of MD calculation.

3 Results and Discussion

As it is said in the previous section, the main goal of the research was the classification of changes occurred in the considered complex system, based on the analysis of short data sets of specially selected two, three characteristics. Generally, one of the most important steps in such tasks of the classification is the systems' feature extraction [4]. This implies the transformation of original data sets into new ones, enabling to focus on targeted features of the investigated process.

As mentioned above, we have started from the multivariate analysis of seismic process based on Southern California earthquakes catalogue [5]. From this database consisting of 155 events from 1929 to 1952, when strong earthquake M7.5 occurred, we compiled IET, IED, and DM data sequences. These data sequences are typical examples of short, unevenly sampled data sets. Thus, because of short catalogue (due to high representative threshold M4.7), we could not analyze systems behavior in the equal time intervals and have decided to compare similarity or dissimilarity of seismic process in the consecutive windows of 40 data, shifted by 10 data. We compared groups which contained IET, IED, DM sequences. In Fig. 1a, we show results of comparison between the first window (which contains first 40 data in catalogue and started in 1929) with other windows shifted by 10 data. As we see, seismic process assessed by the variability of IET, IED, DM sequences, in most of these windows, looks similar like in the first window (MD value is lower than level of significance). Only two windows (4th started in 1940 and 11th started in 1950) reveal statistically significant difference comparing to first window. At the same time, is important to say that these two windows are quite different by the amount of released seismic energy. Indeed, as we see in Fig. 2, in the 4th window almost the maximal amount of seismic energy was released while in the 11th window amount of released seismic energy is close to minimum among all considered 40 data windows. This fact apparently, once again points that dynamics of complex seismic process in whole, can not be correctly understood based only on the knowledge of features of behavior in one of its energetic, temporal or spatial domains. All the said underlines the importance of different multivariate analyses in this field. In present work, we

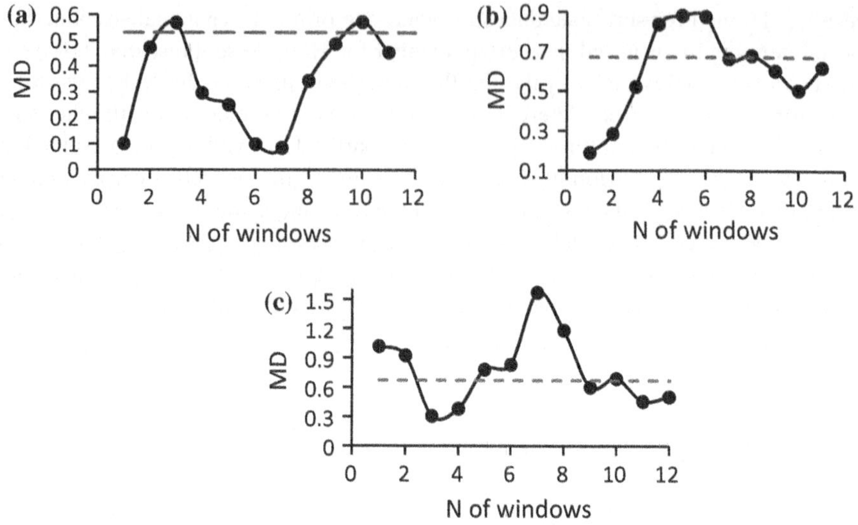

Fig. 1 MD values calculated by IET, IED, and DM sequences for the consecutive 40 data windows, shifted by 10 data. **a** Data obtained from original Southern California earthquake catalogue compared with the first window, **b** data obtained from randomized catalogues compared with the first window, **c** windows contained data from the original catalogue are compared with the corresponding windows contained data from randomized catalogues. Dotted line corresponds to significant difference between windows at $p = 0.05$

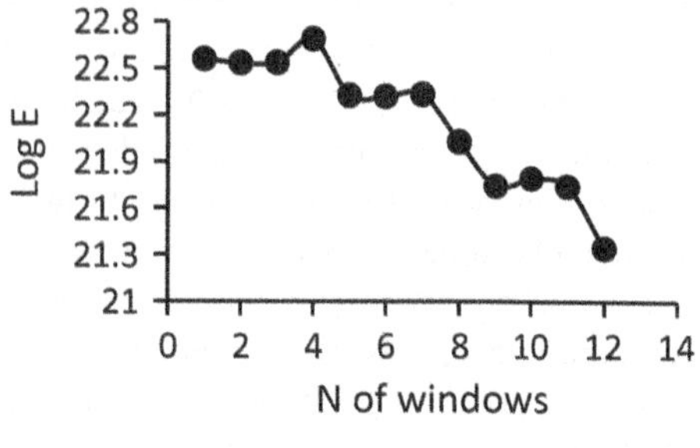

Fig. 2 Seismic energy released in consecutive 40 data windows, shifted by 10 data. Southern California earthquakes catalogue, 1929–1952

demonstrate interesting possibilities provided by one of such approaches and we plan to continue similar researches in future.

Next, to exclude possible mistakes in our analysis, which can be caused by different short- and long-term correlations, we decided to use shuffling procedure. Exactly, as it was described in the methods section, we compared the natural seismic process with the process in which the original temporal, spatial, and energetical structure was intentionally distorted. In other words, we compared original IED, IET, and DM data sets with ones compiled from shuffled catalogues. It is interesting that according to our results (Fig. 1a, b), differences between the first window of original catalogue and next windows is not by chance and apparently are caused by the different character of earthquakes' temporal, spatial, and energetical distributions in different windows.

Further analysis supported this supposition. Indeed, as we see in Fig. 1c, comparison of original IET, IED, and DM data sets with ones compiled from catalogues where natural dynamical structure was distorted, shows that there are windows (third and fourth as well as last two windows) in which dynamical features of seismic process is indistinguishable from random process. At the same time, in other windows (e.g., in first and second and especially in 7th and 8th windows, Fig. 1c) we observe that by its features of temporal, spatial, and energetic distributions, original seismic process is significantly different from randomness. This apparently points out that in these windows seismic process behaves more regular by features of the space, time and energetic distributions comparing to other windows. Taking into consideration that strongest earthquakes (M7.0 and larger) occurred at the beginning and at the end of considered period (in 1934 and 1952) obtained results can be considered in favor of our earlier suggestion that the seismic process looks more regular at periods of decreased local seismic activity. It also not possible to exclude that observed changes are related with the features of aftershocks distribution. It is necessary to underline that we dealt with cleaned catalogue [5] at high magnitude threshold level (M4.7), what definitely decrease possibility of appearance of aftershocks in the catalogue, but to exclude such possibility completely seems impossible.

Anyway, presented analysis convinces that even in the case of short data sets for which using of common dynamical data analysis tools is questionable or impossible, we may draw important conclusions about dynamics of complex seismic process based on the procedure of multivariate MD calculation combined with surrogate data testing.

Second type of data sets, for which we have tested ability of approach to access changes in the dynamical features of complex process based on MD calculation of short data sets, came from climatology. Namely, as it was mentioned above we used WD and CD data sequences derived from data bases of daily max air temperatures in Tbilisi, Georgia (1915–2013). We underline again that these sequences of warmer and colder days stem from the same—daily max air temperature data sets. Additional analysis described elsewhere [6] convinces us that changes occurred in the local climate should be much more complicated than a simple shift toward an increase of mean max air temperatures many times described in literature (see e.g., http://www.ipcc.ch/publications_and_data/ar4/wg1/en/ch3s3-8-2.html). At the same time, it becomes clear that a separate analysis of two processes (i.e., variability of warm and cold days), based only on monovariate considerations, will not give a correct view of changes that occurred as a result of increase in max daily air temperatures.

Therefore, in order to learn more about changes in the local climate in the sense of variability of warmer and colder days, we accomplished simultaneous multivariate analysis based on both (WD and CD) data sets. We compared two groups, each of which contained sequences of yearly number of warmer days as the first column and colder days as the second column. Analysis was accomplished in the consecutive 33-year windows, shifted by 11 years.

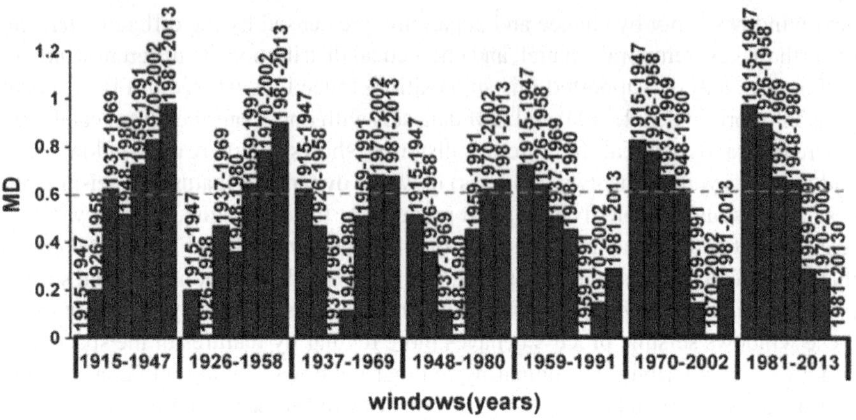

Fig. 3 Mahalanobis distances calculated by sequences of yearly number of warmer and colder days in consecutive 33-year windows. At the bottom are shown sections—windows (of 33-year span), to which other 6 windows (in columns) were compared

In Fig. 3, we show MD values calculated for different sliding windows; dotted line in these figure corresponds to the level of significant difference ($F > 4$, at critical $Fc = 3.14$ for $p = 0.05$). According to obtained results, significant separation was observed for mainly distant (in time) windows. Generally, significant separation between compared groups means that features of variability of yearly number of warmer and colder days have been strongly changed. Thus, assessed by features of the variability of yearly number of warm and cold days the state of the local climate change process in Tbilisi, has significantly changed during last 99 years.

At the same time changes happening with variation of yearly number of warmer and colder days in Tbilisi look regular because they occurred gradually (MD values in later windows compared to the first window almost always increase significantly) during a whole period of observation. MD for windows in the middle of observation period are less different from the first and last window, sometimes the difference is insignificant.

Thus, we can conclude that over the period of our analysis local climate underwent strong changes in the features of variability of yearly number of warmer and colder days. Further, we performed analysis of changes in the local climate over analyzed period of observation, in sense of time correlation features in variation of yearly number of warmer and colder days. For this, as it was described above, it was necessary to compare original data sets with ones, where original temporal structure was intentionally distorted.

In general, time series of daily max air temperature anomalies used in this research, represent a certain type of data sets with removed yearly trends, i.e., the time structure of the original process of daily max air temperature variation is already essentially changed. At the same time, there are different kinds of short- and long-term correlations, which obviously still exist in such data sets even after removal of yearly trends.

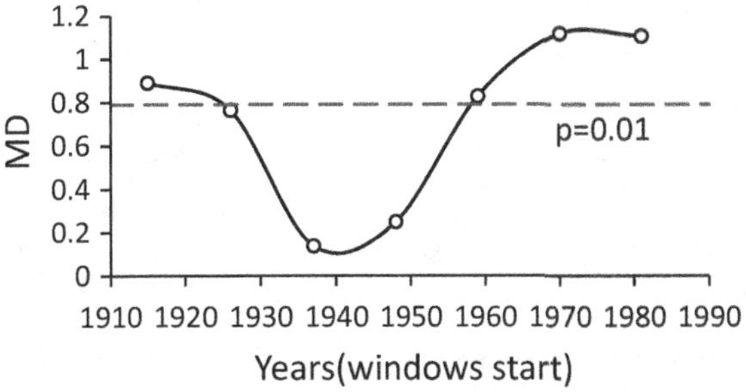

Fig. 4 Mahalanobis distance values between original and time structure distorted sequences of yearly number of warmer and colder days in Tbilisi, calculated for consecutive 33-years windows (11-year step, 1915–2013). Dotted line corresponds to significant difference at $p = 0.05$

This is why we further destroyed all internal time structures of the original process by random shuffling. Exactly, we constructed time structure distorted sequences of yearly number of warmer and colder days, based on randomized data sets of anomalies of max daily temperatures, which in turn have been compiled from averaged rows of several tens of shuffled original max air temperature data sets. As it was expected, time structure distorted data sets of number of warmer and colder days almost do not reveal changes in consecutive 33-years windows (not shown here), in contrast to original (WD and CD) sequences (presented in Fig. 3).

Results of Mahalanobis distance calculation when original and dynamical structure distorted data sets have been compared is presented in Fig. 4. From this figure, it can be concluded that the extent of regularity in the local climate variability, in sense of changes in the amount of warmer and colder days, has significantly changed in different 33-year windows of considered 99 year period. The situation with climate change looked most dysregulated in the middle of the last century. What presently can be said for sure is that changes in the local climate in Tbilisi should be related with changes in large- and small-scale atmospheric dynamics. In the present work, we do not go deeper into these discussions, because the main goal was to demonstrate the effectiveness of used approach for short data sets from different complex natural processes.

Next, for the same purpose of the analysis of short data from complex processes, we proceeded to the analysis of physiological data sets (from the database of Institute of Clinical Cardiology, Tbilisi, Georgia). Namely, we analyzed arterial systolic and diastolic blood pressure data sets (SBP and DBP) of persons falling into optimal arterial hypertension grade, according to guidelines of European Society of Hypertension (ESH) and the European Society of Cardiology (ESC) [7]. Generally, blood pressure variability is one of the most often discussed in the special scientific literature questions. Most attention usually is paid to differences in the blood pressure variability of patients from different hypertensive groups (see e.g., [8, 9]).

Knowledge of the character of variability of blood pressure characteristics is very important because may provide unique information on human physiological system's

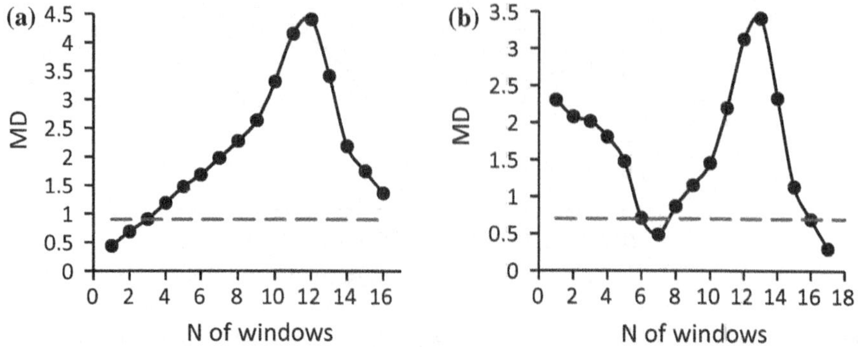

Fig. 5 Mahalanobis distance values calculated for consecutive, 8 h, windows shifted by 1 h. **a** Compared to first window sequences of original systolic and diastolic blood pressures of persons in optimal arterial hypertension group, **b** comparing each window of original data to corresponding window of shuffled time series. Dotted line corresponds to significant difference between windows at p = 0.05

behavior and may serve as characteristic fingerprints for health condition identification as well as for comparison with other systems and models. Among others, is very important to further clear up the character of changes in blood pressure typical for certain hypertensive groups (here in the optimal arterial hypertension grade, defined according to ESH and ESC guidelines (e.g., [7])). Exactly, in this research, we aimed to assess changes occurred during different periods of a day in persons falling in the optimal arterial hypertension group. We accomplished multivariate analysis of averaged (over whole group) values of systolic and diastolic blood pressures which were measured in each 15 min during 24 h observation period.

We see in Fig. 5 (left curve) that by the features of variability of arterial systolic and diastolic blood pressures, all consecutive (8 h long) windows are clearly different from the first window (12.00–20.00) excluding two neighbor windows at the beginning of a period of observation. The most different from the first window, by the largest MD value, is blood pressure variability in 12th window, corresponding to 23.00 PM–07.00 AM period. After, the extent of difference decreases though remains statistically significant compared to the first window (see Fig. 5a). It is not shown here but we have carried out additional analysis and compared all windows with each other and concluded that in about 60% of considered cases neighboring windows (of 8 h long) are statistically not different. It is interesting that two windows, started at 16.00 and 18.00 accordingly, are statistically similar with just one adjacent window and clearly differ from all other, 8 h long, windows. Also, we observed an interesting fact that sometimes blood pressure variation in certain time window may reveal features looking like a long-range correlation (see e.g., [3, 8–10]). Exactly, in these 8 h periods, features of blood pressure variability (assessed by SDP and DBP data sets) was statistically different from closest windows while remained indistinguishable from the more distant windows.

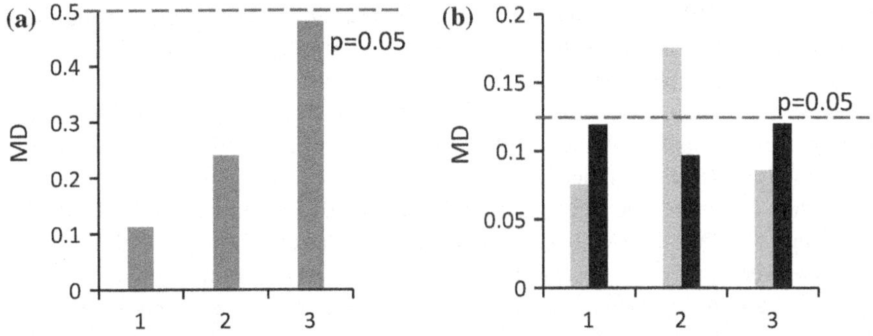

Fig. 6 Mahalanobis distance values, calculated for three southern Caucasian countries in 2014 (1), 2015 (2) and 2016 (3) compared with 2013 year: based on yearly components of IEF (**a**), based on normed currency exchange rates (grey columns in right figure) (**b**), compared with time structure distorted data sets of increments of currency exchange (black columns in right part of figure **b**)

This fact inspired us to compare original time series with data sets in which original structure of systolic and diastolic blood pressure measurements are destroyed by random shuffling procedure. In this case, as we see in Fig. 5b, blood pressure variability as compared with sequences of averaged randomized data sets, reveals interesting features. Exactly, 8 h span windows (6th and 7th in Fig. 5b) starting on 17.00 and 18.00 PM as well as starting on 03.00 and 04.00 AM (16th and 17th) windows, by features of systolic and diastolic blood pressure variability, are indistinguishable from random processes.

As we already mentioned above, in this work, we did not aim to provide exhaustive explanations of the found changes. We just wanted to show that by the presented analysis approach, even having short physiological data sets from the certain (here healthy) group of patients, it is possible to get new and interesting information about changes in the features of the variation of systolic and diastolic blood pressures.

In the next step of our analysis, we have targeted to test economical data sets as an additional example of complex time series. For this purpose, we have selected two types of short data sets for southern Caucasian Countries: Armenia, Azerbaijan, and Georgia. Namely, we started from the comparison of economical characteristics of these countries by their Index of Economic Freedom (IEF), as it is presented at Heritage Foundation Report (http://www.heritage.org/index/) for last 4 years. We have observed that situation in southern Caucasus as whole, according to the yearly IEF components of three countries, tends to be slightly changed and in 2016 change was rather stronger and close to being significant in comparison with 2013 (left part of Fig. 6).

At the same time, comparison with Georgia for 4 years of analysis indicates that economical situation as assessed by IEF components is not different from Armenia (MD value 0.45) while situation is close to being significantly different (MD value 0.51, $F = 2.32$ at critical $Fc = 2.5$ for $p = 0.05$), if we compare Georgia with Azerbaijan (not shown here). Because they are very short and specific, for data sets of yearly components of IEF, we could not use surrogate testing. So, we carried out additional multivariate analysis for southern Caucasian countries based on normed increments of daily exchange rates time series (used data available from official sites of: Central Bank of Azerbaijan, https://en.cbar.az; National Bank of Georgia, https://www.nbg.gov.ge; Central Bank of Armenia, https://www.cba.am). As follows from our results (Fig. 6b), by variability of increments of national currencies, the situation of southern Caucasian countries was significantly different in 2015 compared to 2013, while in 2014 and 2016, we do not observe significant changes comparing to 2013.

Moreover, apparently observed in 2015, changes have serious grounds rooted in the processes taking place in the domestic economies of these countries as far as they are not related to random factors which could be occurred by chance. Indeed, when compared with surrogates, where original structure of exchange rate was distorted we did not find significant difference. This fact convinces that mentioned changes cannot be regarded as random (see black columns in Fig. 6b).

Thus, in present work, for four data sets of different origin, we demonstrated the ability of used approach to indicate interesting dynamical features of complex natural processes even in the case of short available data sets.

4 Summary

In the present work, we aimed to test method which was developed for the purpose to assess the character of changes in the complex dynamical systems when we have just short data sets. Approach is based on the Mahalanobis distance calculation, combined it with the surrogate data testing procedure.

We used an approach for the analysis of data sets from different natural processes. Exactly in the frame of present research, we used seismological, meteorological, physiological, and economic data sets.

For all used data sets, it was shown that combination of Mahalanobis distance calculation with the surrogate data testing enables to have new understanding about changes occurred in complex systems even in the case of relatively short time series.

Acknowledgements This work was supported by Shota Rustaveli National Science Foundation (SRNSF), grant 217838 “Investigation of dynamics of earthquake’s temporal distribution”.

References

1. Tsallis, C.: Introduction to Nonextensive Statistical Mechanics, Approaching a Complex World. Springer, New York, NY, USA (2009)
2. Abarbanel, H.D.I., Brown, R., Sidorowich, J.J., Tsimring, L.S.: The analysis of observed chaotic data in physical systems. Rev. Mod. Phys. **65**(4), 1331–1392 (1993)

3. Matcharashvili, T., Chelidze, T., Janiashvili, M.: Identification of complex processes based on analysis of phase space structures. In: Byrnes, J.S. (ed.), Imaging for Detection and Identification, pp. 207–243. Springer, Dordrecht (2007)
4. McLachlan, G.J.: Mahalanobis distance. Resonance **6**, 20–26 (1999)
5. Kagan, Y.Y., Jackson, D.D., Rong, Y.F.: New catalog of southern California earthquakes, 1800–2005. Seismol. Res. Lett. **77**(1), 30–38 (2006)
6. Matcharashvili, T., Zhukova, N., Chelidze, T., Founda, D., Gerasopoulos, E.: Analysis of long-term variation of the annual number of warmer and colder days using Mahalanobis distance metrics—a case study for Athens. Phys. A **487**, 22–31 (2017) (accepted for publication)
7. Joint National Committee on the Detection: Evaluation, and treatment of high blood pressure. The sixth report of the joint national committee on prevention, detection, evaluation, and treatment of high blood pressure. Arch. Intern. Med. **157**, 2413–2446 (1997)
8. Janiashvili, M., Jibladze, N., Matcharashvili, T., Topchishvili, A.: Comparison of statistical and distributional characteristics of blood pressure and heart rate variation of patients with different blood pressure categories. Model Assist. Stat. Appl. **8**, 177–184 (2013). https://doi.org/10.3233/MAS-130262
9. Peng, C.K., Havlin, S., Stanley, H.E., Goldberger, A.L.: Quantification of scaling exponents and crossover phenomena in nonstationary heartbeat time series. Chaos **5**, 82–87 (1995). https://doi.org/10.1063/1.166141
10. Pastor-Barriuso, R., Banegas, J.R., Damian, J., Appel, L.J., Guallar, E.: Systolic blood pressure, diastolic blood pressure, and pulse pressure: an evaluation of their joint effect on mortality. Ann. Intern. Med. **139**, 731–739 (2003)

On Generalized Additive Models with Dependent Time Series Covariates

Márton Ispány, Valdério A. Reisen, Glaura C. Franco, Pascal Bondon, Higor H. A. Cotta, Paulo R. P. Filho and Faradiba S. Serpa

Abstract The generalized additive model (GAM) is a standard statistical methodology and is frequently used in various fields of applied data analysis where the response variable is non-normal, e.g., integer-valued, and the explanatory variables are continuous, typically normally distributed. Standard assumptions of this model, among others, are that the explanatory variables are independent and identically distributed vectors which are not multicollinear. To handle the multicollinearity and serial dependence together a new hybrid model, called GAM-PCA-VAR model, was proposed in [17] (de Souza et al., J Roy Stat Soc C-Appl 2018) which is the

---

This paper is based on the talk "An application of the GAM-PCA-VAR model to respiratory disease and air pollution data" given by the first author.

M. Ispány (✉)
University of Debrecen, Debrecen, Hungary
e-mail: ispany.marton@inf.unideb.hu
URL: https://www.inf.unideb.hu/en/ispanymarton

V. A. Reisen · P. R. P. Filho · F. S. Serpa
Federal University of Espírito Santo, Vitória, Brazil
e-mail: valderioanselmoreisen@gmail.com

P. R. P. Filho
e-mail: pauloprezotti@hotmail.com

F. S. Serpa
e-mail: faradibasarquis@uol.com.br

G. C. Franco
Federal University of Minas Gerais, Belo Horizonte, Brazil
e-mail: glaura@est.ufmg.br

V. A. Reisen · P. Bondon · H. H. A. Cotta · P. R. P. Filho
Laboratoire des Signaux et Systèmes (L2S), CNRS-CentraleSupélec-Université Paris-Sud, Gif-sur-Yvette, France
e-mail: pascal.bondon@l2s.centralesupelec.fr

H. H. A. Cotta
e-mail: cotta.higor@gmail.com

P. R. P. Filho
e-mail: pauloprezotti@hotmail.com

I. Rojas et al. (eds.), *Time Series Analysis and Forecasting*,
Contributions to Statistics, https://doi.org/10.1007/978-3-319-96944-2_20

combination of GAM with the principal component analysis (PCA) and the vector autoregressive (VAR) model. In this paper, some properties of the GAM-PCA-VAR model are discussed theoretically and verified by simulation. A real data set is also analyzed with the aim to describe the association between respiratory disease and air pollution concentrations.

Keywords Air pollution · Generalized additive model · Multicollinearity Principal component analysis · Time series · Vector autoregressive model

1 Introduction

In the recent literature of time series, there has been an outstanding growth in models proposed for data that do not satisfy the Gaussian assumption. This is mainly the case when the response variable under study is a count series or an integer-valued series. Procedures developed to analyze this kind of data include, for example, observation-driven models, see [3, 6], integer-valued autoregressive (INAR) processes, see [1, 2], or non-Gaussian state space models, see [8, 10].

Particularly in health and environmental studies, where the response variable is typically a count time series, the GAM has been widely used to associate the dependent series, such as the number of respiratory or cardiovascular diseases to some pollutant or climate variables, see, for example, [5, 13, 14, 16–18] among others. In general, the researches related to the study of the association between pollution and adverse health effects usually consider only one pollutant. This simple model choice may be due to the fact that the pollutants are linearly time-correlated variables, see the discussion and references in [17].

Recently, it has become a common practice to use PCA in regression models to reduce the dimensionality of an independent set of data, especially the pollutants, which in some instances can include a large number of variables. The PCA is highly indicated to this purpose, as it can handle the multicollinearity problem that can cause biased regression estimates, see for example, [19].

Nevertheless, use of PCA in the time series context can bring some misspecifications in the fit of the GAM model, as this technique requires that the data should be independent. This problem arises due to the fact that the principal components (PCs) are linear combinations of the variables. In this context, as the covariates are time series, the autocorrelation present in the observations are transferred to the PCs, see [21].

One solution to this issue was recently proposed by [17] and [18], who introduced a model which combines GAM, PCA, and the VAR process. The authors suggest to apply the VAR model to the covariates, in order to eliminate the serial correlation and produce white noise processes, which in turn will be used to build the PCs in the PCA. The new variables obtained in the PCA are finally used as covariates in the

GAM model, originating the so-called GAM-PCA-VAR model. In their work, the authors have focused on presenting the model and showing its superiority compared to the sole use of GAM or the GAM-PCA procedures, but have not deepened on the theoretical properties of the model.

This work aims to state and prove some properties of the GAM-PCA-VAR model, as well as to perform some simulation study to check the results for small samples.

The paper is organized as follows. Section 2 presents the GAM-PCA-VAR model, and its related models as GAM, PCA and VAR, in some detail. In Sect. 3, the theoretical results are proved for the main model. Section 4 discusses the simulation results and Sect. 5 is devoted to the analysis of a real data set. Section 6 concludes the work.

2 The GAM-PCA-VAR Model

The GAM, see [11, 20], with a Poisson marginal distribution is typically used to relate a nonnegative integer-valued response variable Y with a set of covariates or explanatory variables $X_1, \ldots, X_p$, see [11, 20]. In a GAM the expected value μ of the response variable Y, $\mu = \mathsf{E}(Y)$, depends on the covariates via the formula

$$g(\mu) = \beta_0 + \sum_{i=1}^{p} f_i(X_i),$$

where g denotes the link function, β_0 is the intercept parameter and f_i's are functions with a specified parametric form, e.g., they are linear functions $f_i(x) = \beta_i x$, $\beta_i \in \mathbb{R}$, $i = 1, \ldots, p$, or nonparametric, e.g., they are simple smoothing functions like splines or moving averages. The unknown parameters β_0 and f_i, $i = 1, \ldots, p$ can be estimated by various algorithms, e.g., backfitting or restricted maximum likelihood (REML) method. However, if the data observed for variables Y and $X_i, i = 1, \ldots, p$, form a time series, the observations cannot be considered as a result of independent experiments and the covariates present strong interdependence, e.g., multicollinearity or concurvity, the standard fitting methods result in remarkable bias, see, e.g., [7, 17].

Let $\{Y_t\} \equiv \{Y_t\}_{t\in\mathbb{Z}}$ be a count time series, i.e., it is composed of nonnegative integer valued random variables. We suppose that the explanatory variables form a zero-mean stationary vector time series $\{\boldsymbol{X}_t\} \equiv \{\boldsymbol{X}_t\}_{t\in\mathbb{Z}}$ of dimension p, i.e., $\boldsymbol{X}_t = (X_{1t}, \ldots, X_{pt})^\top$ where $\top$ denotes the transpose, with the covariance matrix $\Sigma_X = \mathsf{E}(\boldsymbol{X}_t\boldsymbol{X}_t^\top)$. Let $\mathcal{F}_t$ denote the σ-algebra which contains the available information up to time t for all $t \in \mathbb{Z}$ from the point of view of the response variable, e.g., $\boldsymbol{X}_t$ is $\mathcal{F}_{t-1}$-measurable. The GAM-PCA-VAR model is introduced in [17] as a probabilistic latent variable model. In this paper, we define this model in a more general form as

$$Y_t \mid \mathcal{F}_{t-1} \sim \text{Poi}(\mu_t), \tag{1}$$

$$\boldsymbol{X}_t = \boldsymbol{\Phi}\boldsymbol{X}_{t-1} + A\boldsymbol{Z}_t \tag{2}$$

with link

$$g(\mu_t) = \beta_0 + \sum_{i=1}^{p}\sum_{j=0}^{\infty} f_{ij}(Z_{i(t-j)}), \tag{3}$$

where Poi$(\cdot)$ denotes the Poisson distribution, the latent variables $\{\boldsymbol{Z}_t\}$, $\boldsymbol{Z}_t = (Z_{1t}, \ldots, Z_{pt})^\top$, form a zero-mean Gaussian vector white noise process of dimension p with diagonal variance matrix $\Lambda = \text{diag}\{\lambda_1, \ldots, \lambda_p\}$, where $\lambda_1 \geq \lambda_2 \geq \cdots \geq \lambda_p$, A is an orthogonal matrix of dimension $p \times p$, $\boldsymbol{\Phi}$ is a matrix of dimension $p \times p$, g is a known link function, β_0 denotes the intercept, and f_{ij}'s are unknown functions. For a zero-mean Gaussian vector white noise process $\{\boldsymbol{Z}_t\}$ with covariance matrix Σ we shall use the notation $\{\boldsymbol{Z}_t\} \sim \text{GWN}(\Sigma)$, see also [4, Definition 11.1.2]. Clearly, for all i, the univariate time series $\{Z_{it}\} \sim \text{GWN}(\lambda_i)$, and $\{Z_{it}\}$ is mutually independent from $\{Z_{jt}\}$ for all $j \neq i$. We assume that all the eigenvalues of $\boldsymbol{\Phi}$ are less than 1 in modulus which implies that (2) has a unique stationary causal solution. In the case of a Poisson distributed response variable, the two widely used link functions are the identity link, $g(z) = z$, and the canonical logarithmic link, $g(z) = \log z$. The set $(\beta_0, \{f_{ij}\}, A, \Lambda, \boldsymbol{\Phi})$ forms the parameters of the GAM-PCA-VAR model to be estimated. We remark that in the case of canonical logarithmic link function no additional assumption is needed for the parameters, while in the case of identity link function all the parameters in (3), i.e., β_0 and f_{ij}'s, have to be nonnegative. It should be also emphasized that the underlying intensity process $\{\mu_t\}$ of $\{Y_t\}$ is also a time series with a complex dependence structure, and μ_t is $\mathcal{F}_{t-1}$-measurable for all $t \in \mathbb{Z}$. One can see that the time series $\{\boldsymbol{X}_t\}$ of covariates depends on $\{\boldsymbol{Z}_t\}$ by formula $\boldsymbol{X}_t = \sum_{k=0}^{\infty} \boldsymbol{\Phi}^k A\boldsymbol{Z}_{t-k}$ for all t, see [4, Example 11.3.1].

The dependence of the response time series $\{Y_t\}$ from the explanatory vector time series $\{\boldsymbol{X}_t\}$ in the GAM-PCA-VAR model can be described by three transformation steps. Clearly, by (2), the latent variable can be expressed as $\boldsymbol{Z}_t = A^\top \boldsymbol{U}_t$, where $\boldsymbol{U}_t := \boldsymbol{X}_t - \boldsymbol{\Phi}\boldsymbol{X}_{t-1}$ for all t. Thus, as the first step, the intermediate vector times series $\{\boldsymbol{U}_t\}$ is derived from filtering $\{\boldsymbol{X}_t\}$ by a VAR(1) filter. One can see that $\{\boldsymbol{U}_t\} \sim \text{GWN}(\Sigma_U)$ where $\Sigma_U := A\Lambda A^\top$. Then, as the second step, the latent vector time series $\{\boldsymbol{Z}_t\}$ as PC vecta is derived by instantaneous linear transformation of the intermediate vector white noise $\{\boldsymbol{U}_t\}$. The transformation matrix of the PCA is given by the spectral decomposition of Σ_U. Finally, as the third step, the standard GAM with link (3) is fitting for the response time series $\{Y_t\}$ using the latent vector time series $\{\boldsymbol{Z}_t\}$. The impact of the VAR(1) filter in the first step is to eliminate the serial correlation present in the original covariates. On the other hand, the impact of the PCA in the second step is to eliminate the correlation in the state space of the original covariates. Hence, the result of these two consecutive transformations is the latent vector time series $\{\boldsymbol{Z}_t\}$ whose components, Z_{it}, $i = 1, \ldots, p$, $t \in \mathbb{Z}$, are independent Gaussian variables both in space and time. In the case of logarithmic link function, large positive values in a coordinate of the latent variable indicate locally high influence according to this latent factor. On the contrary, large negative values indicate negligible influence on the response, see, for example, [21]. The order of models in the acronym GAM-PCA-VAR corresponds to these steps starting with the third one and finishing with the first one.

The GAM-PCA-VAR model contains several submodels with particular dependence structure. If $\Phi = 0$ then (2) is simplified to a PC transformation. In this case, we suppose that there is no serial correlation and we only have to handle the correlation in the state space of covariates. We have two transformation steps: PCA and GAM. This kind of models is called GAM-PCA model and has been intensively studied, see, e.g., [15, 22]. Beside the full PCA when all PCs are involved into the GAM, we can fit a restricted PCA model by defining $f_{ij} = 0$ for all $i > r$ and $j \geq 0$ where $r < p$. In this case, the first rth PCs are applied as covariates in the GAM step. If the matrices in VAR(1) model (2) have the following block structures:

$$\Phi = \begin{bmatrix} \Phi_q & 0 \\ 0 & 0 \end{bmatrix}, \quad A = \begin{bmatrix} A_q & 0 \\ 0 & I_{p-q} \end{bmatrix},$$

where the eigenvalues of the $q \times q$ matrix Φ_q are less than one in modulus, A_q is an orthogonal matrix of dimension $q \times q (q \leq p)$, and $f_{i1}(z) = \beta_i z$ with $\beta_i \in \mathbb{R}$ for $i = 1, \ldots, r(r \leq q)$, f_{i1} is a general smoothing function for $i = q+1, \ldots, p$, $f_{ij} = 0$ otherwise, then we obtain the model that was studied in [17] and applied in the data analysis of Sect. 5. In this model, it is supposed that the set of covariates can be partitioned into two sets: $(X_1, \ldots, X_q)$ are normal covariates, e.g., the pollutant variables in the terminology of Sect. 5, while $(X_{q+1}, \ldots, X_p)$ are so-called confounding variables as trend, seasonality, etc. The normal covariates satisfy a q-dimensional VAR(1) model, however, instead of all coordinates of the innovation, only its first rth PCs are involved into the GAM taking into consideration that the covariates present strong inter-correlation. Finally, we note that our model can be further generalized by replacing (2) by the more general VARMA or VARIMA or their seasonal variants (SVARMA or SVARIMA) models.

Since the latent variables $\{\boldsymbol{Z}_t\}$ form a Gaussian vector time series, given a sample $(\boldsymbol{X}_1, Y_1), \ldots, (\boldsymbol{X}_n, Y_n)$, the log-likelihood can be expressed in an explicit form, see [17] for a particular case. Because this log-likelihood is rather complicated a three-stage estimation method is proposed. First, VAR(1) model is fitted to the original covariates by applying standard time series techniques. Second, PCA is applied for the residuals defined by $\widehat{\boldsymbol{Z}}_t = \boldsymbol{X}_t - \widehat{\Phi}\boldsymbol{X}_{t-1}$, $t = 2, \ldots, n$, where $\widehat{\Phi}$ denotes the estimated autoregressive coefficient matrix in the fitted VAR(1) model. Third, GAM model is fitted using the PCs. The approach discussed above is similar to the PC regression, see, e.g., [12, Chap. 8], and it can be considered as a three-stage nonlinear regression method.

The first two steps of the above proposed parameter estimation method for GAM-PCA-VAR model can be interpreted as consecutive orthogonalizations, first in time and then in the state space of covariates. In [17, Remark] we argued that the order of VAR filter and PCA can not be interchanged because the orthogonalization in the state space does not eliminate the serial correlation and, as the necessary next step, the orthogonalization in time by VAR filter bring back the inter-correlation between the covariates. In what follows, we demonstrate this phenomena by giving a simple example. Let $\{\boldsymbol{X}_t\}$ be a zero-mean causal VAR(1) process defined by

$$\boldsymbol{X}_t = \Psi \boldsymbol{X}_{t-1} + \boldsymbol{W}_t,$$

where $\{\boldsymbol{W}_t\}$ is a zero-mean vector white noise process with variance matrix Σ_W. Suppose that the variance matrix Σ_X of $\{\boldsymbol{X}_t\}$ is diagonal, i.e., the coordinates of $\{\boldsymbol{X}_t\}$ can be interpreted as PCs after PCA. Then, Σ_W is not necessarily a diagonal matrix, which implies that a VAR(1) filter may result in an inter-correlated white noise. Namely, consider the following parameters $\Sigma_W = A\Lambda A^\top$ and $\Psi = ASA^\top$, where Λ and S are diagonal matrices and A is an orthogonal matrix. In other words, we suppose that the orthogonal matrix A in the spectral decomposition of Σ_W diagonalizes the autoregressive coefficient matrix as well. Then, we have, by formula (11.1.13) in [4], that

$$\Sigma_X = \sum_{j=0}^{\infty} \Psi^j \Sigma_W (\Psi^\top)^j = \sum_{j=0}^{\infty} AS^j \Lambda S^j A^\top = A\mathrm{diag}\left\{\frac{\lambda_i}{1-s_i^2}\right\} A^\top.$$

Let $\sigma^2 > \max_i\{\lambda_i\}$ arbitrary and define $s_i := \sqrt{1-\lambda_i/\sigma^2}$ for all i. Clearly, Ψ is a causal matrix since all its eigenvalues are less than 1 in modulus and $\Sigma_X = \sigma^2 I$, i.e., the coordinates of $\{\boldsymbol{X}_t\}$ are uncorrelated. However, the innovation variance matrix Σ_W can be arbitrary proving that the application of VAR filter for a non-intercorrelated vector time series can give inter-correlated vector white noise in its coordinates.

Now, we present some particular examples of GAM-PCA-VAR models.

Example 1 One of the simplest GAM-PCA-VAR models is the model with dimension $p = 1$ and log-linear link function. In this case, there is only one covariate $\{X_t\}$, and (2) is an AR(1) model

$$X_t = \phi X_{t-1} + Z_t, \tag{4}$$

where $|\phi| < 1$ which guarantees the existence of a unique stationary causal solution, $\{Z_t\} \sim \mathrm{GWN}(\lambda)$, $\lambda > 0$. We remark that $A = 1$ in (2) in order for the model to be identifiable. The link is log-linear, expressed as

$$\log \mu_t = \beta_0 + \beta_1 Z_t. \tag{5}$$

The parameter set of this model is $(\beta_0, \beta_1, \lambda, \phi)$ with parameter space $\mathbb{R}^2 \times \mathbb{R}_+ \times (-1, 1)$. In this model, there is no dimension reduction. Clearly, $Z_t = X_t - \phi X_{t-1}$, thus the response depends on the covariate through the link

$$\log \mu_t = \gamma_0 + \gamma_1 X_t + \gamma_2 X_{t-1}, \tag{6}$$

where there is a one-to-one correspondence between the parameter sets (β_0, β_1, ϕ) and $(\gamma_0, \gamma_1, \gamma_2)$ defined by the equations $\gamma_0 = \beta_0, \gamma_1 = \beta_1$ and $\gamma_2 = -\phi\beta_1$ provided $\phi \neq 0$. However, if we fit the standard GAM by using the link (6) with covariates X_t and X_{t-1} at time t, we take no count of the interdependence in time series $\{X_t\}$ which can result in biased and inconsistent estimators of the GAM parameters.

Example 2 Define a particular two-dimensional ($p = 2$) GAM-PCA-VAR model with logarithmic link function in the following way. The two-dimensional covariate vector process $\{\boldsymbol{X}_t\}$, $\boldsymbol{X}_t = (X_{1t}, X_{2t})^\top$, satisfies the VAR(1) model

$$\begin{bmatrix} X_{1t} \\ X_{2t} \end{bmatrix} = \begin{bmatrix} \phi_1 & 0 \\ 0 & \phi_2 \end{bmatrix} \begin{bmatrix} X_{1(t-1)} \\ X_{2(t-1)} \end{bmatrix} + \begin{bmatrix} \cos\varphi & -\sin\varphi \\ \sin\varphi & \cos\varphi \end{bmatrix} \begin{bmatrix} Z_{1t} \\ Z_{2t} \end{bmatrix},$$

where $|\phi_1| < 1$, $|\phi_2| < 1$ and $\{Z_{it}\} \sim \text{GWN}(\lambda_i)$ with $\lambda_i > 0$, $i = 1, 2$, which are independent from each other. Note that the set of two-dimensional orthogonal matrices, A, can be parametrized by an angle parameter $\varphi \in [0, 2\pi)$. We assume that the link is

$$\log \mu_t = \beta_0 + \beta_1 Z_{1t}.$$

The parameter set of this model is $(\beta_0, \beta_1, \varphi, \lambda_1, \lambda_2, \phi_1, \phi_2)$ and the parameter space is $\mathbb{R}^2 \times [0, 2\pi) \times \mathbb{R}_+^2 \times (-1, 1)^2$. Note that, in this model, there is a PCA step as a dimension reduction since only the first coordinate $\{Z_{1t}\}$ of the vector innovation is involved into the GAM as covariate. One can see that the response depends on the covariates through the link

$$\log \mu_t = \gamma_0 + \gamma_1 X_{1t} + \gamma_2 X_{2t} + \gamma_3 X_{1(t-1)} + \gamma_4 X_{2(t-1)},$$

where $\gamma_0 = \beta_0$, $\gamma_1 = \beta_1 \cos\varphi$, $\gamma_2 = \beta_1 \sin\varphi$, $\gamma_3 = -\beta_1\phi_1 \cos\varphi$ and $\gamma_4 = -\beta_1\phi_2 \sin\varphi$. Thus, the intensity process $\{\mu_t\}$ depends on all coordinates of $\boldsymbol{X}_t$ and $\boldsymbol{X}_{t-1}$. Clearly, there is a one-to-one correspondence between the two parameter sets $(\beta_0, \beta_1, \varphi, \phi_1, \phi_2)$ and $(\gamma_0, \gamma_1, \gamma_2, \gamma_3, \gamma_4)$.

Example 3 A seasonal one-dimensional GAM-PCA-VAR model with linear link function can be defined in the following way. Suppose that the one-dimensional covariate process $\{X_t\}$ satisfies the $\text{SAR}_s(1)$ model:

$$X_t = \phi X_{t-s} + Z_t,$$

where $|\phi| < 1$, $\{Z_t\} \sim \text{GWN}(\lambda)$ with $\lambda > 0$ and $s \in \mathbb{Z}_+$ denotes the seasonal period. The link is linear and is given by

$$\mu_t = \beta_0 + \beta_1 f(Z_t),$$

where $f : \mathbb{R} \to \mathbb{R}_+$ is a known function and $\beta_0, \beta_1 \in \mathbb{R}_+$ are parameters. The parameter set of this model is $(\beta_0, \beta_1, \lambda, \phi)$ with parameter space $\mathbb{R}_+^3 \times (-1, 1)$. The response variable depends on the original covariates through the link

$$\mu_t = \beta_0 + \beta_1 f(X_t - \phi X_{t-s}).$$

If the function f is sufficiently smooth we have the approximation $f(X_t - \phi X_{t-s}) \approx f(X_t) - \phi f'(X_t)X_{t-s}$, and then

$$\mu_t = \gamma_0 + \gamma_1 f_1(X_t) + \gamma_2 f_2(X_t, X_{t-s}), \tag{7}$$

where f_1, f_2 are known functions and $\gamma_0 = \beta_0$, $\gamma_1 = \beta_1$ and $\gamma_2 = -\beta_1\phi$. Thus, the response depends on the original covariate and its s-step lagged series through the standard GAM. However, the covariates in (7) are clearly dependent.

3 Theoretical Results

In this section, we prove some theoretical results for particular classes of GAM-PCA-VAR models. Consider the log-linear model defined by the link

$$\log \mu_t = \beta_0 + \sum_{i=1}^{p}\sum_{j=0}^{\infty} \beta_{ij} Z_{i(t-j)}, \tag{8}$$

where $\beta_0, \beta_{ij} \in \mathbb{R}, i = 1, \ldots, p, j \in \mathbb{Z}_+$. The first proposition is about the existence of log-linear GAM-PCA-VAR models.

Proposition 1 *Suppose that* $\sigma^2 := \sum_{i=1}^{p} \lambda_i \sum_{j=0}^{\infty} \beta_{ij}^2$ *is finite. Then the GAM-PCA-VAR model with log-linear link* (8) *has solution* $\{(Y_t, \boldsymbol{X}_t)\}$ *which is a strictly stationary process and* $\mathsf{E}(Y_t) = \mathsf{E}(\mu_t) = \exp(\beta_0 + \sigma^2/2)$ *for all* $t \in \mathbb{Z}$.

Proof By conditioning we have that

$$\mathsf{E}(Y_t) = \mathsf{E}(\mathsf{E}(Y_t \mid \mathcal{F}_{t-1})) = \mathsf{E}(\mu_t) = \mathsf{E}(\exp(\log \mu_t)) = \exp(\beta_0 + \sigma^2/2) \tag{9}$$

is finite since, by (8), $\log \mu_t \sim \mathcal{N}(\beta_0, \sigma^2)$, i.e., μ_t has a lognormal distribution, and the moment generating function of $\xi \sim \mathcal{N}(\beta_0, \sigma^2)$ is given by $M_\xi(t) := \mathsf{E}(\exp(t\xi)) = \exp(\beta_0 t + (\sigma t)^2/2)$. Thus, the nonnegative integer valued random variable Y_t is finite with probability one for all $t \in \mathbb{Z}$. The vector time series $\{\boldsymbol{Z}_t\}$ forms a Gaussian white noise. Hence it is strictly stationary process with backshift operator $B(\boldsymbol{Z}_t) = \boldsymbol{Z}_{t-1}$ for all $t \in \mathbb{Z}$. Since both stochastic processes $\{Y_t\}$ and $\{\boldsymbol{X}_t\}$ depend on $\{\boldsymbol{Z}_t\}$ through time-invariant functionals, we have the strict stationarity of $\{(Y_t, \boldsymbol{X}_t)\}$ and $B(\boldsymbol{X}_t) = \boldsymbol{X}_{t-1}$, $B(Y_t) = Y_{t-1}$ for all $t \in \mathbb{Z}$. □

In the next proposition, we prove that all moments of the log-linear GAM-PCA-VAR model are finite.

Proposition 2 *Suppose that* σ^2 *defined in Proposition 1 is finite. Then, all moments of the stochastic process* $\{(Y_t, \boldsymbol{X}_t)\}$ *are finite. In particular, we have, for all* $t \in \mathbb{Z}$,

$$\mathrm{Var}(Y_t) = \exp(2\beta_0 + \sigma^2)(\exp(\sigma^2) - 1 + \exp(-\beta_0 - \sigma^2/2)),$$
$$\mathrm{Var}(\mu_t) = \exp(2\beta_0 + \sigma^2)(\exp(\sigma^2) - 1).$$

Proof Let $r \in \mathbb{N}$. Define the rth factorial of a nonnegative integer k as $k^{[r]} := k(k-1)\cdots(k-r+1)$ and let $k^{[0]} := 1$. For the rth factorial moment of Y_t we have by conditioning that

$$\begin{aligned}\mathsf{E}(Y_t^{[r]}) &= \sum_{k=0}^{\infty} k^{[r]}\mathsf{P}(Y_t = k) = \mathsf{E}\sum_{k=0}^{\infty} k^{[r]}\mathsf{P}(Y_t = k \mid \mathcal{F}_{t-1})\\ &= \mathsf{E}\sum_{k=r}^{\infty} \frac{\mu_t^k}{(k-r)!}e^{-\mu_t} = \mathsf{E}(\mu_t^r)\end{aligned}$$

for all $t \in \mathbb{Z}$. Similarly to (9), we have that the factorial moments are finite, since

$$\mathsf{E}(Y_t^{[r]}) = \mathsf{E}(\mu_t^r) = \mathsf{E}(\exp(r\log\mu_t)) = \exp\{\beta_0 r + (\sigma r)^2/2\}. \tag{10}$$

Since the higher order moments can be expressed by the factorial moment via the formula

$$\mathsf{E}(Y^r) = \sum_{j=0}^{r} S(r, j)\mathsf{E}(Y^{[j]}),$$

where $S(r, j)$'s denotes Stirling numbers of the second kind, the finiteness of all higher order moments follows easily. Since $\{\boldsymbol{X}_t\}$ is a Gaussian process all its moments are finite. Finally, the existence of mixed moments follows by the Cauchy–Schwarz inequality.

From (10), we have

$$\begin{aligned}\mathrm{Var}(\mu_t) &= \mathsf{E}(\mu_t^2) - \mathsf{E}^2(\mu_t) = \exp(2\beta_0 + (2\sigma)^2/2) - \exp(2\beta_0 + \sigma^2)\\ &= \exp(2\beta_0 + \sigma^2)(\exp(\sigma^2) - 1).\end{aligned}$$

Finally, the formula for $\mathrm{Var}(Y_t)$ can be derived by

$$\mathrm{Var}(Y_t) = \mathsf{E}(\mathrm{Var}(Y_t \mid \mathcal{F}_{t-1})) + \mathrm{Var}(\mathsf{E}(Y_t \mid \mathcal{F}_{t-1})) = \mathsf{E}(\mu_t) + \mathrm{Var}(\mu_t).$$

□

The existence of all moments for the log-linear GAM-PCA-VAR process is to be compared with the same result for the integer-valued GARCH, so-called INGARCH, process, see [9, Proposition 6]. This implies that the log-linear GAM-PCA-VAR process possesses second and higher order structures, e.g., the autocorrelation function, the spectral density function, the cumulants and the higher order spectra exist. Let ρ_Y denotes the autocorrelation function of the time series $\{Y_t\}$.

Proposition 3 *For the auto- and cross-correlation functions of the GAM-PCA-VAR process* $\{(Y_t, \boldsymbol{X}_t)\}$ *with intensity process* $\{\mu_t\}$, *we have* $\rho_Y(h) = c_Y\rho(h)$, $\rho_\mu(h) = c_\mu\rho(h)$ *and* $\rho_{Y\mu}(h) = c_{Y\mu}\rho(h)$ *where*

$$\rho(h) := \exp\left(\sum_{i=1}^{p}\lambda_i\sum_{j=0}^{\infty}\beta_{i(j+|h|)}\beta_{ij}\right) - 1, \quad h \in \mathbb{Z}\setminus\{0\},$$

and the constants $c_Y, c_\mu, c_{Y\mu}$ *are defined by*

$$c_Y := (\exp(\sigma^2) - 1 + \exp(-\beta_0 - \sigma^2/2))^{-1}, \quad c_\mu := (\exp(\sigma^2) - 1)^{-1}, \quad c_{Y\mu} := \sqrt{c_Y c_\mu}.$$

Moreover, $\mathrm{Cov}(Y_{t+h}, \boldsymbol{X}_t) = \mathrm{Cov}(\mu_{t+h}, \boldsymbol{X}_t) = \mathsf{E}(Y_{t+h}\boldsymbol{X}_t) = \mathsf{E}(\mu_{t+h}\boldsymbol{X}_t) = C(h)$ *with*

$$C(h) := \exp(\beta_0 + \sigma^2/2) \times \begin{cases} \sum_{k=0}^{\infty} \Phi^k A(\boldsymbol{\lambda} \circ \boldsymbol{\beta}_{h+k}) & \text{if } h \geq 0, \\ \sum_{k=0}^{\infty} \Phi^{k-h} A(\boldsymbol{\lambda} \circ \boldsymbol{\beta}_k) & \text{if } h \leq 0, \end{cases} \tag{11}$$

where $\boldsymbol{\lambda} := (\lambda_1, \ldots, \lambda_p)^\top$, $\boldsymbol{\beta}_j := (\beta_{1j}, \ldots, \beta_{pj})^\top$, $j \in \mathbb{Z}_+$, *and* $\circ$ *denotes the entrywise (Hadamard) product.*

Proof Let $h \in \mathbb{N}$. One can see that for the intensity process we have $\mu_{t+h} = \mu_{th}^{(1)}\mu_{th}^{(2)}$ where

$$\log\mu_{th}^{(1)} := \beta_0 + \sum_{i=1}^{p}\sum_{j=1}^{h}\beta_{i(h-j)}Z_{i(t+j)}, \quad \log\mu_{th}^{(2)} := \sum_{i=1}^{p}\sum_{j=0}^{\infty}\beta_{i(j+h)}Z_{i(t-j)}.$$

Clearly, $\mu_{th}^{(1)}$ is independent of $\mathcal{F}_{t-1}$ and Y_t, while $\mu_{th}^{(2)}$ is $\mathcal{F}_{t-1}$-measurable. Hence, we have by conditioning that

$$\begin{aligned} \mathsf{E}(Y_{t+h}Y_t) &= \mathsf{E}(Y_t\mathsf{E}(Y_{t+h} \mid \mathcal{F}_{t+h-1})) = \mathsf{E}(\mu_{t+h}Y_t) = \mathsf{E}(\mu_{th}^{(1)}\mu_{th}^{(2)}Y_t) \\ &= \mathsf{E}(\mu_{th}^{(1)})\mathsf{E}(\mu_{th}^{(2)}\mathsf{E}(Y_t \mid \mathcal{F}_{t-1})) = \mathsf{E}(\mu_{th}^{(1)})\mathsf{E}(\mu_{th}^{(2)}\mu_t) = \mathsf{E}(\mu_{t+h}\mu_t) \end{aligned}$$

since μ_t is independent of $\mu_{th}^{(1)}$. This gives the result for $h > 0$. On the other hand, for all $h > 0$, again by conditioning, $\mathsf{E}(Y_{t+h}\mu_t) = \mathsf{E}(\mu_{t+h}\mu_t)$. Thus

$$\mathrm{Cov}(Y_{t+h}, Y_t) = \mathrm{Cov}(\mu_{t+h}, \mu_t) = \mathrm{Cov}(Y_{t+h}, \mu_t), \quad h \in \mathbb{Z}\setminus\{0\}.$$

Since

$$\mathsf{E}(\mu_{t+h}\mu_t) = \mathsf{E}(\mu_{th}^{(1)}\mu_{th}^{(2)}\mu_t) = \mathsf{E}(\mu_{th}^{(1)})\mathsf{E}(\mu_{th}^{(2)}\mu_t)$$

similarly to (9) we have

$$\begin{aligned}\mathsf{E}(\mu_{t+h}\mu_t) &= \exp\left(2\beta_0 + \frac{1}{2}\sum_{i=1}^{p}\lambda_i\left(\sum_{j=0}^{h-1}\beta_{ij}^2 + \sum_{j=0}^{\infty}(\beta_{i(j+h)} + \beta_{ij})^2\right)\right) \\ &= \exp\left(\sum_{i=1}^{p}\lambda_i\sum_{j=0}^{\infty}\beta_{i(j+h)}\beta_{ij}\right)\mathsf{E}(\mu_{t+h})\mathsf{E}(\mu_t).\end{aligned}$$

Thus, the first part of the proposition follows by Proposition 2.

Next we prove (11) for the cross-correlations of response and covariate variables. Clearly, by conditioning, $\mathsf{E}(Y_{t+h}\boldsymbol{X}_t) = \mathsf{E}(\mu_{t+h}\boldsymbol{X}_t)$ for all $h \in \mathbb{Z}_+$. On the other hand, for all $t \in \mathbb{Z}$, $h \in \mathbb{Z}_+$, we have $\boldsymbol{X}_{t+h} = \boldsymbol{X}_{th}^{(1)} + \boldsymbol{X}_{th}^{(2)}$ where

$$\boldsymbol{X}_{th}^{(1)} := \sum_{k=1}^{h}\Phi^{h-k}A\boldsymbol{Z}_{t+k}, \quad \boldsymbol{X}_{th}^{(2)} := \sum_{k=0}^{\infty}\Phi^{h+k}A\boldsymbol{Z}_{t-k}.$$

One can see that $\boldsymbol{X}_{th}^{(1)}$ is independent of $\mathcal{F}_{t-1}$ and Y_t, while $\boldsymbol{X}_{th}^{(2)}$ is $\mathcal{F}_{t-1}$-measurable. Thus, we have that

$$\begin{aligned}\mathsf{E}(\boldsymbol{X}_{t+h}Y_t) &= \mathsf{E}((\boldsymbol{X}_{th}^{(1)} + \boldsymbol{X}_{th}^{(2)})Y_t) = \mathsf{E}(\boldsymbol{X}_{th}^{(1)})\mathsf{E}(Y_t) + \mathsf{E}(\boldsymbol{X}_{th}^{(2)}\mathsf{E}(Y_t \mid \mathcal{F}_{t-1})) \\ &= \mathsf{E}(\boldsymbol{X}_{th}^{(1)})\mathsf{E}(\mu_t) + \mathsf{E}(\boldsymbol{X}_{th}^{(2)}\mu_t) = \mathsf{E}(\boldsymbol{X}_{t+h}\mu_t).\end{aligned}$$

Hence $\mathsf{E}(Y_{t+h}\boldsymbol{X}_t) = \mathsf{E}(\mu_{t+h}\boldsymbol{X}_t)$ for all $h \in \mathbb{Z}$ and it is enough to compute the cross-correlation between $\{\boldsymbol{X}_t\}$ and $\{\mu_t\}$. Let $h \geq 0$. For all $\ell \in \{1, \ldots, p\}$, $k \in \mathbb{Z}_+$ let $\mathcal{I}_{\ell k}^h := \{1, \ldots, p\} \times \mathbb{Z}_+ \setminus (\ell, k+h)$ and define the random variables

$$\log \xi_{\ell k}^{th} := \beta_0 + \sum_{(i,j)\in\mathcal{I}_{\ell k}^h}\beta_{ij}Z_{i(t+h-j)}, \quad \log \eta_{\ell k}^{th} := \beta_{\ell(k+h)}Z_{\ell(t-k)}.$$

Then $\mu_{t+h} = \xi_{\ell k}^{th}\eta_{\ell k}^{th}$, where the factors in this decompostion are independent. Since $\mathsf{E}(\mu_{t+h}\boldsymbol{X}_t) = \sum_{k=0}^{\infty}\Phi^k A\mathsf{E}(\mu_{t+h}\boldsymbol{Z}_{t-k})$ and, using the fact that for $Z \sim \mathcal{N}(0, \lambda)$ and $\beta \in \mathbb{R}$ we have $\mathsf{E}(Z\exp(\beta Z)) = \beta\lambda\exp(\lambda\beta^2/2)$,

$$\mathsf{E}(\mu_{t+h}Z_{\ell(t-k)}) = \mathsf{E}(\xi_{\ell k}^{th}\eta_{\ell k}^{th}Z_{\ell(t-k)}) = \mathsf{E}(\xi_{\ell k}^{th})\mathsf{E}(\eta_{\ell k}^{th}Z_{\ell(t-k)}) = \mathsf{E}(\mu_{t+h})\beta_{\ell(k+h)}\lambda_\ell,$$

we obtain the formula (11). The proof is similar in the case of $h < 0$. □

Remark 1 It is easy to see that if $\beta_{ij} = \beta_i^j$ for all i, j, then the function ρ is given by $\rho(h) = \exp(\sum_{i=1}^{p}\lambda_i\beta_i^{|h|}/(1-\beta_i^2)) - 1$, $h \in \mathbb{Z}$. If β_i's are all positive then ρ is positive everywhere and we have autocorrelation functions which are similar to what is displayed in Fig. 1. For the one-dimensional model in Example 1 we have the cross-correlation function (CCF) $C(h) = \exp(\beta_0 + \lambda\beta_1^2/2)\lambda\beta_1\phi^{-h}$ for $h \leq 0$ and

$C(h) = 0$ for $h > 0$. If $\phi > 0$ then, according to positive or negative β_1, we obtain everywhere positive or negative CCFs. For example, see the CCFs in Fig. 2 between the response (Admissions) and pollutants CO, NO_2 that are positive and the CCFs between the response (Admissions) and O_3, SO_2 that are negative at every lag, respectively.

Consider another widely used link function, the linear one, and define the linear GAM-PCA-VAR model by the link

$$\mu_t = \beta_0 + \sum_{i=1}^{p} \sum_{j=0}^{\infty} \beta_{ij} f(Z_{i(t-j)}), \tag{12}$$

where $\beta_0, \beta_{ij} \in \mathbb{R}_+$, $i = 1, \ldots, p$, $j \in \mathbb{Z}_+$, are parameters and $f : \mathbb{R} \to \mathbb{R}_+$ is a known function, e.g., $f(z) = \exp(z)$. Let $\varphi(x \mid \lambda)$ denote the probability density function of the normal distribution with mean 0 and variance λ.

Proposition 4 *Suppose that, for all* $i = 1, \ldots, p$, $\sum_{j=0}^{\infty} \beta_{ij} < \infty$ *and* $\tau_i := \int_{-\infty}^{\infty} f(x)\varphi(x \mid \lambda_i)\mathrm{d}x < \infty$. *Then the GAM-PCA-VAR model with linear link* (12) *has a strictly stationary solution* $\{(Y_t, \boldsymbol{X}_t)\}$. *Moreover,* $\mathsf{E}(Y_t) = \mathsf{E}(\mu_t) = \beta_0 + \sum_{i=1}^{p} \tau_i \sum_{j=0}^{\infty} \beta_{ij}$.

Proof The proof is similar to the proof of Proposition 1. □

Clearly, the assumptions of Proposition 4 do not necessarily garantee the existence of higher order moments of linear GAM-PCA-VAR process. Indeed, the rth order moment $\mathsf{E}(Y_t^r)$ is finite if and only if $\int_{-\infty}^{\infty} f^r(x)\varphi(x \mid \lambda_i)\mathrm{d}x < \infty$ for all i where $r \geq 1$.

4 Simulation Study

In order to evaluate the effect on the parameter estimation of a GAM model in the presence of temporal correlation in the covariate $\{X_t\}$, a simulation study was conducted. The data were generated according to the model discussed in Example 1. Three estimation methods were considered: the standard GAM with only one covariate where the estimated parameters were β_0 and β_1 (M1); the standard GAM with two covariates, the original one and its 1-step lagged series, where the estimated parameters were β_0, β_1, β_2 and $\phi = -\beta_2/\beta_1$ (M2); the full GAM-PCA-VAR model by the procedure described in Sect. 2 where all parameters β_0, β_1, ϕ, λ were estimated (M3).

For the model discussed in Example 1 the data were generated under $\beta_0 = 0.2$, $\beta_1 = 1$, $\lambda = 2$ and three scenarios were considered as $\phi = -0.7, 0.3, 0.9$ to model strong negative, small positive and strong positive correlations, respectively. In order to model the impact due to some unobservable variables, e.g., environmental ones in

Table 1 Simulation results for model in Example 1

Estimation method	ϕ	Parameter	Mean	Bias	MSE
M1: GAM with X_t	−0.7	$\beta_0 = 0.2$	0.699	0.499	0.253
		$\beta_1 = 1$	0.507	−0.492	0.244
M2: GAM with X_t, X_{t-1}		$\beta_0 = 0.2$	0.204	0.004	0.001
		$\beta_1 = 1$	0.999	−0.001	0.0002
		ϕ =-0.7	−0.7	0	0.0001
M3: GAM-PCA-VAR		$\beta_0 = 0.2$	0.205	0.005	0.001
		$\beta_1 = 1$	0.999	−0.001	0.0002
		$\phi = -0.7$	−0.695	0.004	0.0005
		$\lambda = 2$	2.003	0.003	0.008
M1: GAM with X_t	0.3	$\beta_0 = 0.2$	0.302	0.102	0.012
		$\beta_1 = 1$	0.905	−0.095	0.009
M2: GAM with X_t, X_{t-1}		$\beta_0 = 0.2$	0.209	0.009	0.001
		$\beta_1 = 1$	0.998	−0.002	0.0002
		$\phi = 0.3$	0.3	0	0.0002
M3: GAM-PCA-VAR		$\beta_0 = 0.2$	0.209	0.009	0.001
		$\beta_1 = 1$	0.999	−0.001	0.0002
		$\phi = 0.3$	0.306	0.006	0.0008
		$\lambda = 2$	1.995	−0.005	0.009
M1: GAM with X_t	0.9	$\beta_0 = 0.2$	1.002	0.802	0.651
		$\beta_1 = 1$	0.191	−0.809	0.655
M2: GAM with X_t, X_{t-1}		$\beta_0 = 0.2$	0.2	0	0.001
		$\beta_1 = 1$	1	0	0.0002
		$\phi = 0.9$	0.899	−0.001	0
M3: GAM-PCA-VAR		$\beta_0 = 0.2$	0.203	0.003	0.001
		$\beta_1 = 1$	1	0	0.0002
		$\phi = 0.9$	0.899	−0.001	0.0001
		$\lambda = 2$	2.007	0.007	0.0086

the context of the next section, independent $\mathcal{N}(0, 0.1)$ distributed random variables were added to the predictor of $\log \mu_t$ for all $t \in \mathbb{Z}$. The sample size $n = 1000$ and the number of Monte Carlo simulations was equal to 100. The empirical values of mean, bias and mean square error (MSE) are displayed in Table 1. All results were obtained by using R-code.

In the case of standard GAM estimation (M1) it can be seen that the estimate of β_1 is heavily affected by the autocorrelation structure present in the covariate, by presenting a negative bias which increases in absolute value as $|\varphi|$ increases. The estimated MSE also increases substantially with $|\varphi|$. On the other hand, it can also be seen that the fitted standard GAM model tends to severely overestimate β_0. Contrarily, the estimation methods M2 and M3 work equally well, the estimates of the parameters are very close to the true values with noticeably small MSE. The undoubted advantage of method M3 against M2 is that an AR(1) model is also fitted for the covariate where the innovation variance λ is estimated and which can be applied later in the prediction. In this procedure, first, the covariate variable is predicted by (4), and then the response variable is predicted by the GAM using the link (5).

5 Application to Air Pollution Data

In this study, the number of hospital admissions (Admissions) for respiratory diseases (RD) as response variable was obtained from the main children's emergency department in the Vitória Metropolitan Area (called Hospital Infantil Nossa Senhora da Gloria), ES, Brazil. The following atmospheric pollutants as covariates were studied: particulate material (PM_{10}), sulphur dioxide (SO_2), nitrogen dioxide (NO_2), ozone (O_3) and carbon monoxide (CO). For details, e.g., descriptive statistics and basic time series plots, see [17]. The data analyzed in this section can be obtained from http://wileyonlinelibrary.com/journal/rss-datasets.

The graphs of the sampling functions of the autocorrelations and partial autocorrelations in Fig. 1 show that the series of the number of hospital admissions for RD possesses seasonal behavior, which was to be expected for this phenomena. Another characteristic observed in the series was an apparently weak stationarity. Similar graphs for the pollutant series can be found in [17].

Figure 2 shows the sample cross-correlation functions (CCF) between the response and pollutant covariates. As we discussed in Remark 1 four CCF's among them present similar behavior: the impact of pollutants CO and NO_2 is positive while the impact of SO_2 and O_3 are negative to the response variable at every lag. This observation is consistent with the PCA result presented in [17], see Table 5, where CO and NO_2 form a joint cluster for PC1. On the other hand, all CCF's possess seasonal behavior as well.

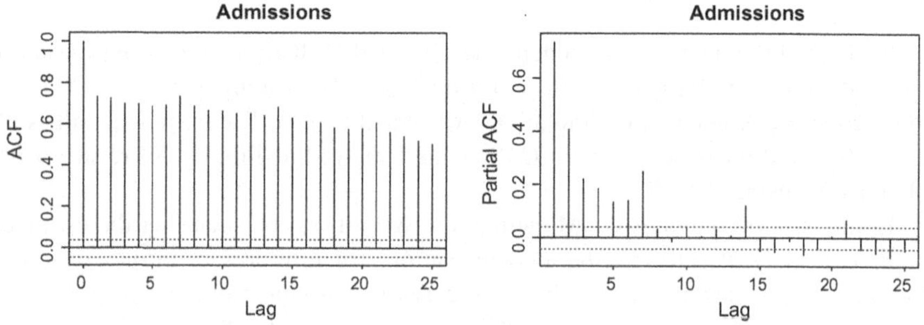

Fig. 1 Sample autocorrelation function (ACF) and partial autocorrelation function (PACF) of the response variable

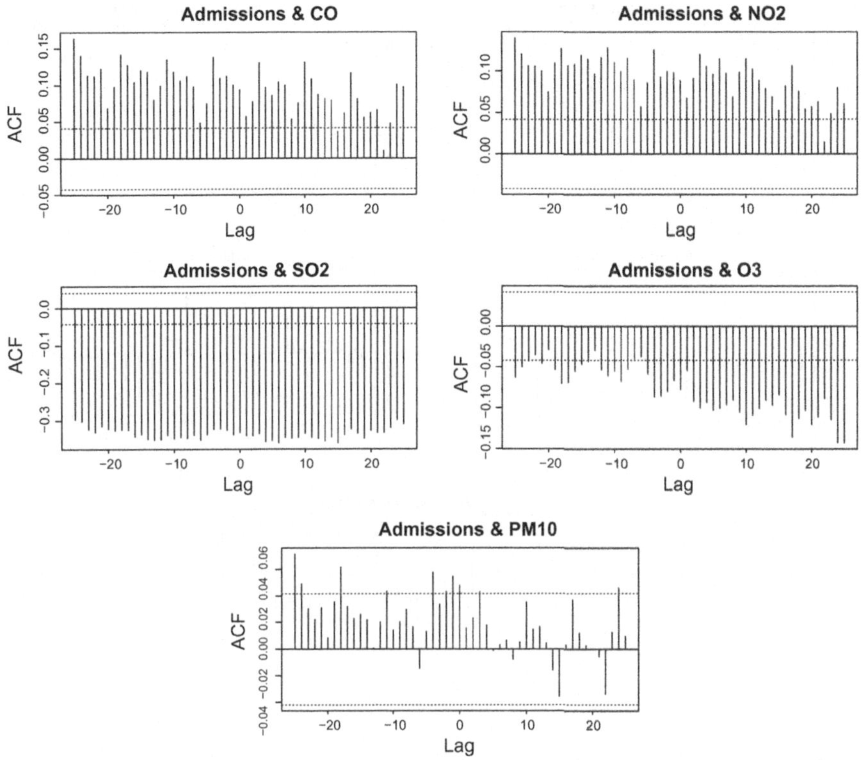

Fig. 2 Sample cross-correlation function (CCF) of the response and pollutant variables

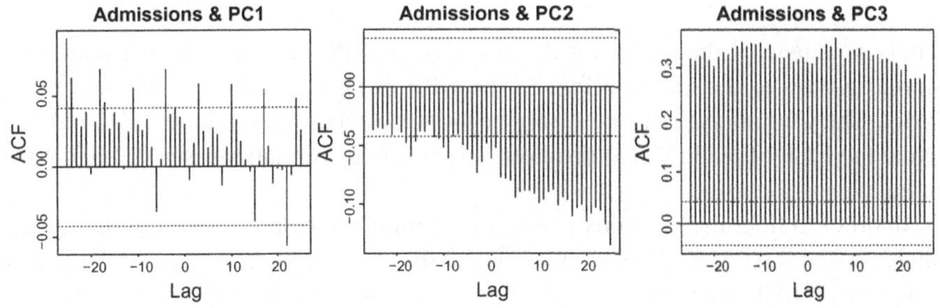

Fig. 3 Sample cross-correlation function (CCF) of the response and first three PCs

Figure 3 shows the sample cross-correlation functions (CCF) between the response variable and the first three PCs derived from applying PCA for the vector of pollutants. In Sect. 3.2 of [17], see Table 5 there, one can see that the first three components correspond to 83.2% of the total variability. The temporal behavior of the PCs is also presented in the autocorrelation plots of [17, Fig. 4]. The autocorrelations and the cross-correlations displayed here presented heavy seasonality as well. On the other

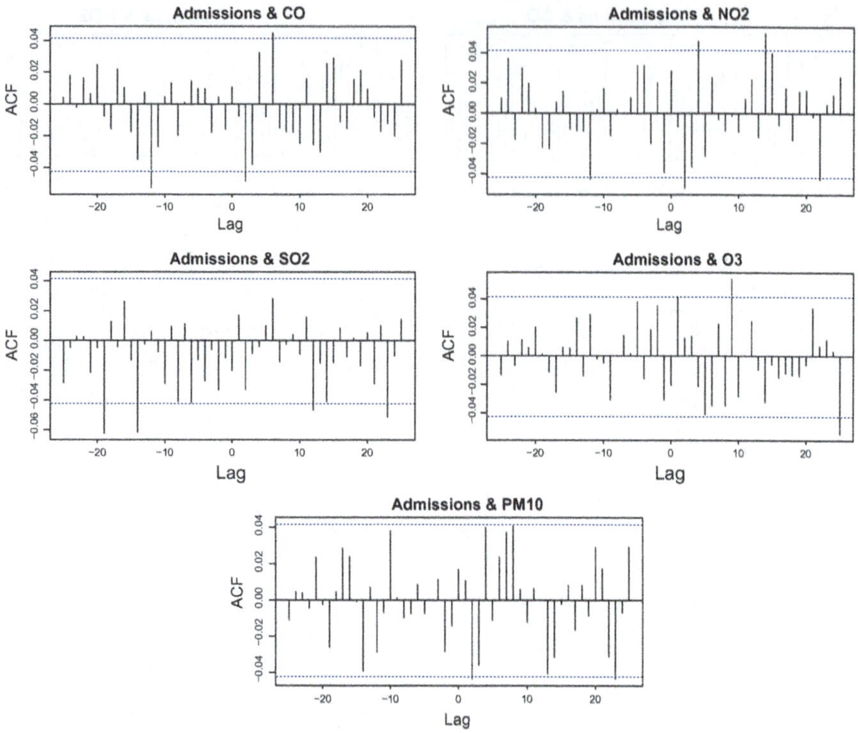

Fig. 4 Sample cross-correlation function (CCF) between the response and pollutant variables after the filtering

hand, the shape of the CCFs for the response and PCs can also be classifed into similar groups to the CCFs in Fig. 2. The CCF of PC1 is similar to the one of the PM_{10}. The CCF of PC2 displays only negative correlations similar to SO_2 and O_3, while the CCF of PC3 (Fig. 3) displays only positive correlations, see CO and NO_2 in Fig. 2.

In order to filter the vigorous seasonality both in the response and pollutant variables, seasonal ARMA filters with a 7-day period were applied. The pollutant vector time series and the one-dimensional response time series were filtered by $SVAR_7(1)$ and $SARMA_7(1, 1)$ processes, respectively. The residuals obtained by these filters indicate remaining significant correlations, see the CCFs between these residuals in Fig. 4. The significant cross-correlations and their respective lags are presented in Table 2. Clearly, the correlations which belong to the negative lags are spurious. However, the correlations which belong to the positive lags measure the true impact of a covariate. For example, there are significant correlations at lag 2 for pollutants PM_{10}, NO_2 and CO equally which could mean that the influence of these pollutants to the response indicates 2 days delay. Contrarily, the influence of the pollutants SO_2 and O_3 presents far delays.

Table 2 Significant cross-correlations and their respective lags between the response and pollutants after the filtering

	RD × SO_2					RD × NO_2					RD × PM_{10}		RD × CO			RD × O_3	
Lag	−19	−14	−6	12	23	−12	2	4	14	22	2	23	−12	2	6	9	25
Value	−0.063	−0.062	−0.042	−0.047	−0.051	−0.044	−0.050	0.048	0.053	−0.044	−0.044	−0.043	−0.053	−0.048	0.045	0.054	−0.055

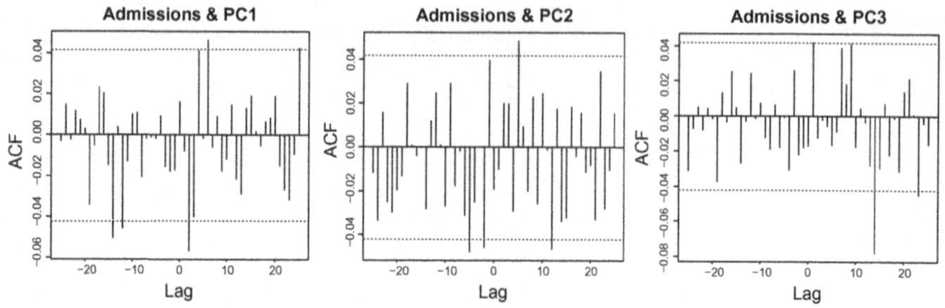

Fig. 5 Sample cross-correlation function (CCF) between the response and PCs after the filtering

Table 3 Significant cross-correlations and their respective lags between the response variable RD and PCs after the filtering

	RD × PC1					RD × PC2				RD × PC3		
Lag	−14	−12	2	6	25	−5	−2	5	12	1	14	23
Value	−0.051	−0.046	−0.057	0.046	0.043	−0.048	−0.046	0.048	−0.047	0.042	−0.078	−0.045

Figure 5 shows the sample CCF between the residuals of the response variable and the first three PCs after the filtering. The significant cross-correlations and its respective lags are presented in Table 3. It should be emphasized that there are strong coincidences in the lags between Tables 2 and 3. For example, the lag 2 in PC1 corresponds to the pollutants PM_{10}, NO_2 and CO, the lag 6 in PC1 corresponds to the pollutant CO, while lag 25 in PC1 corresponds to the pollutant O_3. The lag 12 in PC2 corresponds to the pollutant SO_2. Finally, the lag 14 corresponds to the pollutant NO_2 and the lag 23 to the pollutants SO_2 and NO_2. These correspondences are compatible with the clustering derived in [17, Table 7]. The fitted GAM-PCA-VAR model with its goodness-of-fit measures are reported in [17] as well. We note that in this fitted model $f_{ij} = 0$ was chosen for all $j > 0$. In view of the above results the GAM-PCA-VAR model with link

$$\log \mu_t = \beta_0 + \sum_{i=1}^{p} \sum_{j \in \mathcal{I}_i} f_{ij}(Z_{i(t-j)})$$

can also be a possible candidate, where $\mathcal{I}_i$ denotes the set of lags which belong to the significant cross-correlation between the residuals of the response and the ith PC. This model can be fitted by using the procedure described in Sect. 2.

6 Conclusions

A hybrid model, called GAM-PCA-VAR model, composed by three statistical tools, the VAR model, PCA and the GAM, with Poisson marginal distribution, was developed in a more general framework than in [17]. A three-stage estimation method was proposed and studied by simulation for some examples. Some theoretical properties were also proved. The model was applied to describe the dependence between the number of hospital admissions for respiratory diseases and air pollutant covariates.

An extension of the proposed estimation method for the GAM-PCA-VAR model by a variable selection procedure which ensures that only the significant PCs with their respective lags are involved into the model will be pursued in future works.

Acknowledgements The authors thank the following agencies for their support: the National Council for Scientific and Technological Development (Conselho Nacional de Desenvolvimento Científico e Tecnológico—CNPq), the Brazilian Federal Agency for the Support and Evaluation of Graduate Education (Coordenação de Aperfeiçoamento de Pessoal de Nível Superior—CAPES), Espírito Santo State Research Foundation (Fundação de Amparo à Pesquisa do Espírito Santo—FAPES) and Minas Gerais State Research Foundation (Fundação de Amparo à Pesquisa do estado de Minas Gerais—FAPEMIG). Márton Ispány was supported by the EFOP-3.6.1-16-2016-00022 project. The project is co-financed by the European Union and the European Social Fund. Pascal Bondon thanks to the Institute for Control and Decision of the Université Paris-Saclay.

References

1. Al-Osh, M.A., Alzaid, A.A.: First-order integer valued autoregressive (INAR(1)) process. J. Time Ser. Anal. **8**, 261–275 (1987)
2. Barczy, M., Ispány, M., Pap, G., Scotto, M.G., Silva, M.E.: Additive outliers in INAR(1) models. Stat. Pap. **53**, 935–949 (2012)
3. Benjamin, M.A., Rigby, R.A., Stasinopoulos, D.M.: Generalized autoregressive moving average models. J. Amer. Statist. Assoc. **98**, 214–223 (2003)
4. Brockwell, P.J., Davis, R.A.: Time Series: Theory and Methods. Springer Series in Statistics. Springer-Verlag, New York (1991)
5. Chen, R.J., Chu, C., Tan, J., Cao, J., Song, W., Xu, X., Jiang, C., Ma, W., Yang, C., Chen, B., Gui, Y., Kan, H.: Ambient air pollution and hospital admission in Shanghai. China. J. Hazard. Mater. **181**, 234–240 (2010)
6. Davis, R.A., Dunsmuir, W.T.M., Streett, S.B.: Observation-driven models for Poisson counts. Biometrika **90**, 777–790 (2003)
7. Dionisio, K.L., Chang, H.H., Baxter, L.K.: A simulation study to quantify the impacts of exposure measurement error on air pollution health risk estimates in copollutant time-series models. Environ. Health **15**, 114 (2016)
8. Durbin, J., Koopman, S.J.: Time series analysis of non-Gaussian observations based on state space models from both classical and Bayesian perspectives. J. Roy. Stat. Soc. B **62**, 3–56 (2000)
9. Ferland, R., Latour, A., Oraichi, D.: Integer-valued GARCH process. J. Time Ser. Anal. **27**(6), 923–942 (2006)
10. Gamerman, D., Santos, T.R., Franco, G.C.: A non-Gaussian family of state-space models with exact marginal likelihood. J. Time Ser. Anal. **34**, 625–645 (2013)
11. Hastie, T.J., Tibshirani, R.J.: Generalized Additive Models. Chapman and Hall, London (1990)
12. Jolliffe, I.T.: Principal Component Analysis, 2nd edn. Springer, New York (2002)
13. Nascimento, A.P., Santos, J.M., Mil, J.G., de Souza, J.B., Reis Júnior, N.C., Reisen, V.A.: Association between the concentration of fine particles in the atmosphere and acute respiratory diseases in children. Rev. Saude Publ. **51**, 3 (2017)
14. Ostro, B.D., Eskeland, G.S., Sánchez, J.M., Feyzioglu, T.: Air pollution and health effects: a study of medical visits among children in Santiago. Chile. Environ. Health Persp. **107**, 69–73 (1999)

15. Roberts, S., Martin, M.: Using supervised principal components analysis to assess multiple pollutant effects. Environ. Health Persp. **114**(12), 1877–1882 (2006)
16. Schwartz, J.: Harvesting and long term exposure effects in the relationship between air pollution and mortality. Am. J. Epidemiol. **151**, 440–448 (2000)
17. de Souza, J.B., Reisen, V.A., Franco, G.C., Ispány, M., Bondon, P., Santos, J.M.: Generalized additive models with principal component analysis: an application to time series of respiratory disease and air pollution data. J. Roy. Stat. Soc. C-Appl. **67**(2), 453–480 (2018)
18. de Souza, J.B., Reisen, V.A., Santos, J.M., Franco, G.C.: Principal components and generalized linear modeling in the correlation between hospital admissions and air pollution. Rev. Saude Publ. **48**(3), 451–8 (2014)
19. Wang, Y., Pham, H.: Analyzing the effects of air pollution and mortality by generalized additive models with robust principal components. Int. J. Syst. Assur. Eng. Manag. **2**, 253–259 (2011)
20. Wood, S.N.: Generalized Additive Models: An Introduction with R. 2nd edn. Chapman and Hall/CRC (2017)
21. Zamprogno, B.: PCA in time series with short and long-memory time series. Ph.D. Thesis at the Programa de Pós-Graduação em Engenharia Ambiental do Centro Tecnológico, UFES, Vitória, Brazil (2013)
22. Zhao, J., Cao, J., Tian, S., Chen, Y., Zhang, S., Wang, Z., Zhou, X.: A comparison between two GAM models in quantifying relationships of environmental variables with fish richness and diversity indices. Aquat. Ecol. **48**, 297–312 (2014)

A Bayesian Approach to Astronomical Time Delay Estimations

Mariko Kimura, Hyungsuk Tak and Taichi Kato

Abstract Time delay estimations between two time series data in astronomy have some difficulties due to their sparseness. We propose a fully Bayesian method based on a state–space model for this kind of analyses, and raise one example of the application to astronomical data. Our estimation can deal with heteroskedastic observational errors of astronomical time series and has much smaller errors than the result with a conventional method. This method may be applicable for many kinds of black hole systems and has a potential to derive the information of geometrical structure of astronomical objects after some improvements.

Keywords Bayesian inference · State–space model · Accretion · Black hole physics

1 Introduction

One may think that the brightness of stars is static. However, there are countless stars in the universe whose brightness varies drastically, even though we may hardly recognize such variations with naked eyes. We can see easily the dramatic brightness variations of stars over time by using telescopes. Astronomers call these brightness time series data of a star *a light curve*.

These light curves are sparsely and irregularly observed with heteroskedastic measurements errors. Ground-based telescopes, for example, can monitor stars only in

M. Kimura (✉) · T. Kato
Department of Astronomy, Graduate School of Science, Kyoto University, Oiwakecho, Kitashirakawa, Sakyo-ku, Kyoto 606-8502, Japan
e-mail: mkimura@kusastro.kyoto-u.ac.jp

T. Kato
e-mail: tkato@kusastro.kyoto-u.ac.jp

H. Tak
Department of Applied and Computational Mathematics and Statistics, University of Notre Dame, Notre Dame, IN, USA
e-mail: htak@nd.edu

I. Rojas et al. (eds.), *Time Series Analysis and Forecasting*, Contributions to Statistics, https://doi.org/10.1007/978-3-319-96944-2_21

clear sky at night. Even a fragment of cloud that intervenes between a star and a telescope would prevent us from measuring its brightness. Neither do space telescopes monitor a star regularly because the Sun plays the similar role to the cloud. Consequently, astronomers obtain unevenly-spaced time series data of brightness with large gaps between observations. Also, since the sensitivity of detecters changes according to the brightness of the targeted star, each data point has a heteroskedastic measurement error; brightness of a faint star is estimated with a larger measurement error than that of a bright star.

It is possible to observe a star at several different wavelengths (optical, ultraviolet, X-ray, radio wavelengths, and so on) by using different telescopes and/or detecters. The resulting multiwavelength light curves can be used to better understand physical characteristics of the star. For example, Fig. 1 displays a realistic multiwavelength data set of the black hole binary V404 Cyg; the upper time series is an X-ray light curve and the lower one is an optical light curve. If the fluctuations of these two light curves are similar to each other, we may know the underlying phisics in the black hole binary by investigating their correlation and time lag.

Here, we introduce a novel Bayesian method to estimate the time lag between two light curves. Unlike traditional grid-based estimation methods, such as cross-correlation method (e.g., [1]), the proposed Bayesian method has been physically motivated and accounts for the sparse and heteroskedastic features of the light curves in a fully parametric way.

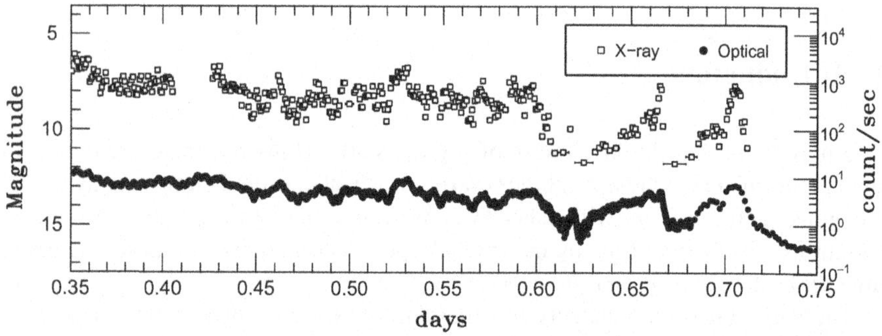

Fig. 1 An example of two light curves of the same star at different wavelengths. They are optical and X-ray short-term variations during the 2015 summer outburst in V404 Cyg, a black hole binary. The vertical axis represents brightness of the object in units of magnitudes (an astronomical logarithmic measure of brightness) or X-ray count rates. The rectangles and circles denote the light curves in the X-ray 25–60 keV band (taken by the *INTEGRAL* satellite) and the optical *V* band (taken by ground-based telescopes), respectively. These data are a part of the light curves shown in [2]

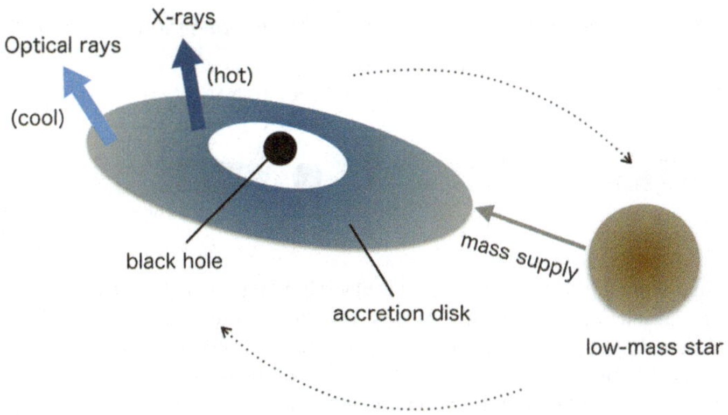

Fig. 2 Schematic figure of a black hole binary

2 Accretion onto Black Holes and Importance of Time Delay Estimations

Before introducing our method, we describe some basic knowledge on black hole binaries, and explain the reason why time lag estimates are important in this kind of objects. Although the Sun which is the most familiar star for us is a single star, about half the stars in the universe are binary systems, in which two stars rotate around each other. Black hole binaries are a subtype of binary systems, and one of them is composed of a central black hole and a low-mass star like the Sun.[1] Since a black hole has a strong gravitational field, the matter of a paired low-mass star is attracted toward a black hole. Then, the matter rotates around the black hole and forms an accretion disk like Fig. 2. Via the accretion disk, the matter falls into the black hole and its energy is released as a lot of light rays. In other words, the light emission from this system is a death cry of the fallen matter. X-rays and optical light rays are emitted from the inner and outer disk, respectively, according to the temperature of the accretion disk (see Fig. 2). Although optical light rays are also emitted from the low-mass star, its contribution is weak in comparison with the brightness of the disk.

If the matter in the accretion disk is absorbed into the black hole constantly, the light emission should be static, but a rapid increase of brightness is sometimes observed [3]. This phenomenon is called *an outburst*. The origin of outbursts is believed to be disk instability which transiently increases the amount of matter accreting onto a black hole [4]. Although the global trend of the outbursts can be explained by the disk instability model, the origin of short-term variability observed in some outbursts like Fig. 1 is not clear. Unraveling this mechanism would contribute to the better understanding of the physics not only in black hole binaries but also in many other black hole systems, for example, active galactic nuclei having accretion disks

[1] There are other black hole binaries having paired high-mass stars, but we regard black hole binaries have low-mass companion stars in this chapter.

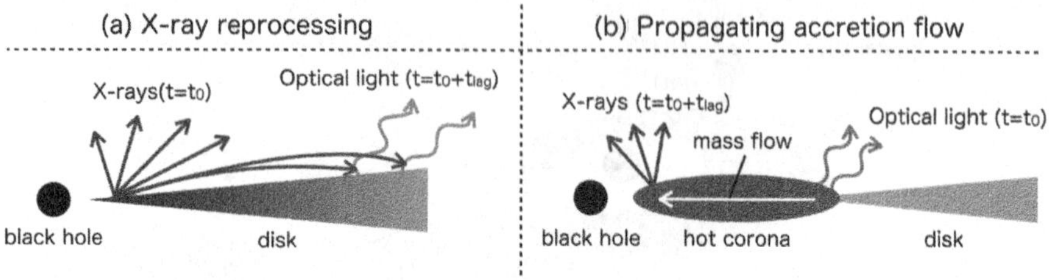

Fig. 3 Schematic picture of the structure of a black hole binary and related radiation processes producing time lags. **a** A part of radiated X-rays from the high-temperature inner disk is absorbed at the outer disk and optical light rays are reemitted from the outer disk. Then, an optical delay on timescales of several tens of seconds is observed [6, 7]. **b** The matter in a disk is transported from the outer region to the inner region. If the inhomogeneous accreting matter propagates inward, we will observe optical emission at first, and X-ray emission some time later

around supermassive black holes because the similar variability has been observed (e.g., [5]). In order to know the clue of the mechanism, the time delay estimations between correlated optical and X-ray light variations are useful because the delays are related to the radiation process of their light variations and the disk structure. We show two different emission mechanisms in black hole binaries in Fig. 3 as examples. The sign of the time delay measurement determines if the X-ray emission is earlier or later than the optical emission ((a) or (b) in Fig. 3), and it helps us figure out which emission mechanism is dominant in the disk.

3 Our Bayesian Approach and Its Application

A Bayesian State–Space Model for Time Delay Estimations

We overview our Bayesian approach that models the observational light curves using a state–space model; see [8] for details. Each observation is recorded in magnitude that is an astronomical measure of brightness on a logarithmic scale. The observed data are n magnitudes of one light curve, $\boldsymbol{x} = \{x_1, x_2, \ldots, x_n\}$, measured at observation times, $\{t_{x_1}, t_{x_2}, \ldots, t_{x_n}\}$, and m magnitudes of the other light curve, $\boldsymbol{y} = \{y_1, y_2, \ldots, y_m\}$ measured at possibly different observation times, $\{t_{y_1}, t_{y_2}, \ldots, t_{y_m}\}$. The number of observations for each light curve is not necessarily the same. We assume that the observed magnitudes are measured around the unknown true magnitudes with heteroskedastic measurement errors. We denote these latent magnitudes by $\boldsymbol{X} = \{X(t_{x_1}), X(t_{x_2}), \ldots, X(t_{x_n})\}$ for $\boldsymbol{x}$ and $\boldsymbol{Y} = \{Y(t_{y_1}), Y(t_{y_2}), \ldots, Y(t_{y_m})\}$ for $\boldsymbol{y}$, and known standard deviations of measurement errors by $\{\delta_1, \delta_2, \ldots, \delta_n\}$ for $\boldsymbol{x}$ and $\{\eta_1, \eta_2, \ldots, \eta_m\}$ for $\boldsymbol{y}$.

A state–space model is composed of (i) an observational equation that defines the relationship between the observed data and unknown states of a latent process, and

(ii) a system equation that specifies the relationship among the unknown states [9]. Assuming that the observed data are measured with heteroskedastic Gaussian errors centered at the unknown true magnitudes, we can specify the observational equation as follows:

$$x_i \mid X(t_{x_i}) \sim \mathrm{N}\big(X(t_{x_i}),\ \delta_i^2\big) \text{ for } i = 1, 2, \ldots, n, \tag{1}$$

$$y_j \mid Y(t_{y_j}) \sim \mathrm{N}\big(Y(t_{y_j}),\ \eta_j^2\big) \text{ for } j = 1, 2, \ldots, m. \tag{2}$$

In addition, we assume that one latent light curve is a shifted version of the other in magnitude and time, i.e., $Y(t) = X(t - \Delta) + \beta$, where β is a magnitude offset between two latent light curves and Δ is the time lag of interest. This is called a curve-shifting assumption [10]. Using this assumption, we can reexpress Eq. (2) as follows:

$$y_j \mid X(t_{y_j} - \Delta), \Delta, \beta \sim \mathrm{N}\big(X(t_{y_j} - \Delta) + \beta,\ \eta_j^2\big) \text{ for } j = 1, 2, \ldots, m. \tag{3}$$

For a system equation, we adopt a continuous-time Ornstein–Uhlenbeck (O–U) process that is also called a damped random walk process among astronomers. This process is empirically proven to describe well the stochastic variations of brightness of active galactic nuclei generated by accretion [11] in that the power spectral density of such light curves are consistent to that of the O–U process. This process is defined by the following stochastic differential equation:

$$dX(t) = -\frac{1}{\tau}(X(t) - \mu)dt + \sigma dB(t). \tag{4}$$

Here, μ denotes the overall mean of the process, and σ represents the short-term variability of the process. Both are on the magnitude scale. Also, τ is a timescale of the process in days, and $B(t)$ is a standard Brownian motion. The solution of this stochastic differential equation defines the relationship among the latent magnitudes by the following Gaussian conditional distributions:

$$X(t_1^{\Delta}) \sim \mathrm{N}\left(\mu,\ \frac{\tau\sigma^2}{2}\right), \text{ and for } i = 2, 3, \ldots, n+m,$$

$$X(t_i^{\Delta}) \mid X(t_{i-1}^{\Delta}) \sim \mathrm{N}\left(\mu + a_i\big(X(t_{i-1}^{\Delta}) - \mu\big),\ \frac{\tau\sigma^2}{2}(1 - a_i^2)\right) \tag{5}$$

where $a_i \equiv \exp(-(t_i^{\Delta} - t_{i-1}^{\Delta})/\tau)$, and $\boldsymbol{t}^{\Delta} = (t_1^{\Delta}, \ldots, t_{n+m}^{\Delta})$ are the sorted vector of $n + m$ observation times among the n observation times, $\{t_{x_1}, \ldots, t_{x_n}\}$, and the m time-lag-shifted observation times, $\{t_{y_1} - \Delta, \ldots, t_{y_m} - \Delta\}$. We denote all of $n + m$ latent magnitudes by $\boldsymbol{X}(\boldsymbol{t}^{\Delta})$. These conditional Gaussian distributions defines the system equation of a state–space model. See Fig. 4 for a diagram that describes the assumed data generation process.

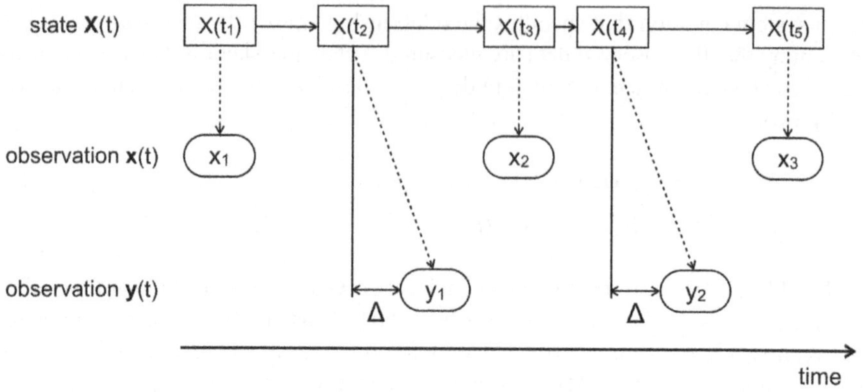

Fig. 4 A diagram that describes the data generation process. The observed data x_i and y_j are generated from a latent light curve $\boldsymbol{X}(t^{\Delta})$ representing the unknown true magnitudes evaluated at discrete observation times

Besides the $n + m$ latent magnitudes, our model has five unknown parameters, i.e., Δ, β, μ, σ, and τ, and we adopt the following prior distributions for these model parameters, considering the knowledge from previous astronomical probes:

$$\Delta \sim \text{Uniform}(-0.04, 0.04), \quad \beta \sim \text{N}(0,\ 10^5), \quad \mu \sim \text{Uniform}(-30,\ 30)$$
$$\sigma^2 \sim \text{inverse Gamma}(1,\ 1), \quad \tau^2 \sim \text{inverse Gamma}(1,\ 2 \times 10^{-7}). \quad (6)$$

See Sects. 2.4 and 2.5 of [8] for motivation and scientific background of these choices.

The resulting full posterior density denoted by $\pi(\Delta, \beta, \mu, \sigma, \tau, \boldsymbol{X}(t^{\Delta}) \mid \boldsymbol{x}, \boldsymbol{y})$ is proportional to the multiplication of the density functions whose distributions are defined in Eqs. (1), (3), (5), and (6), respectively. We sample the full posterior distribution using a Metropolis–Hastings within Gibbs sampler [12]; see Sect. 3 of [8] for details.

Application to Observational Data

We apply our method to the data of the 2015 winter outburst in V404 Cyg, which are given in the right panel of Fig. 5.[2] This figure exhibits correlation between X-ray and optical large-amplitude variations. The X-ray light curve in this data set is scaled to the amplitudes of optical variations by using the result of the logarithmic regression between the X-ray and optical luminosity in order to meet the curve-shifting assumption. We also confirm that the power density spectrum of the X-ray light curve is well expressed by a power law as that of the O-U process is (see Fig. 6).

[2]The data were obtained on new year's day in 2016.

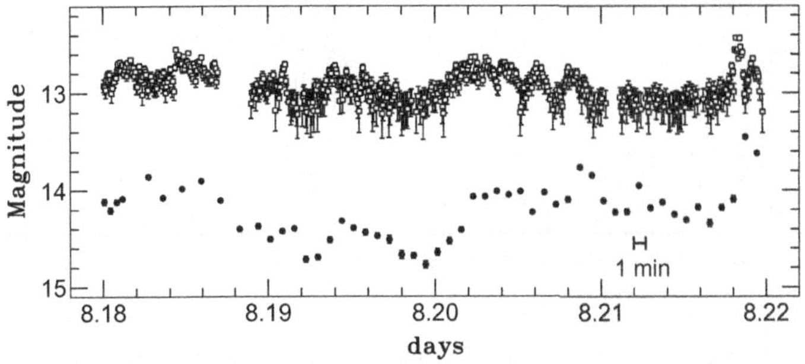

Fig. 5 The observed optical and X-ray light curves during the 2015 winter outburst in V404 Cyg. The length of the time interval between the first and last observations is about an hour. The empty rectangles indicate the X-ray light curve and the filled rectangles denote the optical light curve. The half-length of each vertical line around the rectangles represents the standard deviation of the measurement error

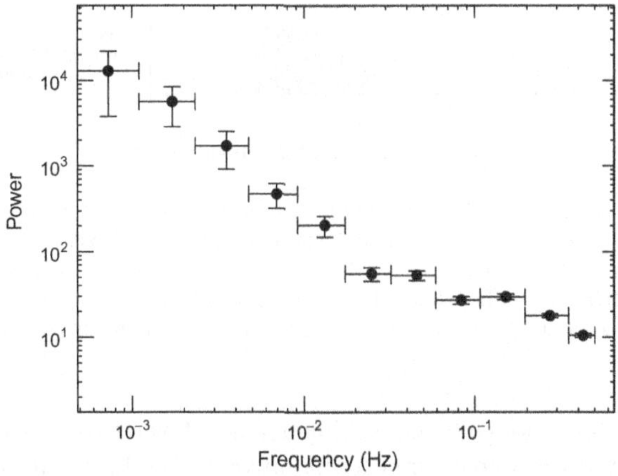

Fig. 6 Power spectral density of the X-ray light curve in Fig. 5. The errors represent the standard deviations

In order to implement our model, we use an R package, `timedelay`, which is publicly available at CRAN.[3] Since the marginal distribution of the time lag often suffers from multimodality, this package adopts a repelling-attracting Metropolis algorithm [13] to better explore a multimodal distribution of Δ; Fig. 7 shows that the autocorrelation time decreases faster with the algorithm. We first compute the profile likelihood of the time lag and check dominant modes of the time lag. Next we prepare initial three Markov chains near the highest mode. We run each Markov chain for 150,000 iterations, discarding the first 50,000 as burn-in.

The left and right panels of Fig. 8 show the histogram and trace plot of the posterior sample of the time lag Δ, respectively. This sampling result indicates that the optical variations are delayed to the X-ray ones by $22.7^{+0.3}_{-0.2}$ s. The Gelman–Rubin convergence diagnostic statistic [14] computed from the three Markov chains is 1.0003,

[3]https://cran.r-project.org/package=timedelay.

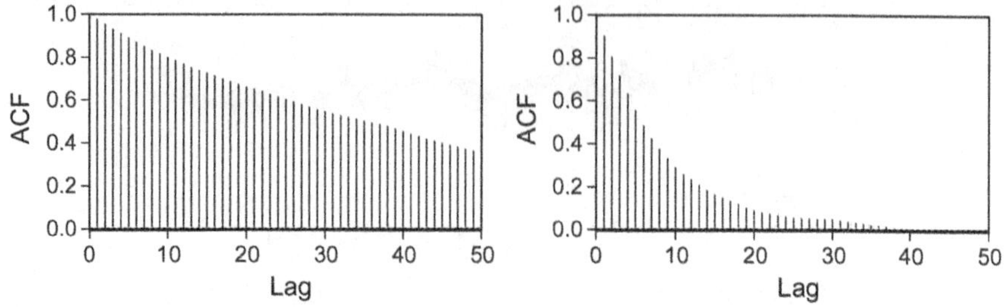

Fig. 7 Autocorrelation functions of posterior sample of Δ for one chain. Left: without the repelling-attracting metropolis algorithm. Right: with the repelling-attracting metropolis algorithm

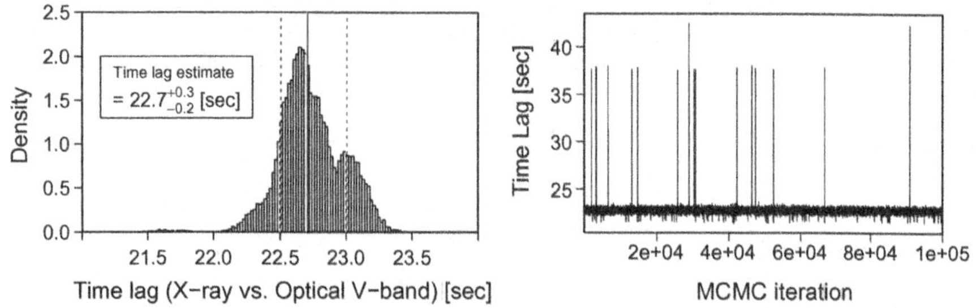

Fig. 8 Left: The resulting posterior distribution of the time lag around the highest mode. In the histogram on the left, the solid blue line indicates the posterior median of Δ and the dashed blue lines represent the 68% quantile-based interval. There are a small mode near 37.7 s and an invisibly mode near 42.4 s, though not shown here. The right panel displays the trace plot of posterior samples of Δ for one chain. We can see the existence of small modes which are not displayed in the histogram

close enough to unity. Although not shown here, we also perform a sensitivity analysis to see that our posterior inference is robust to the choice of the scale parameters of the inverse-Gamma prior distributions for τ and σ^2. We suggest that the ~22-s optical delay originates from the X-ray reprocessing (see (a) of Fig. 3). Figure 5 and the left panel of Fig. 8 represent the preliminary light curves and estimations of the results shown in [15], respectively. The detailed scientific interpretation is also described in that paper.

4 Discussion

Applicability of Our Method for Other Astronomical Systems

As mentioned in Sect. 2, there are many other black hole systems including accretion disks in the universe. Since the brightness variations of these systems are produced by accretion, our method is naturally applicable for them. Actually, it was originally developed for gravitational-lensed light curves [8]. Here, we raise two examples of the application.

Gravitational-Lensed Light Curves
Quasars are a very bright subclass of active galactic nuclei. If there is an object having strong gravitational field between a quasar and the Earth, the light from the quasar is bent and split, and we obtain several images from one quasar because each of the split light rays arrive at the Earth via each different path. Light variations derived from these images delay to each other due to the different length of each path. Since these time delays are proportional to Hubble constant which determines the degree of expansion of the universe, we can estimate the constant by measuring the delays [16, 17]. Please see [8] for the details of the application to this type of phenomena. In gravitational-lensed events, one light curve is offset in brightness to the other due to the different degree of the lensing magnification [18]. This effect is taken into account via a polynomial regression in the paper [8].

Reverberation Mapping
Active galactic nuclei are believed to be surrounded by broad-line regions, lumps of rotating gas around the central black holes. The optical continuum emission from accretion disks illuminates the broad-line regions and ionizes the neutral gas in them. Then, some photons are emitted from the ionized gas. This phenomenon is similar to the X-ray reprocessing explained in (a) of Fig. 3. The light variations of line emission are delayed against those of the continuum, and have smaller amplitudes like Fig. 1. Since the time delay between the continuum and line emission is regarded to reflect the approximate time with which the photons move at the light speed from the vicinity of the central black hole to the broad-line region, we can inference the approximate distance between them by multiplying the delay and the light speed. This method is called *reverberation mapping* [19]. In conjunction with the rotating velocity of the broad-line regions measured by the spectroscopic observations, we can estimate the mass of the central black hole.

This reverberation mapping method is also used to constrain the inner structure of active galactic nuclei. For instance, low-energy X-ray delays have been detected in the delay estimations between low-energy and high-energy X-ray variations. This kind of delays are regarded to be caused by the illumination of the inner disk by high-energy X-rays, i.e., the similar phenomena to X-ray reprocessing shown in (a) of Fig. 3 occur. On the contrary, high-energy X-ray delays against low-energy X-rays have been also observed. They are considered to be the representation of propagating accretion flow in the inner region, and resemble X-ray delays in black hole binaries (see (b) in Fig. 3). Both two phenomena can give key information on the size of the disk and/or the corona [20].

Advantages of Our Method and Future Work

We can inference time delays without any interpolations and their statistic errors with high accuracy by using our Bayesian method. This is an advantage of our method over some conventional methods. Actually, although we also tried the time delay estimation as for our data in Fig. 5 by using a locally normalized discrete correlation function [21], the analysis presented larger errors by more than 35 times than those obtained by our method.

Though our approach is a sophisticated method, there is some room to improve for versatility. In the current version, the difference in amplitudes between optical and X-ray variability is not taken into account. If our model includes a scale parameter as an unknown constant, i.e., the relation between two light curves is assumed to be $Y(t) = CX(t - \Delta) + \beta$ (C: constant), we can inference the difference.

In addition, few photons arrive at the Earth in an exposure time when the system is very faint or when we observe it at extremely high-energy wavelengths. Although we assume that the magnitudes of light curves are expressed by Gaussian distributions, it is better to expand our method to be applicable for Poisson distributions if we need to analyze faint and/or high-energy light curves.

Moreover, as we can see Fig. 1, the optical light curves have more smoother shape. If we model this smoothness, we may obtain more information of the disk structure by the reverberation mapping. For instance, one light curve can be expressed by the convolution of the other light curve and a transfer function as follows:

$$Y(t) = \int \Psi(\Delta)X(t - \Delta)d\Delta, \tag{7}$$

where $\Psi(\Delta)$ is a transfer function. The transfer function in [8] is a delta function of the time delay. Our next step is to develop a method for dealing with some other transfer functions in order to know the width of the time delay depending on the extended shape of the accretion disk. Black hole binaries are tiny systems and active galactic nuclei are too far from the Earth. They always look like point sources on astronomical observational images, and we cannot directly observe their geometrical structures. If it is possible to know the width of time delays, we could estimate the disk size more accurately from it. If the transfer function itself can be estimated, we might be able to estimate the disk structure by the reverberation mapping method.

Diversity of Light Curves in Astronomy

Though we exhibit the light curves of only black hole binaries, there are many other kinds of light variations in astronomy. We display some of them in Fig. 9. The light curves shown in the upper left panel are of 3 C 454, one of blazers whose light variations originate from jet ejections. They seem to be similar to the variability of the price of the stock in spite of the sparseness and large errors. Quasi-periodic oscillations, periodic modulations having time-varying phases, are sometimes observed in this kind of systems. Their power spectral density looks like a power law plus a small sharp peak (e.g., [22]).

There are also many pulsating variable stars in the universe. The second example displayed in the upper right panel of Fig. 9 is the light variations of R Sct, one of RV Tau-type pulsating variable stars, which show alternating shallow and deep minima, and astronomers wonder whether this behavior is predictable [23, 24].

The light variations in the lower panel of Fig. 9 are of V344 Lyr which belongs to a subclass of dwarf novae showing occasional long and big outbursts. These light variations were observed in the early stage of the outburst, and we can see that small-amplitude and periodic variability grows up, and that its period and amplitude are gradually varying with time. We sometimes observe a few different types of variations whose periods are close to each other during the same time interval. It is

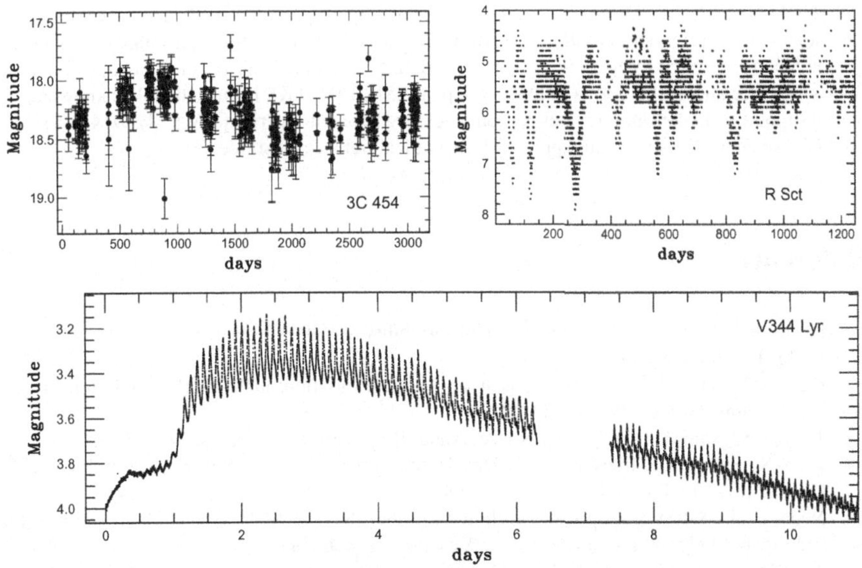

Fig. 9 Light curve collections. Upper left: The optical V-band light curves of 3 C 454 (taken by the Catalina real-time transient survey (CRTS)). Upper right: The visual observations of R Sct (taken by the American association of variable star observers (AAVSO)). Lower: The optical light curves of V344 Lyr (monitored by the Kepler satellite)

demonstrated that these signals are resolved by using the least absolute shrinkage and selection operator (Lasso) method [24, 25].

5 Conclusions

Analyzing astronomical multiwavelength light curves is a challenging problem, since these data usually have each different irregular and sparse samplings. We estimated a time lag between X-ray and optical light curves in a black hole binary by using a fully Bayesian method including state–space model. Our method enables us to estimate the time lag with small errors, and has the advantage of the robust results independent of the choice of the parameters in prior distributions. This method will be applicable for other astronomical systems if the light variations from them are expressed by O-U process. In the near future, enormous observational data will be collected by several upcoming astronomical surveys and big telescopes. Time-domain astronomy is coming soon and further developments of statistical procedures are required in astronomy.

Acknowledgements We are thankful to many amateur observers in the VSNET (Variable Star Network) team and AAVSO team (http://www.aavso.org/data/download/) and the CRTS (http://nesssi.cacr.caltech.edu/DataRelease/) for providing a lot of data used in this study. We also thank the *INTEGRAL* groups for making the products of the ToO data publicly available online at the *INTEGRAL* Science Data Centre. This work was financially supported by the Grant-in-Aid for JSPS Fellows for young researchers (MK) and by the Grant-in-Aid "Initiative for High-Dimensional Data-Driven Science through Deepening of Sparse Modeling" from the Ministry of Education, Culture, Sports, Science and Technology (MEXT) of Japan (25120007, TK). Hyungsuk Tak acknowledges partial support from the United States National Science Foundation grants DMS 1127914 and DMS 1638521 given to Statistical and Applied Mathematical Sciences Institute.

References

1. Hynes, R.I.: Correlated X-Ray and optical variability in V404 Cygni in quiescence. Astrophys. J. **611**, L125–L128 (2004)
2. Kimura, M., et al.: Repetitive patterns in rapid optical variations in the nearby black-hole binary V404 Cygni. Nature **529**, 54–58 (2016)
3. Tanaka, Y., Shibazaki, N.: X-ray Novae. Annu. Rev. Astron. Astrophys. **34**, 607–644
4. Kato, S., Fukue, J., Mineshige, S.: Black-Hole Accretion Disks—Towards a New Paradigm. Kyoto University Press, Kyoto, Japan (2008)
5. Noda, et al.: X-Ray and optical correlation of type I Seyfert NGC 3516 studied with Suzaku and Japanese ground-based telescopes. Astrophys. J. **828**, 78 (2016)
6. van Paradijs, J., McClintock, J.E.: Absolute visual magnitudes of low-mass X-ray binaries. Astron. Astrophys. **290** (1994)
7. Hynes, R.I., O'Brien, K., Horne, K., Chen, W., Haswell, C.A.: Echoes from an irradiated disc in GRO J1655–40. Mon. Not. R. Astron. Soc. **299**, 37 (1998)

8. Tak, H., Mandel, K., van Dyk, D.A., Kashyap, V.L., Meng, X.-L., Siemiginowska, A.: Bayesian Estimates of astronomical time delays between gravitationally lensed stochastic light curves. Ann. Appl. Stat. **11**(3), 1309–1348 (2017)
9. Brockwell, P.J., Davis, R.A.: Introduction to Time Series and Forecasting, 2nd edn. Springer, New York (2002)
10. Pelt, J., Hoff, W., Kayser, R., Refsdal, S., Schramm, T.: Time delay controversy on QSO 0957+561 not yet decided. Astron. Astrophys. **286**, 775–785 (1994)
11. Kelly, B.C., Bechtold, J., Siemiginowska, A.: Are the variations in quasar optical flux driven by thermal fluctuations? Astrophys. J. **698**, 895–910 (2009)
12. Tierney, Luke: Markov chains for exploring posterior distributions. Ann. Stat. **22**, 1701–1728 (1994)
13. Tak, H., Meng, X.-L., van Dyk, D.A.: A repelling-attracting metropolis algorithm for multimodality (2016). arXiv:1601.05633
14. Gelman, A., Rubin, D.B.: Inference from iterative simulation using multiple sequences. Stat. Sci. 457–472 (1992)
15. Kimura, M.: Erratum: rapid optical variations correlated with X-rays in the 2015 second outburst of V404 Cygni (GS 2023+338). Mon. Not. R. Astron. Soc. **475**, 3083–3085 (2018)
16. Refsdal, S.: On the possibility of determining Hubble's parameter and the masses of galaxies from the gravitational lens effect. Mon. Not. R. Astron. Soc. **128**, 307 (1964)
17. Treu, T., Marshall, P.J.: Time delay cosmography. Astron. Astrophys. Rev. **24**, 11 (2016)
18. Refsdal, S., Stabell, R., Pelt, J., Schild, R.: Constraints on source and lens parameters from microlensing variability in QSO 0957+561 A, B. Astron. Astrophys. **360**, 10–14 (2000)
19. Blandford, R.D., McKee, C.F.: Reverberation mapping of the emission line regions of Seyfert galaxies and quasars. Astrophys. J. **225**, 419–439 (1982)
20. Kara, E., et al.: A global look at X-ray time lags in Seyfert galaxies. Mon. Not. R. Astron. Soc. **462**, 511–531 (2016)
21. Lehar, J., Hewitt, J.N., Burke, B.F., Roberts, D.H.: The radio time delay in the double quasar 0957+561. Astrophys. J. **384**, 453–466 (1992)
22. King, O.G.: A quasi-periodic oscillation in the blazar J1359+4011. Mon. Not. R. Astron. Soc. **436**, L114–L117 (2013)
23. Buchler, J.R., Kollath, Z., Serre, T., Mattei, J.: Nonlinear analysis of the light curve of the variable star R scuti. Astrophys. J. **462**, 489 (1996)
24. Kato, T., Uemura, M.: Period analysis using the least absolute shrinkage and selection operator (Lasso). Publ. Astron. Soc. Jpn. **64**, 122 (2012)
25. Osaki, Y., Kato, T.: Study of superoutbursts and superhumps in SU UMa stars by the Kepler light curves of V344 Lyrae and V1504 Cygni. Publ. Astron. Soc. Jpn. **65**, 95 (2013)

Further Results on a Modified EM Algorithm for Parameter Estimation in Linear Models with Time-Dependent Autoregressive and t-Distributed Errors

Boris Kargoll, Mohammad Omidalizarandi, Hamza Alkhatib and Wolf-Dieter Schuh

Abstract In this contribution, we consider an expectation conditional maximization either (ECME) algorithm for the purpose of estimating the parameters of a linear observation model with time-dependent autoregressive (AR) errors. The degree of freedom (d.o.f.) of the underlying family of scaled t-distributions, which is used to account for outliers and heavy-tailedness of the white noise components, is adapted to the data, resulting in a self-tuning robust estimator. The time variability of the AR coefficients is described by a second linear model. We improve the estimation of the d.o.f. in a previous version of the ECME algorithm, which involves a zero search, by using an interval Newton method. We model the transient oscillations of a shaker table measured by a high-accuracy accelerometer, and we analyze various criteria for selecting a simultaneously parsimonious and realistic time-variability model.

Keywords Linear model · Time-dependent AR process · Scaled t-distribution Self-tuning robust estimator · EM algorithm · Vibration analysis

1 Introduction

In many fields of application, observed time series can be approximated by a linear ("deterministic") model, involving desired target quantities in the form of unknown parameters to be estimated. Random deviations ("errors") of the observations from

---

B. Kargoll (✉) · M. Omidalizarandi · H. Alkhatib
Leibniz Universität Hannover, Geodätisches Institut, Nienburger Str. 1, 30167 Hannover, Germany
e-mail: kargoll@gih.uni-hannover.de

M. Omidalizarandi
e-mail: zarandi@gih.uni-hannover.de

H. Alkhatib
e-mail: alkhatib@gih.uni-hannover.de

W.-D. Schuh
Rheinische Friedrich-Wilhelms-Universität Bonn, Institut für Geodäsie und Geoinformation, Nussallee 17, 53115 Bonn, Germany
e-mail: schuh@geod.uni-bonn.de

I. Rojas et al. (eds.), *Time Series Analysis and Forecasting*,
Contributions to Statistics, https://doi.org/10.1007/978-3-319-96944-2_22

the functional model are frequently autocorrelated ("colored noise"), for instance, as a consequence of sampling with small time differences between the measurements (cf. [1]). Colored noise can often be modeled by means of an autoregressive (AR) or by a more general autoregressive moving average (ARMA) process. In geodesy, for instance, linear models based on spherical harmonics have been employed in conjunction with ARMA error processes to model satellite gravity gradiometer data acquired within the Gravity and Gravity Field and Steady-State Ocean Circulation Explorer (GOCE) mission (cf. [2]). Recently, accelerometer measurements of an oscillation were linearly modeled as a sum of sinusoids with additive AR errors [3].

In geodetic applications, colored noise is frequently found to display a time-variable behavior (see, e.g., [4, 5]), mainly as a consequence of temporally changing conditions within the measurement environment. To take such phenomena into account, employment of AR processes with time-dependent coefficients was suggested (cf. [6]). While many different methods for generating time-dependent AR and ARMA processes have been introduced since then (see [7] for a comparative study), the most frequently applied approach appears to have been based on the idea of expressing the values of a particular, temporally changing AR(MA) coefficient as values of a best-fitting linear combination of certain basis functions. Popular choices of basis functions have been polynomials with different orders [8], Legendre polynomials [9], B-splines [10], wavelets [11], sinusoids [12], sigmoid functions [13], and discrete prolate spheroidal sequences [14].

Least squares techniques, which are commonly employed for estimating the parameters of regression ARMA models (see [15] for an early exposition), should be used with caution when outliers are expected to be present in the observations. When reliable prior knowledge about the relative frequencies of outliers is lacking, it makes sense to use a self-tuning robust estimator based on a family of heavy-tailed probability distributions. Such estimators are partially adaptive in the sense that shape parameters which control the tail behavior of the associated probability density function are estimated alongside the main model parameters [16]. In the context of linear regression models, the family of scaled t-distributions has been found useful in this regard [17] since it allows for the application of an expectation maximization (EM) algorithm in the computationally convenient form of iteratively reweighted least squares (IRLS) [18, 19]. To speed up the convergence of this EM algorithm in applied situations, modifications to an expectation conditional maximization (ECM), multicycle ECM or ECM *either* (ECME) algorithm have been proposed (cf. [20]).

The linear regression model with t-distributed errors, as considered within these studies, was recently extended to include a covariance-stationary AR model with t-distributed white noise components [3]. The derived ECME algorithm showed that the parameter estimates concerning the linear regression and AR model can be computed via two separate IRLS schemes. Kargoll et al. [5] demonstrated the extension of this twofold IRLS algorithm from a covariance-stationary to a time-dependent AR process, employing the aforementioned basis function approach with polynomials of different orders. In the current contribution, we improve that algorithm regarding the estimation of the degree of freedom (d.o.f.) of the underlying t-distribution. Furthermore, we substantially extend the numerical analysis of the vibration experiment involving measurements of a high-accuracy accelerometer (considered in the previous two studies), focusing now on the important issue of model selection concerning the time-variable AR process.

2 The Observation Model

We assume that each observable Y_t of a time series is represented by a purely deterministic model $\mathbf{A}_t\boldsymbol{\xi}$ and an additive random deviation or error E_t, i.e.,

$$Y_t = \mathbf{A}_t\boldsymbol{\xi} + E_t \qquad (t = 1, \ldots, n). \tag{1}$$

In the following, we denote by $\mathbf{Y}$ the $(n \times 1)$-random vector of all observables, by $\mathbf{E}$ the $(n \times 1)$-random vector of all random deviations, by $\mathbf{A}$ the $(n \times m)$-design matrix consisting of the known $(1 \times m)$-row vectors $\mathbf{A}_1, \ldots, \mathbf{A}_n$, and by $\boldsymbol{\xi}$ the $(m \times 1)$ column vector of unknown functional parameters. In order for these parameters to be estimable, we assume the number of observations to exceed the number m of functional parameters and the design matrix to be of full rank. Moreover, we allow the random deviations to be autocorrelated according to a time-variable AR process

$$E_t = \alpha_{1,t}E_{t-1} + \cdots + \alpha_{p,t}E_{t-p} + U_t \qquad (t = 1, \ldots, n), \tag{2}$$

in which we assume the random variables $U_1, \ldots, U_n$ to be independently and identically distributed (i.i.d.) with mean 0 and variance σ_0^2. We call the autocorrelated random deviations $E_1, \ldots, E_n$ also the "colored noise components" of the measurement time series $Y_1, \ldots, Y_n$. To fix the initial conditions of the AR(p) model (2), we assume that $E_0, \ldots, E_{1-p}$ take a constant value of 0. Furthermore, for each $j = 1, \ldots, p$ we treat the AR coefficients $\alpha_{j,1}, \ldots, \alpha_{j,n}$ as unknown values of some function on $\{1, \ldots, n\}$, in the sense that

$$\alpha_{j,t} = \mathbf{X}_t\boldsymbol{\beta}_j \qquad (j = 1, \ldots, p; t = 1, \ldots, n). \tag{3}$$

Here, the given $(1 \times q)$-row vectors $\mathbf{X}_1, \ldots, \mathbf{X}_n$ give rise to a second, $(n \times q)$-design matrix $\mathbf{X}$, and the $(q \times 1)$ column vectors $\boldsymbol{\beta}_1, \ldots, \boldsymbol{\beta}_p$ contain the corresponding (unknown) parameters. Stacking these p column vectors then yields the $(pq \times 1)$ column vector $\boldsymbol{\beta}$. Note that the vector $\mathbf{X}_t$ is the same for each of the p AR coefficients $\alpha_{1,t}, \ldots, \alpha_{p,t}$. In other words, we employ the same type of function for all coefficients of one AR model while allowing the functional parameters to be different for the various AR coefficients.

In order to account for heavy-tailed or outlier-afflicted errors, we let the i.i.d. random variables $U_1, \ldots, U_n$ (which we also refer to as the "white noise components" of the time-variable AR process) follow a scaled (Student's) t-distribution $t_\nu(0, \sigma^2)$ with mean 0, unknown d.o.f. ν, and unknown scale parameter σ^2. This family of distributions has the probability density function (pdf)

$$f(u_t) = \frac{\Gamma\left(\frac{\nu+1}{2}\right)}{\sqrt{\nu\pi}\,\sigma\,\Gamma\left(\frac{\nu}{2}\right)}\left[1+\left(\frac{u_t}{\sigma}\right)^2/\nu\right]^{-\frac{\nu+1}{2}} \quad (t = 1,\ldots,n), \tag{4}$$

where Γ is the gamma function. For large d.o.f. (say, $\nu \geq 300$), this pdf closely approximates a Gaussian bell curve, so that short-tailed or outlier-free measurement errors can be represented by this stochastic model as well. As $U_1,\ldots,U_n$ are stochastically independent, their joint pdf is given by the product $\prod_{t=1}^{n} f(u_t)$.

We intend to estimate all of the unknown model parameters $\boldsymbol{\xi}, \boldsymbol{\beta}, \sigma^2, \nu$ by means of ML estimation (all autoregressive coefficients $\alpha_{j,t}$ can be calculated directly from $\boldsymbol{\beta}$ and need, therefore, not be included in the list of parameters to be estimated). To this end, we assume that the AR model (2) can be inverted. Substituting (1)–(3) for the tth realization of U_t, we obtain

$$u_t = e_t - \alpha_{1,t}e_{t-1} - \cdots - \alpha_{p,t}e_{t-p} \tag{5}$$

$$= (y_t - \mathbf{A}_t\boldsymbol{\xi}) - \mathbf{X}_t\boldsymbol{\beta}_1(y_{t-1} - \mathbf{A}_{t-1}\boldsymbol{\xi}) - \cdots - \mathbf{X}_t\boldsymbol{\beta}_p(y_{t-p} - \mathbf{A}_{t-p}\boldsymbol{\xi}), \tag{6}$$

where we set the initial conditions $y_0 = \cdots = y_{1-p} = 0$ and $\mathbf{A}_0 = \cdots = \mathbf{A}_{1-p} = \mathbf{0}_{[1\times m]}$, according to the previous specifications of $e_0 = \cdots = e_{1-p} = 0$. To give the expressions for u_t a form that is easier to manage and interpret, we use the notation $L^j\mathbf{Z}_t := \mathbf{Z}_{t-j}$ in connection with $\boldsymbol{\alpha}_t(L) := 1 - \alpha_{1,t}L - \cdots - \alpha_{p,t}L^p$ and $\overline{\mathbf{Z}}_t = \alpha_t(L)\mathbf{Z}_t$ (where $\mathbf{Z}_t$ is a term of an arbitrary sequence of matrices). We can then write (5) more briefly as

$$u_t = \overline{e}_t = \boldsymbol{\alpha}_t(L)e_t = \boldsymbol{\alpha}_t(L)(y_t - \mathbf{A}_t\boldsymbol{\xi}) = \overline{y}_t - \overline{\mathbf{A}}_t\boldsymbol{\xi}. \tag{7}$$

Here, $\overline{e}_t$, $\overline{y}_t$ and $\overline{\mathbf{A}}_t$ can be interpreted as the outputs of a time-variable digital filter $\boldsymbol{\alpha}_t(L)$ that is applied, respectively, to a certain segment of the error time series $e_1,\ldots,e_n$, of the observation time series $y_1,\ldots,y_n$ and of the vector sequence $\mathbf{A}_1,\ldots,\mathbf{A}_n$. The entire collection of digital filters $\boldsymbol{\alpha}_1(L),\ldots,\boldsymbol{\alpha}_n(L)$ can be viewed as turning the colored noise components $e_1,\ldots,e_n$ progressively into white noise components $u_1,\ldots,u_n$, so that we call it a "decorrelation filter".

Substituting (6) into (4) and forming the product of these univariate pdf, we have a joint pdf $f(\mathbf{u})$ for the white noise components $u_1,\ldots,u_n$, which evidently is a function of all observations $\mathbf{y}$ and of all unknown parameters. Therefore, we could write $f(\mathbf{u})$ as the likelihood function $\mathcal{L}(\boldsymbol{\xi}, \boldsymbol{\beta}, \sigma^2, \nu; \mathbf{y})$. As this function is conditional on the previously fixed initial conditions for $E_0,\ldots,E_{1-p}$, we thus have a conditional likelihood function (in the sense of [21]), which can be expected to give reasonable estimation results in case the number of observations is large enough for the number of distorted values—created by the simple choice of initial conditions (also known as "warm-up effect", cf. [22])—to be negligible in comparison to the total number of observations.

Unfortunately, the previous likelihood function does not allow for closed-form expressions of the parameter estimates. To obtain such expressions at least for some of the parameter groups, we employ a well-known latent-variables approach (described in detail, e.g., in [3]), in which unobservable ("latent") gamma-distributed random variables $W_1, \ldots, W_n$ are introduced in addition to the observables $Y_1, \ldots, Y_n$. Letting ν be the d.o.f. of the original t-distribution error model, we choose the gamma distribution defined by the pdf

$$f(w_t) = \begin{cases} \frac{(\frac{\nu}{2})^{\frac{\nu}{2}}}{\Gamma(\frac{\nu}{2})} \cdot w_t^{\frac{\nu}{2}-1} \cdot e^{-\frac{\nu}{2}w_t} & \text{if } w_t > 0, \\ 0 & \text{if } w_t \leq 0. \end{cases} \quad (8)$$

We assume the latent variables to be stochastically independent, so that their joint pdf $f(\mathbf{w})$ constitutes the product $\prod_{t=1}^{n} f(w_t)$. To obtain an ML estimator which is equivalent to the one constructed before directly from the pdf $f(\mathbf{u})$ based on t-distributions, each white noise component U_t is now assumed to follow a normal distribution conditional on the occurrence of the value w_t of the latent variable W_t. For this purpose, we choose the particular pdf

$$f(u_t|w_t) = \frac{1}{\sqrt{2\pi(\sigma/\sqrt{w_t})^2}} \exp\left\{-\frac{u_t^2}{2(\sigma/\sqrt{w_t})^2}\right\}, \quad (9)$$

and we assume U_t to be conditionally independent from $U_1, W_1, \ldots, U_{t-1}, W_{t-1}, U_{t+1}, W_{t+1}, \ldots, U_n$ and W_n. Thus, the values of the latter random variables do not affect the density of u_t. In a first step, we can combine (8) and (9) for every time instance t to the joint pdf $f(u_t, w_t) = f(w_t)\, f(u_t|w_t)$ of the latent variable W_t and the time-wise associated white noise component U_t. In a second step, we may combine these joint pdfs for all time instances $t = 1, \ldots, n$ to the joint pdf $f(\mathbf{u}, \mathbf{w}) = \prod_{t=1}^{n} f(u_t, w_t)$ by exploiting the previous assumptions of (conditional) independence. We now use this pdf to define a proxy likelihood function $\mathcal{L}(\boldsymbol{\xi}, \boldsymbol{\beta}, \sigma^2, \nu; \mathbf{y}, \mathbf{w})$ in place of the original likelihood function $\mathcal{L}(\boldsymbol{\xi}, \boldsymbol{\beta}, \sigma^2, \nu; \mathbf{y})$. It can be shown that the latter defines the marginal distribution of the former, so that both models may be viewed as being equivalent (cf. [19]). It should also be mentioned that the conditional pdf $f(w_t|u_t)$ defines the gamma distribution $G(a, b)$ with parameters $a = (\nu + 1)/2$ and $b = (\nu + u_t^2/\sigma^2)/2$, given the value u_t. This fact is used in the subsequent derivation of an EM algorithm based on the proxy likelihood function.

3 The Modified EM Algorithm

It is convenient to collect all the unknown model parameters $\boldsymbol{\xi}$, $\boldsymbol{\beta}$, σ^2 and ν within a single column vector $\boldsymbol{\theta}$ and to use the natural logarithm of the proxy likelihood function, which we can write as

$$\log \mathcal{L}(\boldsymbol{\theta}; \mathbf{y}, \mathbf{w}) = -\frac{n}{2}\log(2\pi) - \frac{n}{2}\log(\sigma^2) + \frac{n\nu}{2}\log\left(\frac{\nu}{2}\right) - n\log\Gamma\left(\frac{\nu}{2}\right)$$
$$-\frac{1}{2}\sum_{t=1}^{n}\log w_t - \frac{1}{2\sigma^2}\sum_{t=1}^{n} w_t[\boldsymbol{\alpha}_t(L)(y_t - \mathbf{A}_t\boldsymbol{\xi})]^2 + \frac{\nu}{2}\sum_{t=1}^{n}(\log w_t - w_t). \quad (10)$$

Note that each part $\boldsymbol{\alpha}_t(L)$ of the decorrelation filter is a function of the parameters $\boldsymbol{\beta}$. Further note that the values $w_1, \ldots, w_n$ of the unobservable latent variables are not given, so that (10) cannot be maximized directly. This can be remedied by maximizing instead the conditional expectation

$$Q(\boldsymbol{\theta}|\boldsymbol{\theta}^{(i)}) = E_{\mathbf{W}|\mathbf{y};\boldsymbol{\theta}^{(i)}}\{\log \mathcal{L}(\boldsymbol{\theta}; \mathbf{y}, \mathbf{W})\}. \quad (11)$$

of the log-likelihood function (also known as the "Q-function"). Here, besides the observations $\mathbf{y}$, prior parameter values $\boldsymbol{\theta}^{(i)}$ must also be given. Then, maximizing the Q-function yields a new solution $\boldsymbol{\theta}^{(i+1)}$, which is used in the next iteration step in place of the previous solution $\boldsymbol{\theta}^{(i)}$. Alternating the expectation (E) step and the maximization (M) step yields an EM algorithm in the sense of [18].

Concerning the E step, since we defined the likelihood function in terms of the white noise $\mathbf{U}$ rather than the observables $\mathbf{Y}$, we may condition directly on the values $\mathbf{u}$. We then find for the Q-function the expression

$$Q(\boldsymbol{\theta}|\boldsymbol{\theta}^{(i)}) = c - \frac{n}{2}\log(\sigma^2) - \frac{1}{2\sigma^2}\sum_{t=1}^{n} w_t^{(i)}\left[\boldsymbol{\alpha}_t(L)(y_t - \mathbf{A}_t\boldsymbol{\xi})\right]^2 + \frac{n\nu}{2}\log\nu \quad (12)$$
$$-n\log\Gamma\left(\frac{\nu}{2}\right) + \frac{n\nu}{2}\left[\psi\left(\frac{\nu^{(i)}+1}{2}\right) - \log\left(\nu^{(i)}+1\right) + \frac{1}{n}\sum_{t=1}^{n}\left(\log w_t^{(i)} - w_t^{(i)}\right)\right]$$

(see [5] for details), where c is a constant independent of the unknown parameters, ψ is the digamma function, and

$$w_t^{(i)} = \frac{\nu^{(i)}+1}{\nu^{(i)} + \left(\frac{\boldsymbol{\alpha}_t^{(i)}(L)(y_t - \mathbf{A}_t\boldsymbol{\xi}^{(i)})}{\sigma^{(i)}}\right)^2}. \quad (13)$$

Regarding the M-Step, we determine the first partial derivatives of the Q-function (12) with respect to the individual parameters $\boldsymbol{\xi}$, $\boldsymbol{\beta}$, σ^2 and ν in $\boldsymbol{\theta}$, and we set these derivatives equal to zero. This yields for parameter ξ_j

$$\frac{\partial}{\partial\xi_j}Q(\boldsymbol{\theta}|\boldsymbol{\theta}^{(i)}) = \frac{1}{\sigma^2}\sum_{t=1}^{n} w_t^{(i)}\overline{A}_{t,j}(\overline{y}_t - \overline{\mathbf{A}}_t\boldsymbol{\xi}) = 0, \quad (14)$$

and then for all m of these equations

$$\begin{bmatrix} \overline{A}_{1,1} & \cdots & \overline{A}_{n,1} \\ \vdots & & \vdots \\ \overline{A}_{1,m} & \cdots & \overline{A}_{n,m} \end{bmatrix} \mathbf{W}^{(i)} \begin{bmatrix} \overline{y}_1 - \overline{\mathbf{A}}_1 \boldsymbol{\xi} \\ \vdots \\ \overline{y}_n - \overline{\mathbf{A}}_n \boldsymbol{\xi} \end{bmatrix} =: \overline{\mathbf{A}} \mathbf{W}^{(i)} \left(\overline{\mathbf{y}} - \overline{\mathbf{A}} \boldsymbol{\xi} \right) = \mathbf{0}, \tag{15}$$

where $\mathbf{W}^{(i)}$ denotes a diagonal matrix with elements $w_1^{(i)}, \ldots, w_n^{(i)}$. Since the filtered quantities $\overline{\mathbf{y}}$ and $\overline{\mathbf{A}}$ depend on the unknown parameters $\boldsymbol{\beta}$ through the AR coefficients in model (3), we employ the additional principle of conditional maximization (CM) in the sense of [20], substituting the available values $\boldsymbol{\beta}^{(i)}$ of the preceding iteration step for these unknowns. The procedure accordingly is to compute from $\boldsymbol{\beta}^{(i)}$ the AR coefficients $\alpha_{1,1}^{(i)}, \ldots, \alpha_{p,n}^{(i)}$, then the filtered quantities $\overline{y}_t^{(i)} = \boldsymbol{\alpha}_t^{(i)}(L) y_t$ and $\overline{A}_{t,j}^{(i)} = \boldsymbol{\alpha}_t^{(i)}(L) A_{t,j}$. The normal equations (15) are now solved by the reweighted least squares estimate

$$\boldsymbol{\xi}^{(i+1)} = \left((\overline{\mathbf{A}}^{(i)})^T \mathbf{W}^{(i)} \overline{\mathbf{A}}^{(i)} \right)^{-1} (\overline{\mathbf{A}}^{(i)})^T \mathbf{W}^{(i)} \overline{\mathbf{y}}^{(i)}. \tag{16}$$

The resulting colored noise residuals follow to be $e_t^{(i+1)} = y_t - \mathbf{A}_t \boldsymbol{\xi}^{(i+1)}$. Similarly to $\boldsymbol{\xi}$, we find the normal equations for $\boldsymbol{\beta}$ by setting the partial derivatives of the Q-function with respect to $\boldsymbol{\beta}_1, \ldots, \boldsymbol{\beta}_p$ equal to zero. Substituting now the available estimates $\boldsymbol{\xi}^{(i+1)}$ and $e_t^{(i+1)}$ in the spirit of CM, we thus obtain

$$\begin{bmatrix} e_0^{(i+1)} \mathbf{X}_1^T & \cdots & e_{n-1}^{(i+1)} \mathbf{X}_n^T \\ \vdots & & \vdots \\ e_{1-p}^{(i+1)} \mathbf{X}_1^T & \cdots & e_{n-p}^{(i+1)} \mathbf{X}_n^T \end{bmatrix} \mathbf{W}^{(i)} \begin{bmatrix} e_1^{(i+1)} - e_0^{(i+1)} \mathbf{X}_1 \boldsymbol{\beta}_1 - \cdots - e_{1-p}^{(i+1)} \mathbf{X}_1 \boldsymbol{\beta}_p \\ \vdots \\ e_n^{(i+1)} - e_{n-1}^{(i+1)} \mathbf{X}_n \boldsymbol{\beta}_1 - \cdots - e_{n-p}^{(i+1)} \mathbf{X}_n \boldsymbol{\beta}_p \end{bmatrix}$$
$$=: (\mathbf{E}^{(i+1)})^T \mathbf{W}^{(i)} \left(\mathbf{e}^{(i+1)} - \mathbf{E}^{(i+1)} \boldsymbol{\beta} \right) = \mathbf{0}, \tag{17}$$

using the initial conditions $e_0^{(i+1)} = \cdots = e_{1-p}^{(i+1)} = 0$ and the stacked vector $\boldsymbol{\beta}^T = [\boldsymbol{\beta}_1^T \ldots \boldsymbol{\beta}_p^T]$. Then, the reweighted least squares solution for $\boldsymbol{\beta}$ reads

$$\boldsymbol{\beta}^{(i+1)} = \left((\mathbf{E}^{(i+1)})^T \mathbf{W}^{(i)} \mathbf{E}^{(i+1)} \right)^{-1} (\mathbf{E}^{(i+1)})^T \mathbf{W}^{(i)} \mathbf{e}^{(i+1)}. \tag{18}$$

Applying the resulting decorrelation filter $\boldsymbol{\alpha}_1^{(i+1)}(L), \ldots, \boldsymbol{\alpha}_n^{(i+1)}(L)$ to the previously computed colored noise residuals yields the white noise residuals $u_t^{(i+1)} = \boldsymbol{\alpha}_t^{(i+1)}(L) e_t^{(i+1)}$ $(t = 1, \ldots, n)$, which can now be used to compute the scale factor. Setting the derivative of the Q-function with respect to σ^2 equal to zero and substituting $u_t^{(i+1)}$, we find the average weighted sum of squared residuals

$$(\sigma^2)^{(i+1)} = \frac{1}{n} \sum_{t=1}^{n} w_t^{(i)} \left(u_t^{(i+1)} \right)^2 = \frac{(\mathbf{u}^{(i+1)})^T \mathbf{W}^{(i)} \mathbf{u}^{(i+1)}}{n}. \tag{19}$$

Concerning the estimation of the d.o.f. ν of the t-distribution, it was generally found in previous studies (e.g., [20]) that the convergence of the algorithm can be sped up by maximizing the original log-likelihood $\mathcal{L}(\boldsymbol{\xi}, \boldsymbol{\beta}, \sigma^2, \nu; \mathbf{y})$ instead of the Q-function. As shown in detail in [3], this leads to the equation

$$0 = 1 + \log \nu^{(i+1)} - \psi\left(\frac{\nu^{(i+1)}}{2}\right) + \psi\left(\frac{\nu^{(i+1)}+1}{2}\right) - \log\left(\nu^{(i+1)}+1\right) \tag{20}$$
$$+ \frac{1}{n}\sum_{t=1}^{n}\left[\log\left(\frac{\nu^{(i+1)}+1}{\nu^{(i+1)} + \left(u_t^{(i+1)}/\sigma^{(i+1)}\right)^2}\right) - \frac{\nu^{(i+1)}+1}{\nu^{(i+1)} + \left(u_t^{(i+1)}/\sigma^{(i+1)}\right)^2}\right],$$

which can be solved for $\nu^{(i+1)}$. With this modification and the application of CM, the EM algorithm turns into an ECME algorithm.

When initial parameter values $\boldsymbol{\theta}^{(0)}$ and weights $\mathbf{W}^{(0)}$ are lacking, we start the algorithm with the computation of a least squares solution $\boldsymbol{\xi}^{(0)} = (\mathbf{A}^T\mathbf{A})^{-1}\mathbf{A}^T\mathbf{y}$. Using the unit weight matrix $\mathbf{W}^{(0)} = \mathbf{I}$, solutions $\boldsymbol{\beta}^{(0)}$ and $(\sigma^2)^{(0)}$ can then be determined; we usually choose $\nu^{(0)} = 30$ concerning the d.o.f. Having thus fixed the initial values, the E step and CM(E) steps (16)–(20) are iterated until a sufficient level of convergence is reached. In our numerical example, we check if the greatest absolute value of the differences between the estimates of two subsequent iteration steps is less than 10^{-6} for the parameters $\boldsymbol{\xi}$, $\boldsymbol{\beta}$, σ^2, and less than 10^{-4} for ν. To estimate ν, we carried out a reliable zero search using an interval Newton method according to Algorithm 6.1 in [23].

4 An Application to Vibration Analysis

The measurement and estimation of amplitudes (possibly in conjunction with the identification of the frequencies) of oscillating structures such as bridges is an important task of engineering geodesy, with high relevance to disciplines such as structural health monitoring (see, e.g., [24]). The measurements of such oscillations throughout time can be modeled by means of a sum of sinusoids and additive random deviations, that is,

$$y_t = \frac{a_0}{2} + \sum_{j=1}^{M} a_j \cos(2\pi f_j x_t) + b_j \sin(2\pi f_j x_t) + e_t \qquad (t = 1, \ldots, n). \tag{21}$$

Typically, an oscillation is induced by moving objects on the structure, and relatively stable amplitudes in terms of the coefficients $a_0, a_1, \ldots, a_M$ and $b_1, \ldots, b_M$ are reached after some time of "transient oscillations". The expected frequencies $f_1, \ldots, f_M$ can often be derived through an analysis of the eigenfrequencies of the particular structure. Whereas the stable part of the oscillations can be explained deterministically by laws of physics, the transient oscillations appear to be less regular, suggesting the incorporation of a stochastic model component into (21). One idea (which we investigate in the sequel) is to model the smoothly changing deviations of the transient oscillations from the stable oscillations as the combination of a time-variable AR process and outliers. Modeling outliers stochastically via a scaled t-distribution, we thus have an observation model within the framework of the generic model described in Sect. 2.

To investigate the adequacy of high-accuracy and low-cost accelerometers for the purpose of structural health monitoring of bridges, a vibration experiment was carried out at the Institute of Concrete Construction of the Leibniz Universität Hannover/Germany. This experiment involved several sensors that were mounted on a shaker table, which consisted of a plexiglass plate (fixed between two wooden supports) and two imbalance motors in the center. The experiment lasted for about 45 min, throughout which a constant oscillation frequency of 16 Hz was induced. In the following, we analyze, in particular, the measurements of a high-accuracy single-axis PCB Piezotronics accelerometer, in which data reflect the induced oscillations in terms of the sinusoids (21). The sampling frequency of that sensor is approximately 195 Hz, so that the induced frequency of 16 Hz is well below the Nyquist frequency. The recorded data set includes both a period of transient oscillations (clearly visible within the first 1500 values or 7.7 s) and a period of stable oscillation (maintained until the end of the experiment). The latter period was analyzed by [3], who employed the model (21) with $M = 11$ frequencies (to capture the significant side frequencies that are caused by the sampling of the originally continuous-time phenomenon as well as by the physical properties of the shaker table) in connection with a time-constant AR process and t-distributed white noise components. In the current contribution, we specify $M = 12$ frequencies by $f_j = j \cdot 8$ Hz ($j = 1, \ldots, 12$), so that the vector ξ of functional parameters consists of the 25 unknown coefficients $a_0, a_1, \ldots, a_{12}$ and $b_1, \ldots, b_{12}$.

Concerning the time-variable AR process (3), we used linear combinations of polynomials $x^0, x^1, \ldots, x^k$ up to a maximum degree k, treating the coefficients as unknown parameters. To identify an adequate and parsimonious model for the given data, we applied the described ECME algorithm for different AR orders p and different polynomial degrees k, beginning with the smallest considered AR model ($p = 1$ and $k = 0$). Then, for each $k = 0, \ldots, 5$, we determined the smallest AR order p^* for which the estimated residuals $\hat{u}_1, \ldots, \hat{u}_n$ (resulting from the final iteration step, as indicated by the "hat") passed a white noise test (WNT). The adopted test measures the maximum cumulated periodogram excess $T = \max_i |S_i - i/M|$ with respect to a cumulated, theoretical white noise periodogram (cf. Sect. 7.3.3 in [25]). Here, the normalized cumulated periodogram defined by $S_0 = 0$, $S_i = \sum_{k=1}^{i} I_k / \sum_{k=1}^{m} I_k$ for $i = 1, \ldots, m$ ($m = \text{floor}(n/2)$) involves partial sums of the one-sided periodogram $I_1, \ldots, I_m$, which we computed by means of the MATLAB routine `periodogram`. The significance of T can be evaluated by comparison with critical values based on a specified significance level α. Note that thc choice $k = 0$ corresponds to a time-constant AR process; for this simple case, we also identified the most parsimonious AR model order p^* with successful WNT (see Fig. 1) by executing the ECME algo-

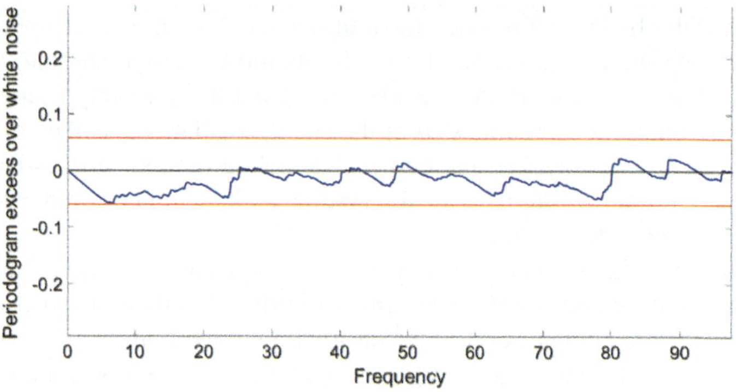

Fig. 1 Acceptance of the white noise test regarding the time-constant AR(20) model ($k = 0$) estimated with fixed degree of freedom ($\nu = 300$), shown as the excess of the periodogram (blue) of the white noise residuals with respect to the theoretical white noise periodogram (black), lying within the 99% significance bounds (red)

Table 1 Model selection criteria based on (1) the maximum polynomial degree (p.d.) k, (2) the lowest AR order p^* for which the periodogram-based WNT is accepted, (3) the total number of AR parameters—determined by $p^*q = p^*(k+1)$, (4) the estimated scale factor, (5) the estimated d.o.f., (6) the log-likelihood of the original observation model, and (7) the colored noise residual sum of squares (RSS). Results are also shown for the largest AR order $p^- = p^* - 1$ for which the model did not pass the WNT

	p.d.:	$k = 0$	$k = 0$	$k = 1$	$k = 2$	$k = 3$	$k = 4$	$k = 5$
		$\nu = 300$	adaptive	adaptive	adaptive	adaptive	adaptive	adaptive
AR orders	p^-	19	29	20	20	5	14	12
	p^*	20	30	21	21	6	15	13
#AR param.	p^*q	20	30	42	63	**24**	75	78
est. scale factor	p^-	11.24	3.44	6.76	6.23	9.50	7.67	7.82
($\hat{\sigma}^2$ [$\cdot 10^{-6}$])	p^*	10.91	3.53	6.12	5.69	**9.48**	6.60	6.90
estimated d.o.f.	p^-	–	2.6	7.1	6.5	4.3	7.3	7.6
($\hat{\nu}$)	p^*	–	2.9	6.1	5.8	**4.8**	4.8	5.0
log-likelihood	p^-	6412	6694	6583	6623	6175	6493	6487
($\log \mathcal{L}(\hat{\boldsymbol{\theta}}; \mathbf{y})$)	p^*	6434	6743	6621	6662	6219	6489	6473
RSS	p^-	11.71	12.38	11.45	11.49	1.95	11.57	11.30
($\Omega = \mathbf{e}^T\mathbf{e}$)	p^*	11.61	11.34	11.53	11.47	**1.82**	1.87	1.88

rithm with the a priori fixed degree of freedom $\nu = 300$ (which closely approximates the standard assumption of normally distributed white noise). A summary of results relevant to model selection is given by Table 1.

Based on these findings, it makes sense to select the adaptively determined solution that involves cubic polynomials ($k = 3$) in connection with an AR(6) model for the following reasons. Besides the accepted WNT test, the colored noise residual sum of squares (RSS) $\Omega = 1.82$ is minimal among all shown solutions, whose

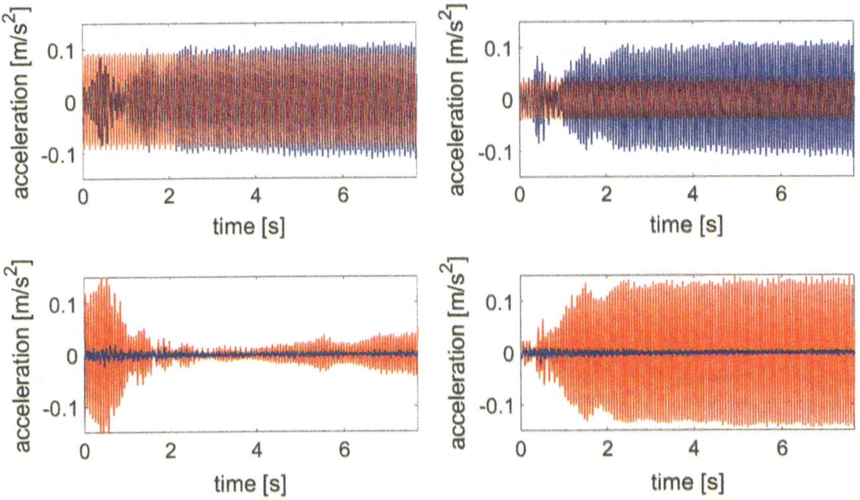

Fig. 2 Top row: oscillation dataset $\mathbf{y}$ (blue), the adjusted observations $\hat{\mathbf{y}} = \mathbf{A}\hat{\boldsymbol{\xi}}$ for the time-variable (tv) AR(6) model based on cubic polynomials (in red on the left subplot), and the adjusted observations for the nonadaptively estimated time-constant (tc) AR(20) model (in red on the right subplot). Bottom row: the corresponding estimated residuals (white noise residuals $\hat{\mathbf{u}}$ in blue, colored noise residuals $\hat{\mathbf{e}}$ in red)

fact indicates the relatively best approximation of the given measurements by the adjusted observations $\hat{\mathbf{y}} = \mathbf{A}\hat{\boldsymbol{\xi}}$. Figure 2 demonstrates this effect in comparison to the nonadapted solution with $k = 0$. Whereas the time-variable AR(6) model reproduces the eventual oscillation amplitude quite accurately, much of the oscillation signal is absorbed into the colored noise of the time-constant model, resulting in the relatively large RSS value $\Omega = 11.61$.

In light of the fact that the estimated d.o.f. for the adaptive solution involving the time-variable AR(6) process takes the rather low value $\hat{\nu} = 4.8$, we see that the estimated white noise residuals display a considerable level of heavy-tailedness and deviation from a Gaussian error distributions. As shown in Table 1, the solutions for $k = 1$ and $k = 2$ give larger $\hat{\nu}$ and smaller $\hat{\sigma}^2$, indicating that the colored noise residuals absorbed outliers (located under the tails of the pdf), besides absorbing vibration signal. Larger polynomial degrees ($k = 4$ and $k = 5$) produce values for the d.o.f. and the RSS (Ω) that are similar to the values resulting from the choice $k = 3$. However, the latter model involves only 24 AR coefficients, which is much less than the 75 and 78 coefficients required for the models $k = 4$ and $k = 5$, respectively. Table 1 also shows that the log-likelihoods of the desirable models (in which the time-variable AR models do not absorb too much of the vibration signal) are smaller than for other solutions (in which vibration signal migrates into the AR model).

Having identified the time-variable AR(6) model as the most reasonable one, we analyze its characteristics in the time and frequency domain. On the one hand, the estimated cubic polynomials describing the time-variability of that model show

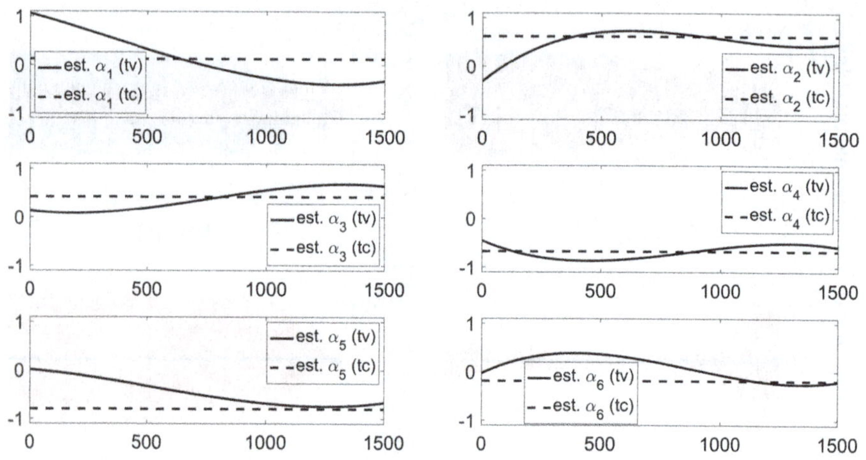

Fig. 3 Estimated coefficients of the time-variable (tv) AR(6) model based on cubic polynomials and of the first six coefficients of the nonadaptively estimated time-constant (tc) AR(20) model

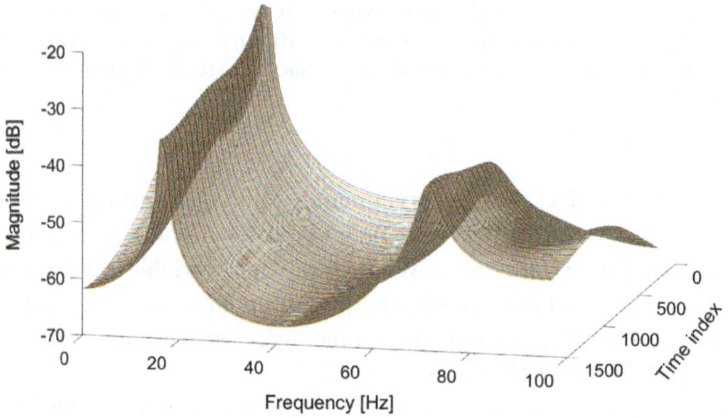

Fig. 4 Power spectral density of the time-variable AR(6) model

rather smooth changes of the individual coefficients (see Fig. 3). On the other hand, we determined the time-variable power spectral density (see Fig. 4), defined by $\mathrm{PSD}(f,t) = \hat{\sigma}_u^2 / \left|1 - \sum_{j=1}^{p} \hat{\alpha}_{j,t} e^{-i2\pi jf}\right|^2$ (cf. [26]), where the standard deviation of the t-distributed white noise components is given by $\hat{\sigma}_u^2 = \frac{\hat{\nu}}{\hat{\nu}-2}\hat{\sigma}^2$. We can observe that also the PSD changes rather smoothly with time. Furthermore, the evident fact that the PSDs have peaks around 16 and 72 Hz demonstrates that the oscillation signal is still partially captured by the colored noise model, possibly as a consequence of nonconstant signal frequency.

5 Summary, Conclusions, and Outlook

We defined an observation time series model consisting of a linear regression model with time-variable autoregressive errors, where each coefficient is described by a second linear regression model throughout time and where the white noise components follow a scaled t-distribution with unknown degree of freedom. To obtain closed-form expressions for the corresponding self-tuning robust maximum likelihood estimator, we derived a modified expectation maximization algorithm, which takes the form of iteratively reweighted least squares for the solution of the two linear models. We applied this algorithm to model transient oscillations measured by a high-accuracy accelerometer in the presence of outliers. We identified a parsimonious observation model that clearly separates the desired vibration signal, the transient oscillation component (described by colored noise residuals), and the outliers (captured by the heavy-tailed white noise residuals). This identification was based on choosing the most parsimonious, time-variable AR model that simultaneously passes the white noise test and produces the least RSS with respect to the colored noise. The adaptive solution for this model was also characterized by the lowest occurring degree of freedom and a relatively large scale factor. Unfortunately, the log-likelihood appears to be an irrelevant quantity in the process of model selection. This finding makes sense since the likelihood function of very complicated observation models, such as the one considered in the analyzed vibration experiment, must be expected to have local maxima. EM algorithms are known for getting stuck in such local maxima, which might correspond to undesirable solutions in our example. To shed further light on this problem, it might be helpful to carry out global optimization not only for the degree of freedom, but for the entire model. Since multidimensional interval methods are very time consuming, we intend to apply heuristic methods such as a genetic algorithm for this task in the future.

Acknowledgements The presented application of the PCB Piezotronics accelerometer within the vibration analysis experiment was performed as a part of the collaborative project "Spatio-temporal monitoring of bridge structures using low cost sensors" with ALLSAT GmbH, which is funded by the German Federal Ministry for Economic Affairs and Energy (BMWi) and the Central Innovation Programme for SMEs (ZIM Kooperationsprojekt, ZF4081803DB6). In addition, the authors acknowledge the Institute of Concrete Construction (Leibniz Universität Hannover) for providing the shaker table and the reference accelerometer.

References

1. Kuhlmann, H.: Importance of autocorrelation for parameter estimation in regression models. In: 10th FIG International Symposium on Deformation Measurements, pp. 354–361. FIG (2001)
2. Brockmann, J.M., Kargoll, B., Krasbutter, I., Schuh, W.D., Wermuth M.: GOCE data analysis: from calibrated measurements to the global Earth gravity field. In: Flechtner, F., Gruber, T., Gntner A., Mandea, M., Rothacher, M., Schne, T., Wickert, J. (eds.) Advanced Technologies in Earth Sciences System Earth via Geodetic-Geophysical Space Techniques, pp. 213–229. Springer, Berlin (2010)

3. Kargoll, B., Omidalizarandi, M., Loth, I., Paffenholz, J.A., Alkhatib, H.: An iteratively reweighted least-squares approach to adaptive robust adjustment of parameters in linear regression models with autoregressive and t-distributed deviations. J. Geod. (2017). https://doi.org/10.1007/s00190-017-1062-6
4. Krasbutter, I., Brockmann, J.M., Kargoll, B., Schuh, W.D., Goiginger, H., Pail, R.: Refinement of the stochastic model of GOCE scientific data in a long time series. In: Ouwehand, L. (ed.) 4th International GOCE User Workshop. ESA Publication SP-696, ESA/ESTEC (2011). ISBN: 978-92-9092-260-5
5. Kargoll, B., Omidalizarandi, M., Alkhatib, H., Schuh, W.D.: A modified EM algorithm for parameter estimation in linear models with time-dependent autoregressive and t-distributed errors. In: Valenzuela, O., Rojas, F., Pomares, H., Rojas, I. (eds.) Proceedings ITISE 2017–International work-conference on Time Series, vol. 2, pp. 1132–1145 (2017). ISBN 978-84-17293-01-7
6. Subba Rao, T.: The fitting of non-stationary time-series models with time-dependent parameters. J. Roy. Stat. Soc. B Met. **32**(2), 312–322 (1970)
7. Azrak, R., Mélard, G.: AR models with time-dependent coefficients—a comparison between several approaches. Technical Report 0642, IAP Statistics Network, Interuniversity Attraction Pole (2006)
8. Dahlhaus, R.: Fitting time series models to nonstationary processes. Ann. Statist. **25**(1), 1–37 (1997). https://doi.org/10.1214/aos/1034276620
9. Rudoy, D., Quatieri, T.F., Wolfe, P.J.: Time-varying autoregressions in speech: detection theory and applications. IEEE Trans. Audio Speech Lang. Process. **19**(4), 977–989 (2011). https://doi.org/10.1109/TASL.2010.2073704
10. Schuh, W.D., Brockmann, J.M.: Refinement of the stochastic model for GOCE gravity gradients by non-stationary decorrelation filters. In: ESA Living Planet Symposium 2016, Prag, Poster 2382 (2016)
11. Tsatsanis, M.K., Giannakis, G.B.: Time-varying system identification and model validation using wavelets. IEEE Trans. Sign. Process. **41**, 3512–3523 (1993). https://doi.org/10.1109/78.258089
12. Eom, K.B.: Analysis of acoustic signatures from moving vehicles using time-varying autoregressive models. Multidim. Sys. Sign. Process. **10**(4), 357–378 (1999). https://doi.org/10.1023/A:1008475713345
13. Härmä, A., Juntunen, M., Kaipio, J.P.: Time-varying autoregressive modeling of audio and speech signals. In: 10th European Signal Processing Conference. IEEE (2000)
14. Grenier, Y.: Time-dependent ARMA modeling of nonstationary signals. IEEE Trans. Acoust. Speech Sign. Process. **31**(4), 899–911 (1983)
15. Pierce, D.A.: Least squares estimation in the regression model with autoregressive-moving average errors. Biometrika **58**(2), 299–312 (1971). https://doi.org/10.2307/2334518
16. McDonald, J.B., Newey, W.K.: Partially adaptive estimation of regression models via the generalized t distribution. Econom. Theor. **4**(3), 428–457 (1988)
17. Parzen, E.: A density-quantile function perspective on robust estimation. In: Launer, L., Wilkinson, G.N. (eds.) Robustness in Statistics, pp. 237–258. Academic Press, New York (1979). https://doi.org/10.1016/B978-0-12-438150-6.50019-4
18. Dempster, A.P., Laird, N.M., Rubin, D.B.: Maximum likelihood from incomplete data via the EM algorithm. J. Roy. Stat. Soc. B Met. **39**(1), 1–38 (1977)
19. Lange, K.L., Little, R.J.A., Taylor, J.M.G.: Robust statistical modeling using the t-distribution. J. Am. Stat. Assoc. **84**, 881–896 (1989). https://doi.org/10.2307/2290063
20. Liu, C.H., Rubin, D.B.: ML estimation of the t distribution using EM and its extensions. ECM ECME. Stat. Sin. **5**, 19–39 (1995)
21. Hamilton, J.D.: Time Series Analysis. Princeton University Press, Princeton (1994)
22. Schuh, W.D.: The processing of band-limited measurements; filtering techniques in the least squares context and in the presence of data gaps. Space Sci. Rev. **108**(1), 67–78 (2003). https://doi.org/10.1023/A:1026121814042

23. Hargreaves, G.I.: Interval Analysis in MATLAB. Numerical Analysis Report No. 416, Manchester Centre for Computational Mathematics, The University of Manchester (2002). ISSN 1360-1725
24. Neitzel, F., Niemeier, W., Weisbrich, S., Lehmann, M.: Investigation of low-cost accelerometer, terrestrial laser scanner and ground-based radar interferometer for vibration monitoring of bridges. In: 6th European Workshop on Structural Health Monitoring (2012). http://www.ndt.net/?id=14063
25. Schlittgen, R., Streitberg, B.H.J.: Zeitreihenanalyse, 9th edn. R. Oldenbourg Verlag, Munich (2001)
26. Tary, J.B., Herrera, R.H., van der Baan, M.: Time-varying autoregressive model for spectral analysis of microseismic experiments and long-period volcanic events. Geophys. J. Int. **196**(1), 600–611 (2014). https://doi.org/10.1093/gji/ggt400

Author Index

I. Rojas et al. (eds.), *Time Series Analysis and Forecasting*,
Contributions to Statistics, https://doi.org/10.1007/978-3-319-96944-2

The manufacturer's authorised representative in the EU is Springer Nature Customer Service Centre GmbH, Europaplatz 3, 69115 Heidelberg, Germany. If you have any concerns regarding our products, please contact ProductSafety@springernature.com

Printed and bound by CPI Group (UK) Ltd, Croydon, CR0 4YY
30/09/2026
02225655-0009